Power Systems Conceptualization with MATLAB

NIPA® GENX ELECTRONIC RESOURCES & SOLUTIONS P. LTD.
New Delhi-110 034

About the Authors

Naveen Jain has obtained his B.E. (Electrical Engineering) and M. E. (Power System) from Malaviya Regional Engineering College (presently known as Malaviya National Institute of Technology), Jaipur (India) in 1997 and 1999, respectively. Further, he obtained Ph.D. from IIT Kanpur in 2013. He was awarded POSOCO Award for his excellent Ph.D. research work on March 06, 2013 (The POSOCO is a subsidiary of POWERGRID and FITT, IIT Delhi).

Dr. Jain has been working with Rajasthan Government Engineering Institute, "College of Technology and Engineering (CTAE), Udaipur", as assistant professor since March 2005. He has served to the college in different capacity as Hostel Warden, Placement in-Charge (EE), faculty in-charge (Library), Editor-in-Chief (College Magazine: Tech-View). He has published several research papers in International/national journals/conferences. He co-authored a book on "Basic Electrical Engineering". In 2012, his biography was also published with Marquis Who's Who. He has visited USA and Singapore many times for presenting his research work in international conferences. In addition, he is reviewer and member of several journals/profession organization of good repute. His research interests include planning of Distributed Generation (DG), reliability issues in distribution system, technical issues in electricity energy market and application of heuristic methods in electrical power system, etc.

Dr. Jain is a Fellow of Institution of Electronics and Telecommunication Engineers (IETE) India, life Member of Indian Society of Technical Education (ISTE), life Member of Institute of Engineers (IE India), Member of International Association of Engineers and Senior Member of IEEE (USA).

Umesh Agarwal graduated in Electrical Engineering from the College of Technology and Engineering (CTAE, Udaipur) India, in 2012. He received his M. Tech. degree in Power Systems in 2015 from Rajasthan Technical University (RTU, Kota), India. Further, he has completed his Doctoral degree from College of Technology and Engineering (CTAE, Udaipur) India, in 2022. Currently, he is working as an Associate Professor at JIET, Jodhpur. He has over 09 years of teaching and research experience. He has published many research papers in reputed journals and National/International conferences. He is reviewer of many reputed journals and conferences. His areas of interest are power system restructuring, AI application to the power system, reliability analysis, and planning of distributed energy. He is a member of Soft Computing Research Society.

Dedicated
To
Almighty GOD
and
Our parents

Power Systems Conceptualization with MATLAB

Naveen Jain, B.E. (Electrical Engineering), M.E. (Power System)
Ph.D. (Electrical Engineering), SMIEEE, FIETE, LMISTE, MIE
Associate Professor, Department of Electrical Engineering
College of Technology and Engineering
Maharana Pratap University of Agriculture and Technology
Udaipur-313 001, Rajasthan, India

Umesh Agarwal, B.E. (Electrical Engineering), M.E. (Power System)
Ph.D. (Electrical Engineering)
Member SCRS
Associate Professor, Department of Electrical Engineering
Jodhpur Institute of Engineering and Technology
Jodhpur, Rajasthan India

NIPA® GENX ELECTRONIC RESOURCES & SOLUTIONS P. LTD.
New Delhi-110 034

NIPA® GENX ELECTRONIC RESOURCES & SOLUTIONS P. LTD.

101,103, Vikas Surya Plaza, CU Block
L.S.C. Market, Pitam Pura, New Delhi-110 034
Ph : +91 11 27341616, 27341717, 27341718
E-mail: newindiapublishingagency@gmail.com
www: www.nipabooks.com

For customer assistance, please contact
Phone: + 91-11-27 34 17 17
Fax: + 91-11-27 34 16 16
E-Mail: feedbacks@nipabooks.com

Print ISBN: 978-93-58876-20-8

ebook ISBN: 978-93-58875-82-9

Composed and Designed by NIPA®.

Acknowledgement

Writing a book is not a solitary endeavour; it's a collaborative effort that involves the support, encouragement, and assistance of many people. I would like to express my deepest gratitude to those who have contributed to the creation of this work:

First and foremost, I am immensely thankful to almighty god for making me capable to write this book. Further, I would like to thanks my family for their unwavering love, understanding, and patience throughout this journey. Your support has been my rock, and I am endlessly grateful for your encouragement and belief in me.

I extend my heartfelt appreciation to my mentor and Co-author of this book, Dr. Naveen Jain, whose guidance, wisdom, and expertise have been invaluable in shaping this book. Your insights have been instrumental in refining my ideas and sharpening my writing.

I am indebted to my colleagues and friends for their encouragement, feedback, and support. Your enthusiasm for this project has kept me motivated during the challenging moments and inspired me to push through obstacles.

I extend my gratitude to the team at NIPA for their professionalism, dedication, and passion in bringing this book to life. Your expertise and attention to detail have elevated this work beyond my expectations.

Lastly, I want to express my appreciation to the readers who have supported and embraced this book. Your interest and engagement are the ultimate reward for my efforts, and I am humbled by your enthusiasm.

Thank you, each and every one of you, for being a part of this journey. Your contributions have made this book possible, and I am profoundly grateful for your presence in my life.

Warm regards

Authors

Preface

This book on "Power Systems Conceptualization with MATLAB" deals with the MATLAB exposure in power system for postgraduate as well as undergraduate students of Electrical Engineering. Electric Power System is the most important subject in electrical engineering and the most complex subject ever studied by the students. This book covers all the prominent topics like per unit systems, unsymmetrical fault calculations, fuse/relay/circuit breakers etc., and the more advanced topic like symmetrical components and arcing phenomena in power systems. The book covers a very wide spectrum of electrical power system studies which is normally not available in one single book. The book is so comprehensibly written that MATLAB programme on power systems are ready to use with the data of various standard distribution test systems.

It has been the constant endeavour of the author to understand the difficulties of the students in the laboratory and accordingly prepare the laboratory notes after consulting various books on electrical power systems.

An extensive bibliography will help the students to locate detailed information on various topics of his interest. Any constructive suggestions for the improvement of the book will be gratefully acknowledged.

Authors

Contents

Acknowledgement .... *vii*

Preface .... *ix*

Laboratory Practices .... *xiii*

1. Basics of MATLAB 1
2. MATLAB Application for Transmission Line Modelling 13
3. Per Unit System 29
4. Overhead Insulators 40
5. Symmetrical Components 42
6. Symmetrical and Unsymmetrical Components 53
7. Protection of Generator and Transformer 68
8. Fundamentals of Fault Clearing and Switching Phenomena 80
9. Fundamentals of Fuse and Relays 93
10. Fundamentals of Neutral Grounding 107
11. MATLAB Application of Renewable Energy Resources 111

Experiments 124

- Experiment Number-1 124
- Experiment Number-2 127
- Experiment Number-3 130
- Experiment Number-4 133
- Experiment Number-5 136
- Viva Questions Based on Laboratory 138

Appendices 144
Appendix A1.1: Power system one liners.. 144
Appendix A1.2: Power system protection one liners 158
Appendix A2: Line and load data for various test systems 187

Laboratory Practices

Safety Rules

1. Safety is of paramount importance in the electrical engineering laboratories.
2. Electricity never excuses careless persons. So, exercise enough care and attention in handling electrical equipment and follow safety practices in the laboratory.
3. Avoid direct contact with any voltage source and power line voltages. (Otherwise, any such contact may subject you to electrical shock).
4. Wear rubber-soled shoes. (To insulate you from earth so that even if you accidentally contact a live point, current will not flow through your body to earth and hence you will be protected from electrical shock).
5. Wear laboratory-coat and avoid loose clothing. (Loose clothing may get caught on an equipment/instrument and this may lead to an accident particularly if the equipment happens to be a rotating machine).
6. Girl students should have their hair tucked under their coat or have it in a knot.
7. Do not wear any metallic rings, bangles, bracelets, wristwatches and neck chains. (When you move your hand/body, such conducting items may create a short circuit or may touch a live point and thereby subject you to electrical shock).
8. Be certain that your hands are dry and that you are not standing on wet floor. (Wet parts of the body reduce the contact resistance thereby increasing the severity of the shock).
9. Ensure that the power is off before you start connecting up the circuit. (Otherwise, you will be touching the live parts in the circuit).
10. Get your circuit diagram approved by the staff member and connect up the circuit strictly as per the approved circuit diagram.
11. Check power chords for any sign of damage and be certain that the chords use safety plugs and do not defeat the safety feature of these plugs by using ungrounded plugs.

12. When using connection leads, check for any insulation damage in the leads and avoid such defective leads.
13. Do not defeat any safety devices such as fuse or circuit breaker by shorting across it. Safety devices protect you and your equipment.
14. Switch on the power to your circuit and equipment only after getting them checked up and approved by the staff member.
15. Take the measurement with one hand in your pocket. (To avoid shock in case you accidentally touch two points at different potentials with your two hands).
16. Do not make any change in the connection without the approval of the staff member.
17. In case you notice any abnormal condition in your circuit (like insulation heating up, resistor heating up etc), switch off the power to your circuit immediately and inform the staff member.
18. Keep hot soldering iron in the holder when not in use.
19. After completing the experiment show your readings to the staff member and switch off the power to your circuit after getting approval from the staff member.
20. While performing load-tests in the electrical machine's laboratory using the brake-drums:
    i. Avoid the brake-drum from getting too hot by putting just enough water into the brake-drum at intervals; use the plastic bottle with a nozzle (available in the laboratory) to pour the water. (When the drum gets too hot, it will burn out the braking belts)
    ii. Do not stand in front of the brake-drum when the supply to the load-test circuit is switched off. (Otherwise, the hot water in the brake-drum will splash out on you)
    iii. After completing the load-test, suck out the water in the brake-drum using the plastic bottle with nozzle and then dry off the drum with a sponge which is available in the laboratory (the water, if allowed to remain in the brake-drum, will corrode it).
21. Determine the correct rating of the fuse/s to be connected in the circuit after understanding correctly the type of the experiment to be performed: no-load test or full-load test, the maximum current expected in

the circuit and accordingly use that fuse-rating. (While an over-rated fuse will damage the equipment and other instruments like ammeters and watt-meters in case of over load, an under-rated fuse may not allow one even to start the experiment).

22. At the time of starting a motor, the ammeter connected in the armature circuit overshoots, as the starting current is around 5 times the full load rating of the motor. Moving coil ammeters being very delicate may get damaged due to high starting current. A switch has been provided on such meters to disconnect the moving coil of the meter during starting. This switch should be closed after the motor attains full speed. Moving iron ammeters and current coils of watt meters are not so delicate and hence these can stand short time overload due to high starting current. No such switch is therefore provided on these meters. Moving iron meters are cheaper and more rugged compared to moving coil meters. Moving iron meters can be used for both a.c. And d.c. Measurement. Moving coil instruments are however more sensitive and more accurate as compared to their moving iron counterparts and these can be used for d.c. Measurements only. Good features of moving coil instruments are not of much consequence for you as other sources of errors in the experiments are many times more than those caused by these meters.

23. Some students have been found to damage meters by mishandling in the following ways:

    i. Keeping unnecessary material like books, lab records, unused meters etc. Causing meters to fall down the table.

    ii. Putting pressure on the meter (especially glass) while making connections or while talking or listing somebody.

Students Are Strictly Warned That Full Cost of the Measuring Instrument Will Be Recovered From The Individual Who Will Damaged It.

Copy these rules in your Lab Record.

Observe this yourself and help your friends to observe.

I have read and understand these rules and procedures. I agree to abide by these rules and procedures at all times while using these facilities. I understand that failure to follow these rules and procedures will result in my immediate dismissal from the laboratory and additional disciplinary action may be taken.

Guidelines for Preparing Laboratory Notebook

The laboratory notebook is a record of all work pertaining to the experiment. This record should be sufficiently complete so that you or anyone else of similar technical background can duplicate the experiment and data by simply following your laboratory notebook. Record everything directly into the notebook during the experiment. Do not use scratch paper for recording data. Do not trust your memory to fill in the details at a later time.

Organization in your notebook is important. Descriptive headings should be used to separate and identify the various parts of the experiment. Record data in chronological order. A neat, organized and complete record of an experiment is just as important as the experimental work.

1. **Heading:** The experiment identification (number) should be at the top of first page. Your name and date should be mentioned at the top of each page of days experimental.
2. **Object:** A brief complete statement of what you intend to find out or verify in the experiment should be at the beginning of each experiment.
3. **Equipment List:** List those items of equipment which have a direct effect on the accuracy of the data. It may be necessary later to locate specific items of equipment for rechecks if discrepancies develop in the results.
4. **Circuit Diagram:** A circuit diagram should be drawn and labeled so that the actual experiment circuitry could be easily duplicated at any time in the future. Be especially careful to record all circuit changes made during the experiment.
5. **Theory:** As per experiment.
6. **Procedure:** In general, lengthy explanations of procedures are unnecessary so be in brief, short commentaries alongside the corresponding data may be used. Keep in mind the fact that the experiment must be reproducible from the information given in your notebook.
7. **Data:** Think carefully about what data is required and prepare suitable data tables. Record instruments reading directly. Do not use calculated results in place of direct data however; calculated results may be recorded in the same table with the direct data. Data tables should be clearly identified and each data column labelled and headed by the proper units of measure.

8. **Calculations:** Not always necessary but equations and sample calculations are often given to illustrate the treatment of the experimental data in obtaining the results.

9. **Results:** The results should be presented in a form which makes the interpretation easy. Large amounts of numerical results are generally presented in graphical form. Tables are generally used for small amounts of results. Theoretical and experimental results should be on the same graph or arrange in the same table in a way for easy correlation of these results.

10. **Graphs:** Graphs are used to present large amounts of data in a concise visual form. Data to be presented in graphical form should be plotted in the laboratory so that any questionable data points can be checked while the experiment is still set up. The grid lines in the notebook can be used for most graphs. If special graph paper is required, affix the graph permanently into the notebook. Give all graphs a short descriptive title. Label and scale the axes. Use units of measure. Label each curve if more than one on a graph.

11. **Conclusion:** This is your interpretation of the results of the experiment as an engineer. Be brief and specific. Give reasons for important discrepancies.

12. **Precautions:**

Trouble Shooting Hints

1. Be Sure that the power is turned ON.
2. Be sure the ground connections are common.
3. Be sure the circuit you build is identical to your circuit diagram (Do a node-by-node check).
4. Be sure that the supply voltages are correct.
5. Be sure that the equipment is set up correctly and you are measuring the correct parameters.
6. If steps 1 to 5 are correct then you probably have used a component with the wrong value or one that doesn't work. It is also possible that the equipment does not work (although this is not probable or the protoboard you are using may have some unwanted paths between nodes. To find your problem you must trace through the voltages in your circuit

node by node and compare the signal you expect to have. Then if they are different use your engineering judgment to decide what is causing the different or ask your lab assistant.

1

Basics of MATLAB

Introduction

Electrical Engineering generally uses a consistent, structured approach to the solving of various practical problems. An acceptable problem-solving approach and method can be defined, although variations can be seen in specific problems. Problems need to be approached methodically, applying an algorithm by which engineer arrives at an acceptable solution.

Problem-Solving Process in Engineering: (i) Define the engineering problem (ii) Create a mathematical model (iii) Develop a computational method for solving the problem (iv) Implement the computational method (v) Test and assess the solution.

Towards problem solving in Engineering, MATLAB* provides a technical computing environment designed to support the implementation of computational tasks.

The MATLAB is an interactive computing environment that enables numerical computation and data visualization. It has thousands of built-in functions and can be used to solve complex problems. It is an effective tool to do some simple numerical/statistical calculations, solve simultaneous equations. It can run entire simulation program and produce result in various formats viz. text, graph, etc. It has proven to be extraordinarily versatile and capable in its ability to help solve problems in applied science, engineering, and almost in any application area that deals with complex numerical calculations. It starts by double-click on the MATLAB icon or select MATLAB from Start or Programs.

Command Window & Workspace

The command window is the most active window. As commands are executed, appropriate windows will automatically appear. Any window can be activated by clicking the mouse on it. In this window, short instructions can be executed. One can check any syntax and use 'help' command to learn about any inbuilt

---

*MATLAB[R] is a Licensed Software. It is popularly used in solving Engineering Problems.

function. For example: type ‘help while’ and enter, it provides importance guidance about using ‘while’ as mentioned below:

> 'while', repeat statements an indefinite number of times. The general form of a while statement is:
>
> **while expression**
>
> **statements**
>
> **End**
>
> The statements are executed while the real part of the expression has all non-zero elements. The expression is usually the result of ==, <, >, <=, >=, or ~=. The BREAK statement can be used to terminate the loop prematurely.

Students can use ‘help’ command towards statements as ‘for’, ‘if ‘, ‘switch’, etc. Further, ‘help’ can also be used to understand any inbuilt function as ‘fuzzy’, ‘nntool’, ‘optimtool’, etc. A similar important command ‘lookfor text’ looks for the string text in the first comment line in all m-files found on MATLAB path. Wherever a match occurs, lookfor displays it.

In addition, another important syntax ‘demo’ can be run on this window, which opens:

Document Center: (i) Getting started with MATLAB (ii) Examples (iii) Release Notes (iv) MATLAB Functions (v) Language fundamentals (vi) Mathematics (vii) Graphics (viii) Programming Scripts and Functions (ix) Data and File Management (x) GUI building (xi) Advance Software Development (xii) Desktop Environment.

The ‘demo’ help is a great way to learn MATLAB in proper steps as it covers key part of MATLAB.

Ctrl-C (pressing the Ctrl and c keys simultaneously) can be used to interrupt (abort) processing, but does not terminate MATLAB. To exit MATLAB, either quit or terminate can be used. The command window can be cleared by the instruction ‘clc’.

The MATLAB workspace variables created in the ‘Command Window’ are said to reside in the workspace. The workspace retains the values of these variables and allowing them to be used in subsequent expressions. It can be cleared by the command ‘clear all’.

Important Command and Advance Features in MATLAB

Important commands for beginners are given in Table 1., which can be used to solve many engineering problems. In addition, ‘help’ can be a great help to learn other commands. The MATLAB adds several features in its new version. Therefore, students can see its website **‘https://in.mathworks.com/’** for advance features.

Table 1.1: General Purpose MATLAB Commands for Beginners

Operators and Special Characters			
*	Scalar and matrix multiplication operator.	.*	Array multiplication operator.
+	Plus; addition operator.		Minus; subtraction operator.
^	Scalar and matrix exponentiation operator.	.^	Array exponentiation operator.
\	Left-division operator.	/	Right-division operator.
.\	Array left-division operator.	[ ]	Enclosures array elements.
…	Line-continuation operator.		
,	Separates elements in a row.		
	Separates columns and suppresses display.		
:	Colon; generates regularly spaced elements and represents an entire row or column.		
( )	Parentheses; encloses function arguments and array indices; overrides precedence.		
%	Designates a comment and specifies formatting.		

Special Commands	
help	Searches for a help topic.
clc	Clears Command window.
clear	Removes variables from memory.
lookfor	Searches help entries for a keyword.
who	Lists current variables.
whos	Lists current variables.
quit	Stops MATLAB.
exist	Checks for existence of file or variable.

Special Variables and Constants	
ans	Most recent answer.
eps	Accuracy of floating-point precision.
i or φ	The imaginary unit
Inf	Infinity.
NaN	Undefined numerical result (not a number).

System and File Commands	
disp	Displays contents of an array or string.
fscanf	Read formatted data from a file.
fprintf	Performs formatted writes to screen or file.
input	Displays prompts and waits for input.
load	Loads workspace variables from a file.
save	Saves workspace variables in a file.

Format Codes for fprintf and fscanf	
%s	Format as a string.
%d	Format as an integer.
%f	Format as a floating point value.
%e	Format as a floating point value in scientific notation.
%g	Format in the most compact form: %f or %e.
\n	Insert a new line in the output string.
\t	Insert a tab in the output string.
Vector, Matrix and Array Commands	
cat	Concatenates arrays.
find	Finds indices of nonzero elements.
length	Computers number of elements.
linspace	Creates regularly spaced vector.
logspace	Creates logarithmically spaced vector.
max	Returns largest element.
min	Returns smallest element.
prod	Product of each column.
reshape	Change size
size	Computes array size.
sort	Sorts each column.
sum	Sums each column.
Special Matrices	
eye	Creates an identity matrix.
ones	Creates an array of ones.
zeros	Creates an array of zeros.
Matrix Arithmetic	
cross	Computes cross products.
dot	Computes dot products.
Matrix Commands for Solving Linear Equations	
det	Computes determinant of an array.
inv	Computes inverse of a matrix.
rank	Computes rank of a matrix.
Cell Array Functions	
cell	Creates cell array.
celldisp	Displays cell array.
cellplot	Displays graphical representation of cell array.
num2cell	Converts numeric array to cell array.

deal	Matches input and output lists.
iscell	Identifies cell array.
Plotting Commands	
axis	Sets axis limits.
fplot	Intelligent plotting of functions.
grid	Displays gridlines.
plot	Generates xy plot.
print	Prints plot or saves plot to a file
title	Puts text at top of plot.
xlabel	Adds text label to x-axis.
ylabel	Adds text label to y-axis.
Advance Plot Commands	
axes	Creates axes obφects.
close	Closes the current plot.
close all	Closes all plots.
figure	Opens a new figure window.
gtext	Enables label placement by mouse.
hold	Freezes current plot.
legend	Legend placement by mouse.
refresh	Redraws current figure window.
set	Specifies properties of obφects such as axes.
subplot	Creates plots in sub windows.
text	Places string in figure.
Other Plot Commands	
bar	Creates bar chart.
loglog	Creates log-log plot.
polar	Creates polar plot.
semilogx	Creates semilog plot (logarithmic abscissa).
semilogy	Creates semilog plot (logarithmic ordinate).
stairs	Creates stairs pot.
stem	Creates stem plot.

Frequently used Colors, Symbols and Line Types					
Colour		Line		Symbol	
b	blue	--	dashed	o	circle
c	cyan	-*	dash star	*	star
y	yellow	-	solid	+	Plus

Important Three-Dimensional Plotting Commands	
contour	Creates contour plot.
plot3	Creates three-dimensional plots from lines and points.
surf	Creates shaded three-dimensional mesh surface plot.

Few Histogram Functions	
bar	Creates a bar chart.
hist	Aggregates the data into equally spaced bins.

Some Logical and Relational Operators							
==				Equal to.			
~=				Not equal to.			
<	>			Less than,	Greater than.		
>=	<=			greater than or equal to	Less than or equal to.		
&	|	~	xor	AND	OR	NOT	EXCLUSIVE OR

Flow control command	
break	Terminates execution of a loop.
case	Alternate execution paths within switch structure.
else	Delineates alternate block of statements.
elseif	Conditionally executes statements.
end	Terminates for, while, and if statements.
error	Display error messages.
for	Repeats statements a specific number of times
if	Executes statements conditionally.
otherwise	Default part of switch statement.
return	Return to the invoking function.
switch	Directs program execution by comparing point with case expressions.
while	Repeats statements an indefinite number of times.

Logical Functions	
any	True 'logical 1' if any elements are nonzero.
all	True 'logical 1' if all elements are nonzero.
find	Finds indices of nonzero elements.
finite	True 'logical 1' if elements are finite.
isnan	True 'logical 1' if elements are undefined.
isinf	True 'logical 1' if elements are infinite.
isempty	True 'logical 1'if matrix is empty.
isreal	True 'logical 1' if all elements are real.

M-Files	
eval	Execute string with MATLAB expression.
feval	Execute the specified function.

function	Creates a user-defined subroutine M-file.
global	Define global variables.
nargin	Number of function input arguments.
nargout	Number of function output arguments.
script	MATLAB scripts and M-files.
Timing	
cputime	CPU time in seconds.
clock	Current date and time as date vector.
tic / toc	Start / stop a stopwatch timer
Some Mathematical Functions	
exp(t)	Exponential; et.
log(t)	Natural logarithm; ln(t).
log10(t)	Common (base 10) logarithm; log(x)= log10(t).
sqrt(t)	Square root of 't'
Complex Functions	
abs(t)	Absolute value of 't'
angle(t)	Angle of a complex number 't'
conφ(t)	Complex conφugate of 't'
imag(t)	Imaginary part of a complex number 't'
real(t)	Real part of a complex number 't'
Some Trigonometric Functions	
acos(t)	Inverse cosine of 't'
acsc(t)	Inverse cosecant of 't'
asec(t)	Inverse secant of 't'
asin(t)	Inverse sine of 't'
atan(t)	Inverse tangent of 't'
cos(t)	Cosine of 't'
cot(t)	Cot of 't'
csc(t)	Cosecant of 't'
sec(t)	Secant of 't'
sin(t)	Sine of 't'
acosh(t)	Inverse hyperbolic cosine of 't'
acoth(t)	Inverse hyperbolic cotangent of 't'
acsch(t)	Inverse hyperbolic cosecant of 't'
asech(t)	Inverse hyperbolic secant of 't'
cosh(t)	hyperbolic cosine of 't'
coth(t)	hyperbolic cotangent of 't'
csch(t)	hyperbolic cosecant of 't'
sech(t)	hyperbolic secant of 't'

Statistical Functions	
erf	Computes the error function
mean	Calculates the average.
median	Calculates the median.
std	Calculates the standard deviation.
rand	Generates uniformly distributed random numbers between 0 and 1.
randn	Generates normally distributed random numbers.
ceil	Rounds to the nearest integer toward plus infinity.
fix	Rounds to the nearest integer toward zero.
floor	Rounds to the nearest integer toward negative infinity.
round	Rounds towards the nearest integer.
sign	Signum function.
String Functions	
findstr	Finds occurrences of a string.
strcmp	Compares strings, and returns logical 1 or 0.
char	Creates character string array
Direct Functions	
conv	Computes product of two polynomials
deconv	Computes ratio of polynomials.
eig	Computes the eigenvalues of a matrix.
poly	Computes polynomial from roots.
polyfit	Fits a polynomial to data.
polyval	Evaluates polynomial and generates error estimates.
roots	Computes polynomial roots.
Minimization	
fminsearch	Finds minimum of a function.
Numerical Integration and Differentiation Functions	
quad	Numerically evaluate integral (Adaptive Simpson quadrature)
diff(t)	Computes the difference between adφacent elements in the vector 't'
Differential Equation Solver	
ode23	Nonstiff, low-order solver.
ode45	Nonstiff, medium-order solver.
Predefined Input Functions	
gensig	Generates a periodic sine, square, or pulse input.
sawtooth	Sawtooth and triangle wave generation.
square	Square wave generation.
stepfun	Unit step function

Other Functions	
limit	Limit of an expression.
taylor	Taylor series expansion.
solve	Solves symbolic equations.
ilaplace	Inverse Laplace transform.
laplace	Laplace transform.
det	Determinant of a matrix.
eig	Eigenvalues (characteristic roots) of a matrix.
inv	Inverse of a matrix.
poly	Characteristic polynomial of a matrix.
fminbnd	Returns a value of variable that is the local minimum.

Screenshot of MATLAB website is shown in Figure 1.1. At this website, one can search about product detail, solutions, technical support, community, events and many more.

It is important to remember that operators are sometimes updated in latest version. Hence, old program may require some changes. Therefore, it should be a practice to regularly visit the MATLAB website at least once in a year to know about the updates.

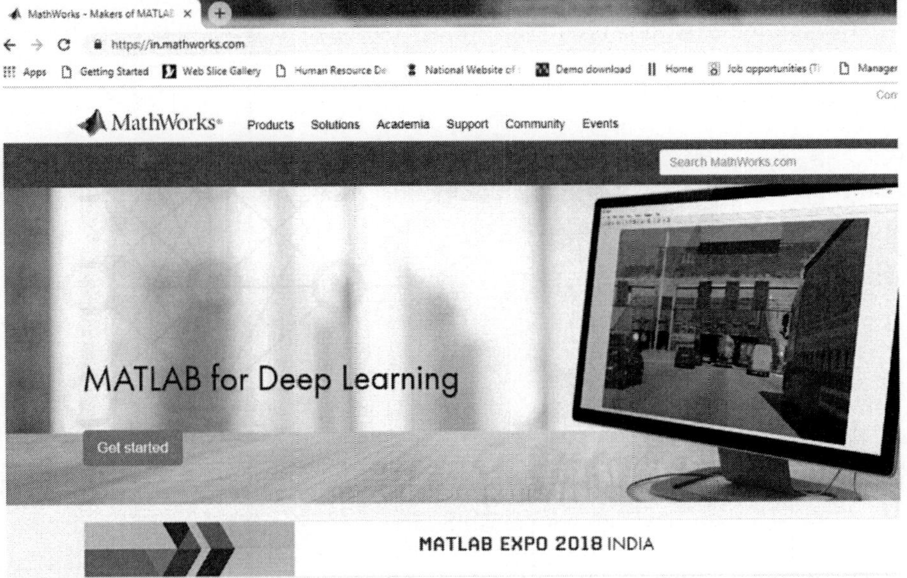

Figure 1.1: MATLAB webpage

Syntax

The MATLAB cannot make sense out of φust any command. Since, commands must be written using the correct syntax (rules for forming commands). The MATLAB clearly tells the user that he/she has made a mistake and tries its best to tell him/her about the error.

There is clear precedence of operations as many operations can be combined in single expression. Hence, there are rules about the order in which operations are performed: (i) Parentheses '()', innermost part first (ii) Exponentiation '^', left to right (iii) Multiplication '*' and division '/ or \' with equal precedence, left to right (iv) Addition '+ and subtraction –' with equal precedence, left to right.

In case, operators in an expression have equal precedence then the operations are carried out from left to right.

Variables, Assignment and Miscellaneous Statements

Variable names can be assigned to represent numerical values in the MATLAB. These variable names are required to follow certain rules as:

(i) They must start with a letter (ii) They can consist only of the letters from 'a' to 'z', digits '0' to '9', and underscore '_' (iii) They can have desired length of letter but the MATLAB only recognizes the first 31 characters (iv) Name is case sensitive (upper case letter and lower-case letter are not considered same) and treated as different name.

The MATLAB has several special variables, which are 'date', 'eps', 'inf', 'nan', 'pi' etc. One can see their values by φust executing them in the command window.

Let see 'date':

```
>> date

    ans =

    10-Dec-2018
```

It is to be noted that 'ans' is the default variable name. Further, the command 'who' lists the names of defined variables and 'whos' lists the names as well as sizes of defined variables. In addition, the command ' clf ' clears the current figure.

In MATLAB, 'semicolon, percent sign and three periods', all three have great importance to punctuate and comment the instruction. As 'semicolon' at the end of a command suppresses the display of the result, a line starts with 'percent sign' comments all text up to the next line ignored by the MATLAB. 'Commas and semicolons' can be used to place multiple commands on one line and 'three periods (. . .)' at the end of a command shows that the command continues on the next line.

In MATLAB, there is several display format viz. number display (Integers, Short fixed point, Short floating point), displaying values & text (disp, fprintf, etc.) which can be used with the help file. Further, an important instruction to input value to the program is 'input', it prompts user with text in quotes, accept input until "Enter Key" is typed.

File Management

There are a group of commands to manage user files. For additional information, type 'help iofun'. The group of commands include 'pwd' (displays the full path of the present working directory), 'cd' (change to directory folder given by path), 'dir' (display the names of the directories and files in the present working directory), 'what' (display the names of the M-files and MAT-files in the current director), 'delete file' (delete file from current directory) and 'type file' (display contents of file), 'diary' (saving the session output without >>), 'save' (stores workspace values viz. variable names, sizes, and values in the binary file matlab.mat) and 'load' (read the information into an array).

Script M-Files

A group of MATLAB commands placed in a text file with a text editor. The MATLAB can open and execute the commands exactly as they are entered at the MATLAB command window. The term "script" indicates that Matlab reads from the "script" found in the file. It is also called "m-files," as the filenames must end with the extension '.m'.

Minimization Function

It is important to determine the minimum value of the function. To find the minimum of a function of a single variable, 'fminbnd' is a simple command.

- v = fminbnd('F',v_1,v_2) returns a value of v that is the local minimizer of F(v) in the interval [v_1,v_2], where F is a string containing the name of the function.
- [v,fval] = fminbnd(...) Also returns the value of the obφective function, fval, computed in F at v.

Let us take an example: fminbnd('sec',0,5) gives answer as 1.5708. In addition, another example can be taken as u= $0.025v^5-0.0625v^4-0.333v^3+v^2$

Defining the function file test.m:

function u = test(v)

% test, higher degree polynomial function

u= $5v^5-5v^4-3v^3+v^2$ -v

Now, using vmin = fminbnd('test',-0.1,3) will give the minimum value of the function 'u' in the selected range -10 and 4. The minimum value is 1.0930.

2

MATLAB Application for Transmission Line Modeling

Short Transmission Line

A **short transmission line** is defined as a transmission line with an effective length less than 80 km (50 miles), or with a voltage less than 69 kV. Unlike medium transmission lines and long transmission lines, the line charging current is negligible, and hence the shunt capacitance can be ignored.

For short length, the shunt capacitance of this type of line is neglected and other parameters like electrical resistance and inductor of these short lines are lumped, hence the equivalent circuit is represented as given below. Let's draw the vector diagram for this equivalent circuit, taking receiving end current I_r as reference. The sending end and receiving end voltages make angle with that reference receiving end current, of φ_s and φ_r, respectively.

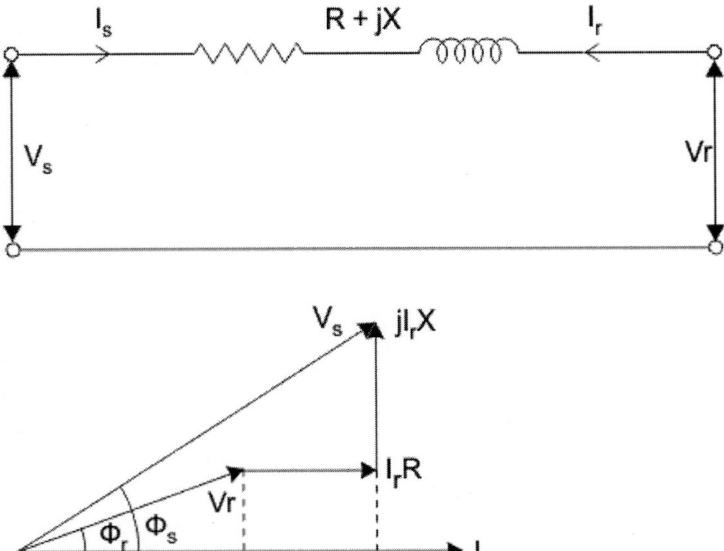

Figure 2.1: Short Transmission Line system with Phasor Diagram

As the shunt capacitance of the line is neglected, hence the sending end current and the receiving end current is same, i.e.

$Is = Ir$

We can see from the phasor diagram of short transmission line that V_s is approximately equal to:

$Vr + Ir* R* cos\ \varphi r + Ir*\ X* \sin \varphi r$

That means,

$Vs \cong Vr + Ir* R* cos\ \varphi r + Ir* X*$ Sin φr

As it is assumed that:

$\varphi s = \varphi r$

As there is no capacitance, during no-load condition the current through the line is considered as zero, hence at no load condition, receiving end voltage is the same as sending end voltage.

As per dentition of voltage regulation of power transmission line,

$$\% \text{ Regulation } = \frac{V_s - V_r}{V_r} * 100$$

$$\frac{I_r * R * \cos\varphi_r + I_r * X * \sin\varphi_r}{V_r} * 100$$

Any electrical network generally has two input terminals and two output terminals. If we consider any complex electrical network in a black box, it will have two input terminals and output terminals. This network is called a two-port network. A two-port model of a network simplifies the network solving technique. Mathematically, a two-port network can be solved by 2 by 2 matrixes.

A transmission as it is also an electrical network, and hence the transmission line can be represented as a two-port network.

Hence, two-port network of the transmission line can be represented as 2 by 2 matrix. Here the concept of ABCD parameters comes into play. Voltage and currents of the network can be represented as:

$$Vs = A* Vr + B* Ir \quad (1)$$

$$Is = C* Vr + D* Ir \quad (2)$$

Where, A, B, C and D are the different constants of the transmission network.

Now from the equivalent circuit, it is found that,

$Vs = Vr + Z^* Ir$

$Is = Ir$

Comparing these equations with equation 1 and 2 we get, A = 1, B = Z, C = 0 and D = 1.

Performance of Short Transmission Line

The performance (i.e. the efficiency) of a short transmission line as simple as efficiency equation of any other electrical equipment, that means

$$\%efficiency(\mu) = \frac{Power\ received\ at\ receiving\ end}{Power\ delievered\ from\ sending\ end} * 100$$

$$\%efficiency(\mu) = \frac{Power\ received\ at\ receiving\ end}{Power\ received\ at\ receiving\ end + 3 * I_r \wedge 2 * R} * 100$$

Where R is the per phase electrical resistance of the transmission line.

Medium Transmission Line

A **medium transmission line** is defined as a transmission line with an effective length more than 80 km (50 miles) but less than 250 km (150 miles). Unlike a short transmission line, the line charging current of a medium transmission line is appreciable and hence the shunt capacitance must be considered (this is also the case for long transmission lines). This shunt capacitance is captured within the admittance ("Y") of the ABCD circuit parameters.

The ABCD parameters of a medium length transmission line is calculated using a lumped shunt admittance, along with the lumped impedance in series to the circuit. These lumped parameters of a medium length transmission line can be represented using three different models, namely:

1. Nominal **Π** representation (nominal pi model)
2. Nominal **T** representation (nominal T model)

Let's now go into the detailed discussion of these above-mentioned models, deriving the ABCD parameters for medium transmission lines.

Nominal Π Representation of a Medium Transmission Line

In case of a nominal **Π** representation (i.e. nominal pi model), the lumped series impedance is placed at the middle of the circuit whereas the shunt admittances are at the ends. As we can see from the diagram of the Π network below, the total lumped shunt admittance is divided into 2 equal halves, and each half with value Y/2 is placed at both the sending and the receiving end while the entire circuit impedance is between the two.

The shape of the circuit so formed resembles that of a symbol **Π**, and for this reason, it is known as the nominal Π representation of a **medium transmission line**. It is mainly used for determining the general circuit parameters and performing load flow analysis.

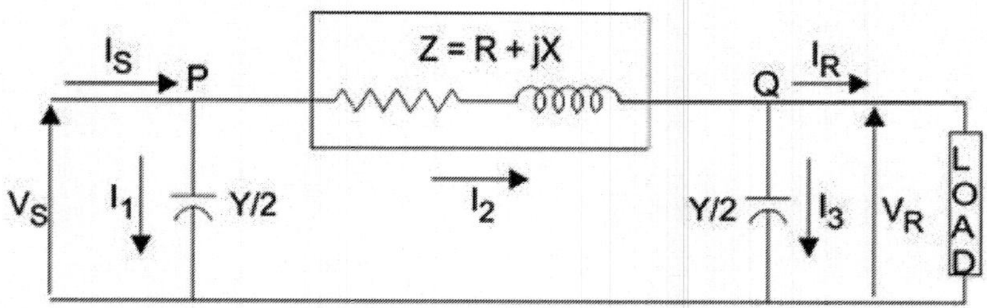

Figure 2.2: Nominal π model of Medium Transmission Line

As we can see here, V_S and V_R is the supply and receiving end voltages respectively, and I_S is the current flowing through the supply end. I_R is the current flowing through the receiving end of the circuit. I_1 and I_3 are the values of currents flowing through the admittances. And I_2 is the current through the impedance Z.

Now applying KCL, at node P, we get.

$$Is = I_1 + I_2 \tag{1}$$

Similarly applying KCL, to node Q.

$$I_2 = I_3 + I_R \tag{2}$$

Now substituting equation (2) to equation (1)

$$Is = I_1 + I_3 + I_R \tag{3}$$

$$Is = \frac{Y}{2}Vs + \frac{Y}{2}VR + IR$$

Now by applying KVL to the circuit,

$Vs = V_R + Z^* I_2$

$$Vs = V_R + Z * (\frac{V_R * Y}{2} + I_R)$$

$$Vs = (1 + \frac{ZY}{2})V_R + Z * I_R \tag{4}$$

Now substitute the value of Vs in equation (3)

$$I_s = Y(1 + \frac{ZY}{4}) * V_R + (1 + \frac{ZY}{2}) * I_R \tag{5}$$

Comparing equation (4) and (5) with the standard ABCD parameter equations

$$Vs = A^* Vr + B^* Ir \tag{6}$$

$$Is = C^* Vr + D^* Ir \tag{7}$$

We derive the ABCD parameters of a medium transmission line as:

$$A = (\frac{ZY}{2} + 1)\Omega \qquad B = Z\Omega$$

$$C = Y * (\frac{ZY}{4} + 1)\Omega \quad D = (\frac{ZY}{2} + 1)\Omega$$

Nominal T Representation of a Medium Transmission Line

In the **nominal T** model of a medium transmission line the lumped shunt admittance is placed in the middle, while the net series impedance is divided into two equal halves and placed on either side of the shunt admittance. The circuit so formed resembles the symbol of a capital **T**, and hence is known as the nominal T network of a medium length transmission line and is shown in the diagram below.

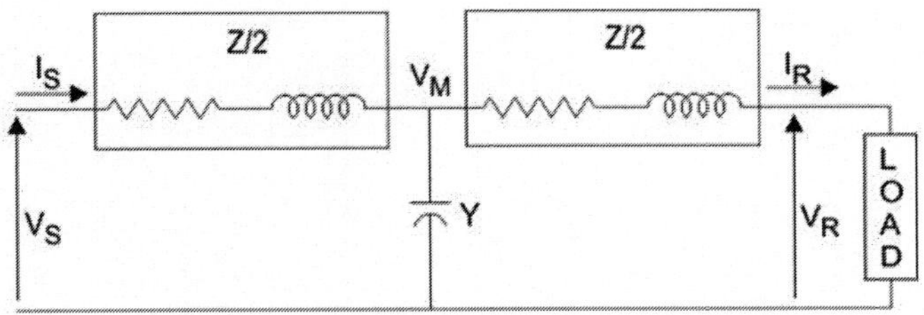

Figure 2.3: Nominal T model of Medium Transmission Line

Here also V_s and V_r is the supply and receiving end voltages respectively, and I_s is the current flowing through the supply end. I_r is the current flowing through the receiving end of the circuit. Let M be a node at the midpoint of the circuit, and the drop at M, be given by V_m.

The sending end voltage is,

$$Vs \quad (1+\frac{ZY}{2})*V_R + Z*(1+\frac{ZY}{4})*I_R \tag{8}$$

The sending end current is,

$$Is = Y*V_R + (1+\frac{ZY}{2})*I_R \tag{9}$$

Again, comparing equation (8) and (9) with the standard ABCD parameter equations,

$Vs = A* Vr + B* Ir$

Is = C* Vr + D* Ir

The parameters of the **T** network of a **medium transmission line** are

$$A = (\frac{ZY}{2}+1)\Omega \quad B = Z*(\frac{ZY}{4}+1)\Omega$$

$$C = Y\ mho \qquad D = (\frac{ZY}{2}+1)\Omega$$

Long Transmission Line

A **long transmission line** is defined as a transmission line with an effective length more than 250 km (150 miles). Unlike short transmission lines and medium transmission lines, it is no longer reasonable to assume that the line parameters are lumped. To accurately model a long transmission line, we must consider the exact effect of the distributed parameters over the entire length of the line. Although this makes the calculation of ABCD parameters of transmission line more complex, it also allows us to derive expressions for the voltage and current at any point along the line.

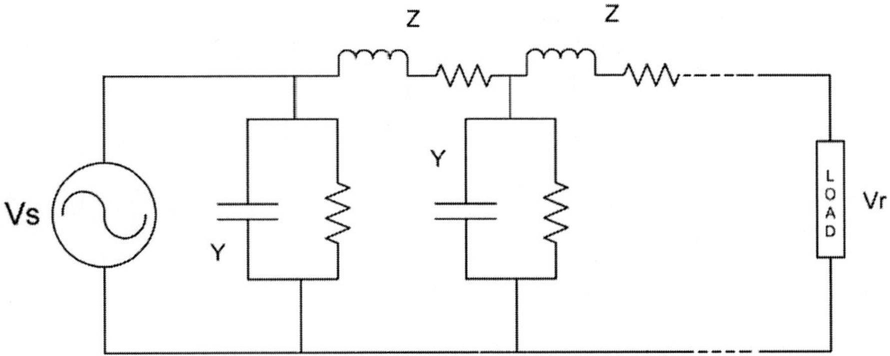

Figure 2.4: Model of Long Transmission Line

In a long transmission line, the line constants are uniformly distributed over the entire length of line. For all practical reasons, we should consider the circuit impedance and admittance being distributed over the entire circuit length as shown in the figure below. The calculations of circuit parameters, for this reason, are going to be slightly more rigorous as we will see here. For accurate modeling to determine circuit parameters let us consider the circuit of the long transmission line as shown in the diagram below.

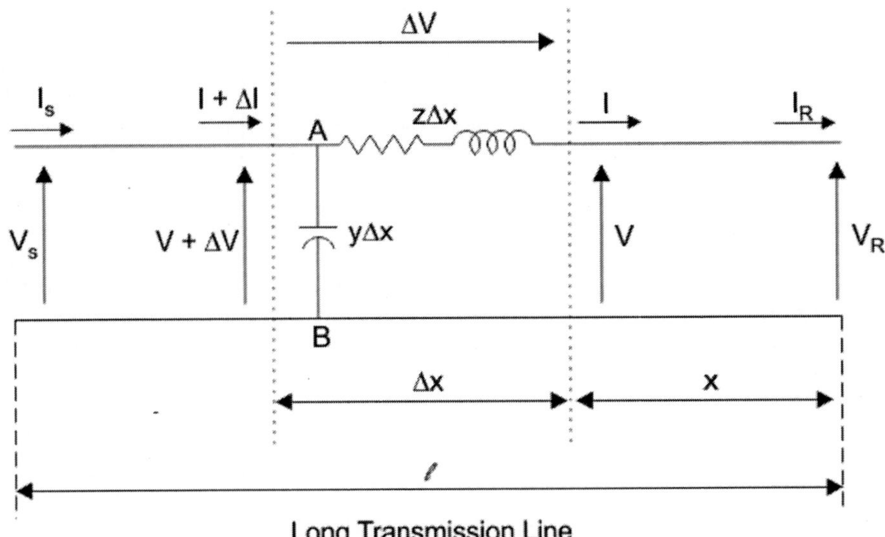

Figure 2.5: Approximate Model of long transmission line

Here a line of length l > 250km is supplied with a sending end voltage and current of V_S and I_S respectively, whereas the V_R and I_R are the values of voltage and current obtained from the receiving end. Let's us now consider an element of infinitely small length x at a distance x from the receiving end as shown in the figure where.

V = voltage φust before entering the element x.

I = current φust before entering the element x.

V+ΔV = voltage leaving the element x.

I+ΔI = current leaving the element x.

ΔV = voltage drop across element x.

zx = series impedance of element x

yx = shunt admittance of element x

Where, Z = z l and Y = y l are the values of total impedance and admittance of the long transmission line.

ABCD parameters of a long transmission line are given as,

$A = cosh\ \delta l \qquad B = Zc\ Sinh\ \delta l$

$C = \dfrac{Sinh \delta l}{Zc} \qquad D = cosh\ \delta$

Where,

$$Zc = \sqrt{\frac{z}{y}}\ and\ \delta = \sqrt{yz}$$

MATLAB Programme

Example 1: A 3-phase overhead transmission line deliver a 1100kw at 11kv at 0.8pf lagging the total resistance and inductive reactance of the line are 8 ohm and 16 ohm respectively. Determine

1) Line current
2) Sending end voltage
3) Sending end power factor
4) Sending end real power
5) Sending end reactive power
6) Transmission efficiency

7) Percentage regulation
8) ABCD parameters.

Program

```
clear;
clc;
R=input('resistance of the line in ohm=');
X=input('reactance of the line in ohm=');
Vr3ph=input('voltage at receiving end in KV=');
Pr=input('real load at receiving end in MW=');
Qr=input('reactive load at receiving end in MVAR=');
Z=R+φ*X;
ABCD=[1 Z;0 1];
Vr=Vr3ph/sqrt(3)+φ*0;
Sr=Pr+φ*Qr;
Ir=conφ(Sr)/(3*conφ(Vr));
VSIS=ABCD*[Vr ; Ir];
Vs=VSIS(1);
Vs3ph=sqrt(3)*abs(Vs);
Is=VSIS(2);
Ism=1000*abs(Is);
pfs=cos(angle(Vs)-angle(Is));
Ss=3*Vs*conφ(Is);
Reg=((Vs3ph-Vr3ph)/(Vr3ph))*100;
Eff=PR/real(Ss)*100;
fprintf('\n IS=%g A',Ism);                fprintf('\n pf=%g',pfs);
fprintf('\n Vs=%g L-L KV',Vs3ph);         fprintf('\n Ps=%g MW',real(Ss));
fprintf('\n Qs=%g MVAR',imag(Ss));        fprintf('\n percentage voltage
                                          regulation = %g',Reg);
```

```
fprintf('\n percentage transmission line efficiency=%g',Eff); fprintf('\n ABCD parameters of transmission line\n'); disp(ABCD);
```

Inputs

resistance of the line in ohm=8

reactance of the line in ohm=16

voltage at receiving end in KV=11

real load at receiving end in MW=1.1

reactive load at receiving end in MVAR=0.825

Result

Is=72.1688 A

pf=0.751626

Vs=13.0384 L-L KV

Ps=1.225 MW

Qs=1.075 MVAR

percentage voltage regulation=18.531

percentage transmission line Efficiency=89.7959

ABCD parameters of transmission line

0.0 8.0000 +16.0000i

0 1.0000

Example 2: A 3-phase 50Hz overhead transmission line delivers 10 MW at 0.8pf lagging at 66 kV. The resistance, inductive reactance and capacitive susceptance 10-ohm, 20 ohm and 4*10^-4 Siemen. Determine

1) Sending end current
2) Sending end voltage
3) Sending end power factor
4) ABCD parameter
5) Regulation
6) Transmission efficiency using nominal T method.

Program

```
clear;
clc;
R=input('resistance of the line in ohm=');
X=input('reactance of the line in ohm=');
B=input('susceptance of shunt line in mho=');
VR3ph=input('voltage at receiving end in KV=');
PR=input('real load at receiving end in MW=');
QR=input('reactive load at receiving end in MVAR=');
Z=R+φ*X;
Y=0+φ*B;
type=input('type(P-Pie/T-t method)=','s');
switch type
case'P'
ABCD=[1+Z*Y/2 Z ; Y*(1+Z*Y/4) 1+Z*Y/2];
case'T'
ABCD=[1+Z*Y/2 Z*(1+Z*Y/4);Y 1+Z*Y/2]; otherwise
Error('Invaid type choosen!!!')
end

VR=VR3ph/sqrt(3)+φ*0;
SR=PR+φ*QR;
IR=conφ(SR)/(3*conφ(VR));
VSIS=ABCD*[VR;IR];
VS=VSIS(1);
VS3ph=sqrt(3)*abs(VS);
IS=VSIS(2);
ISm=1000*abs(IS);
```

```
Pfs=cos(angle(VS)-angle(IS));
SS=3*VS*conφ(IS);
Reg=(VS3ph-VR3ph)/VR3ph*100;
Eff=PR/real(SS)*100;
fprintf('\n IS=%g A',ISm);              fprintf('\n Pfs=%g',Pfs);
fprintf('\n VS=%g L-L KV',VS3ph);      fprintf('\n PS=%g MW',real(SS));
fprintf('\n QS=%g MVAR',imag(SS)); fprintf('\n percentage voltage
                                        regulation=%g',Reg);
fprintf('\n percentage transmission line efficiency=%g',Eff);
fprintf('\n ABCD parameters of transmission line\n');
disp(ABCD);
```

Inputs

resistance of the line in ohm=10

reactance of the line in ohm=20

voltage at receiving end in KV=66

real load at receiving end in MW=10

reactive load at receiving end in MVAR=7.5

susceptance of shunt line in mho=4e-4

type(P-Pie/T-t method)=T

Result

IS=100.533 A

Pfs=0.853122

VS=69.5439 L-L KV

PS=10.331 MW

QS=6.31771 MVAR

percentage voltage regulation=5.36958

percentage transmission line efficiency=96.7965 ABCD parameters of transmission line

0.9960 + 0.0020i 9.9600 +19.9700i

0 + 0.0004i 0.9960 + 0.0020i

Example 3: A 3-phase 50Hz overhead transmission line of 100 km has the following constants.

Resistance/km/phase=0.1 ohm

Inductive reactance/km/phase=0.5 ohm

Capacitive susceptance/km/phase=10*10^-5 Siemen

The line supplies the load of 20MW at 0.9pf lagging at 66kV at receiving end.

Determine 1) Sending end current 2) Sending end voltage 3) Sending end power factor 4) ABCD parameter 5) Regulation 6) Transmission efficiency using nominal pi method.

Input

resistance of the line in ohm=10

reactance of the line in ohm=50

susceptance of shunt line in mho=1e-3

voltage at receiving end in KV=66

real load at receiving end in MW=20

reactive load at receiving end in MVAR=9.67

type(P-Pie/T-t method)=P

Output

IS=176.475 A

Pfs=0.905904

VS=76.0096 L-L KV

PS=21.0471 MW

QS=9.83893 MVAR

Percentage voltage regulation=15.166 percentage transmission line efficiency=95.0248

ABCD parameters of transmission line

0.9750 + 0.0050i   10.0000 +50.0000i

-0.0000 + 0.0010i   0.9750 + 0.0050i

Example 4: Determine the efficiency and regulation of 3-phase, 50Hz, 120km long transmission line delivering 40MW at 132kV at 0.8 lagging pf with following details.

Resistance/km/phase=0.2 ohm

Inductive reactance/km/phase=1.3mH

Capacitive susceptance/km/phase=0.01 μf

Program

```
clear;
clc;
R=input('resistance of the line in ohm=');
X=input('reactance of the line in ohm=');
B=input('susceptance of shunt line in mho=');
VR3ph=input('voltage at receiving end in KV=');
PR=input('real load at receiving end in MW=');
QR=input('reactive load at receiving end in MVAR=');
Z=R+φ*X;
Y=0+φ*B;
gamma=sqrt(Z*Y);
Zc=sqrt(Z/Y);
A=cosh(gamma);
B=Zc*sinh(gamma);
C=1/Zc*sinh(gamma);
D=A;
ABCD=[A B;C D];
VR=VR3ph/sqrt(3)+φ*0;
SR=PR+φ*QR;
```

```
IR=conφ(SR)/(3*conφ(VR));
VSIS=ABCD*[VR;IR];
VS=VSIS(1);
VS3ph=sqrt(3)*abs(VS);
IS=VSIS(2);
ISm=1000*abs(IS);
Pfs=cos(angle(VS)-angle(IS));
SS=3*VS*conφ(IS);
Reg=(VS3ph-VR3ph)/VR3ph*100;
Eff=PR/real(SS)*100;
Val=(A*D)-(B*C);
fprintf('\n IS=%g A',ISm);
fprintf('\n Pfs=%g',Pfs);
fprintf('\n VS=%g L-L KV',VS3ph);
fprintf('\n PS=%g MW',real(SS));
fprintf('\n Qs=%g MVAR',imag(SS));
fprintf('\n percentage voltage regulation=%g',Reg); fprintf('\n Efficiency=%g',Eff);
fprintf('\n A*D-B*C=%g',Val);
fprintf('\n ABCd parameters of transmission line\n'); disp(ABCD);
```

Inputs

resistance of the line in ohm=24

reactance of the line in ohm=49.0088

susceptance of shunt line in mho=3.76e-4

voltage at receiving end in KV=132

real load at receiving end in MW=40

reactive load at receiving end in MVAR=30

Output

IS=200.922 A

Pfs=0.830052

VS=149.46 L-L KV

PS=43.1736 MW

Qs=29.007 MVAR

percentage voltage regulation=13.2272

Efficiency=92.6493

A*D-B*C=1

ABCD parameters of transmission line

```
0.9908 + 0.0045i     23.8528 +48.8944i
-0.0000 + 0.0004i    0.9908 + 0.0045i
```

3

Per Unit System

Introduction

For the analysis of electrical power system, different values are required, thus, per unit system provides the value for voltage, current, power, impendence, and admittance. The Per Unit System also makes the calculation easier as all the values are taken in the same unit. The per-unit system is mainly used in the circuit where variation in voltage occurs. The per-unit value of any quantity is defined as the ratio of actual value in any unit to the base or reference value in the same unit. Any quantity is converted into per unit quantity by dividing the numeral value by the chosen base value of the same dimension. The per-unit value is dimensionless.

$$\text{Per Unit Value} = \frac{\text{Actual value in any unit}}{\text{Base or reference value in the same unit}}$$

The base values can be selected arbitrarily. It is usual to assume the base values as given below

- Base voltage = rated voltage of the machine
- Base current = rated current of the machine
- Base impedance = base voltage /base current
- Base power = base voltage x base current

Firstly, the value of base power and the base voltage are selected, and their choice automatically fixes the other base values. As

$$\text{Per Unit kV} = \frac{\text{Actual kV}}{\text{Base kV}} = \frac{kV_{actual}}{kV_B}$$

So,

$$\text{Base current } (I_B) = \frac{\text{Base kVA}}{\text{Base kV}} = \frac{kVA_B}{kV_B} \quad (1)$$

$$\text{Per Unit current } (I_{pu}) = \frac{\text{Actual value of current}}{\text{Base current}}$$

$$\text{Per Unit current } (I_{pu}) = \frac{\text{Actual value of current}}{\frac{kVA_B}{kV_B}} \tag{2}$$

$$\text{Per Unit current } (I_{pu}) = \frac{\text{Actual value of current} \times kV_B}{kVA_B}$$

$$\text{Base impedance } (Z_B) = \frac{\text{Base kV} \times 1000}{\text{Base current}} \tag{3}$$

Putting the value of base current from the equation (1) in the equation (3) we get

$$\text{Base impedance } (Z_B) = \frac{kV_B \times 1000}{\frac{kVA_B}{kV_B}}$$

$$\text{Base impedance } (Z_B) = \frac{(kV_B)^2 \times 1000}{kVA_B}$$

$$\text{Base power} = kVA_B \tag{4}$$

Now,

$$Z_{pu} = \frac{\text{Actual value of impedance}}{\text{Base impedance}} \tag{5}$$

Putting the value of base impedance from the equation (4) in the equation (5) we will get the value of impedance per unit

$$Z_{pu} = \frac{\text{Actual value of impedance} \times kVA_B}{(kV_B)^2 \times 1000}$$

Advantages of Per Unit System

There are mainly two advantages of using the Per Unit System.

1. The parameters of the rotating electrical machines and the transformer lie roughly in the same range of numerical values, irrespective of their ratings if expressed in a per-unit system of ratings.

2. It relieves the analyst of the need to refer circuit quantities to one or the other side of the transformer, making the calculations easy.

Taking the example of a transformer having the resistance in the per unit as R_{pu} ohm and the reactance as X_{pu} in ohm with referred to primary then per unit values will be

$$R_{ep_{pu}} = \frac{R_{ep}}{\text{Base impedance}} = \frac{R_{ep}}{\frac{V_1}{I_1}}$$

Similarly

$$X_{ep_{pu}} = \frac{I_1 \times X_{ep}}{V_1}$$

Where R_{ep} and X_{ep} are resistance and reactance referred to primary and pu means in the per-unit system. The resistance and leakage reactance referred to primary in per unit are the same as referred to as secondary in the per-unit system because

$$R_{ep_{pu}} = \frac{R_{ep} \times I_1}{V_1} = \frac{\frac{R_{ep}}{K^2} \times KI_1}{\frac{V_1}{K}} = \frac{R_{es} \times I_2}{V_2} = R_{es}(pu)$$

$$X_{ep_{pu}} = \frac{X_{ep} \times I_1}{V_1} = \frac{\frac{X_{ep}}{K^2} \times KI_1}{\frac{V_1}{K}} = \frac{X_{es} \times I_2}{V_2} = X_{es}(pu)$$

Where R_{es} and X_{es} are the equivalent resistance and reactance referred to secondary

Therefore, from the above two equations, it is clear that the need for an ideal transformer is eliminated. This is so because the per-unit impedance is the same for the equivalent circuit of the transformer whether computed from the primary or secondary as long as the voltage bases on the two sides are selected in the ratio of transformation.

MATLAB Programs

Example 1: A 230 kV transmission line has a series impedance of (4+φ60) Ω and a shunt admittance of φ2x10^{-3} S. using 100 MVA and the line voltage as base values, calculate per unit impedance and per unit admittance of the line.

Program

```
clear;
clc;
z=complex(4,60);
mva_base=100;
kv_base=230;
zpu=z*mva_base/(kv_base^2);
printf('the per unit impedance is:');
disp(zpu)
ys=(%i)*(2*10^(-3));
ypu=ys*(kv_base^2)/mva_base;
printf("the per unit admittance is:");
disp(ypu);
```

Result

the per unit impedance is:

0.0075614 + 0.1134216i

the per unit admittance is:

1.058i

Example 2: A three phase, star-connected is rated at 50 MVA and 110 kV. Express 40,000 kVA of three-phase apparent power as a per-unit value referred to (a) the three-phase system kVA as base, and (b) the per phase kVA as base.

Program

```
clear;
clc;
sb=50000;
s=40000;
pu=s/sb;
printf("\n The PU value for the three phase is: %.2f\n",pu);
```

```
sb1=sb/(3);
v=110;
vb=v/sqrt(3);
pu_kva=s/(3*sb1);
printf("\n The PU kVA is: %.2f\n",pu_kva);
```

Result

The PU value for the three phase is: 0.80

The PU kVA is: 0.80

Example 3: A 5 kVA 400/200 V, 50 Hz single-phase transformer has primary and secondary leakage reactance each of 2.5 Ω. Determine the total reactance in per unit.

Program

```
clear;
clc;
sb=5000
x1=2.5;
x2=2.5;
n1=400;
n2=200;
x1e=x1+x2*(n1/n2)^2;
vb1=400;
x1epu=x1e*sb/vb1^2;
printf("\n Total per unit resistance referred to primary is: %.2f",x1epu);
x2e=x1+x2*(n2/n1)^2;
vb2=200;
x2epu=x2e*sb/vb2^2;
printf("\n Total per unit resistance refered to secondary is: %.2f",x2epu);
//c)if the primary voltage is used as base
```

```
xpu1=x1*sb/vb1^2;
printf(" \n Primary per unit reactance is: %.2f ",xpu1);
xpu2=x1*sb/vb2^2;
printf("\n Seconadary per unit reactnace is: %.2f Ohm",xpu2);
```

Result

Sb =

5000.

Total per unit resistance referred to primary is: 0.39

Total per unit resistance referred to secondary is: 0.39

Primary per unit reactance is: 0.08

Secondary per unit reactance is: 0.31 Ohm

{\displaystyle Z_{\text{base}}={\frac {V_{\text{base}}}{I_{\text{base}}\times {\sqrt {3}}}}={\frac {V_{\text{base}}^{2}}{S_{\text{base}}}}=1{\text{ pu}}}

Example 4: A portion of a power system consists of two generators in parallel, connected to a step-up transformer that links them with a 230 kV transmission line. The ratings of these components are:

Generator G_1: 100 MVA, 12 percent reactance

Generator G_2: 5 MVA, 8 percent reactance

Transformer: 15 MVA, 6 percent reactance

Transmission Line: (4+φ60) Ω, 230 kV

Where the percent reactance is computed on the basis of individual component ratings. Express the reactance and the impedance with 15 MVA as the base value.

Program

```
clear;
clc;
x1_pu=.12;
sb2=15;
```

```
sb1=10;
xg2=8;
xg1=.12;
sbg2=5;
sbg1=100;
xtf=6;
sbtf=15;
xt=complex(4,60);
kvl=230;
x2_pu=x1_pu*(sb2/sb1);
xg_pu1=x2_pu*sbg1;
printf("\n The per unit reactance of generator 1 is: %.2f",xg_pu1);
xg_pu2=xg2*(sb2/sbg2);
printf("\n The per unit reactance of generator 2 is: %.2f",xg_pu2);
xtf_pu=xtf*(sb2/sbtf);
printf("\n The per unit reactance of transformer is: %.2f",xtf_pu);
xt_pu=xt*sb2/((kvl)^2);
printf("\n The per unit impedance of transmission line is:");
disp(xt_pu)
```

Result

The per unit reactance of generator 1 is: 18.00

The per unit reactance of generator 2 is: 24.00

The per unit reactance of transformer is: 6.00

The per unit impedance of transmission line is: 0.0011342 + 0.0170132i

{\displaystyle Y_{\text{base}}={\frac {1}{Z_{\text{base}}}}=1{\text{ pu}}}
Example 5: A three-phase synchronous generator delivers 10 MVA at a voltage of 10.5 kV. The line impedance is 5 Ω. Determine the voltage drop in the line in per unit and in volts. Use the reference base a s12 MVA at 11 kV.

Program

```
clear;
clc;
s3q=10*10^6;
vl=10.5*1000;
il=s3q/(sqrt(3)*vl);
iph=il;
zph=5;
printf("\n Voltage drop in the line is: %.2f V",zph*iph);
//b).using per unit method to soolve
sb=10;
vlb=11;
vpu=vl/vlb;
sb3q=12;
spu=s3q/sb3q;
ipu=spu/vpu;
zpu=zph*sb3q/vlb^2;
v=zpu*ipu*vlb/sqrt(3);
printf("\n voltage drop in the line per phase is: %.2f V",v);
```

Result

Voltage drop in the line is: 2749.29 V

voltage drop in the line per phase is: 2749.29 V

Example 6: In the figure below shows a two-machine system. The rating are as follows:

Synchronous generator 20 MVA, 11 kV, X"=0.15 pu

Synchronous motor 15 MVA, 11 kV, X"=0.15 pu

Transformer T_1 25 MVA, 12.5/132 (Delta-Star) kV

Transformer T_2 20 MVA, 132/11 (Star- Delta) kV

Line (200+φ500) Ω

Static load 5 MVA 0.8 power factor lagging.

Draw the impedance diagram for the system. Choose a base voltage of 132 kV for the transmission line and a base volt-ampere of 20 MVA.

If the motor is a synchronous machine drawing 15 MVA at 0.9 power factor (leading) and the terminal voltage 1.1 pu find the generator bus voltage.

Program

```
clear;

clc;

function [zpu]=puz(z1pu,sb2,sb1,vb1,vb2)

zpu=z1pu*(sb2/sb1)*(vb1/vb2)^2;

endfunction

previousprot = funcprot(0)

funcprot(0)

z1pu=0.15;

sb2=20;

sb1=20;

vb1=11;

vb2=12.5;

zpu=puz(z1pu,sb2,sb1,vb1,vb2);

printf("\n the per unit reactance of the generator is: %.2f",zpu);

xg=zpu;

zpu=puz(0.15,20,15,11,11);

printf("\n the per unit reactance of the generator is: %.2f",zpu);

xm=zpu;

zpu=puz(0.10,20,25,132,132);

printf("\n the per unit reactance of the generator is: %.2f",zpu);

xt1=zpu;
```

```
zpu=puz(0.10,20,20,132,132);
printf("\n the per unit reactance of the generator is: %.2f",zpu);
xt2=zpu;
xtl=((200+(%i)*500))*20/(132)^2;
pl=5;
pf=0.8;
p=pl*pf;
q=pl*0.6;
vpu=1;
sb=20;
rpu=vpu^2*sb*p/(p*p+q*q);
xpu=vpu^2*sb*q/(p*p+q*q);
//when pf is changed to 0.9
vpu=1.1;
im=15*(0.9+(%i)*sqrt(1-0.9^2))/(sb*vpu);
il=(p-(%i)*q)/(sb*vpu);
i=im+il;
zt=(%i)*(xt1+xt2)+xtl;
vg=vpu+zt*i;
function [mag,theta]=c(r,i)
mag=sqrt(r*r + i*i)
theta=atand(i/r)
endfunction
previousprot = funcprot(0)
funcprot(0)
[mag,theta]=c(real(vg),imag(vg));
printf("\nthe terminal generator voltage is: %.2f kV",mag*11);
```

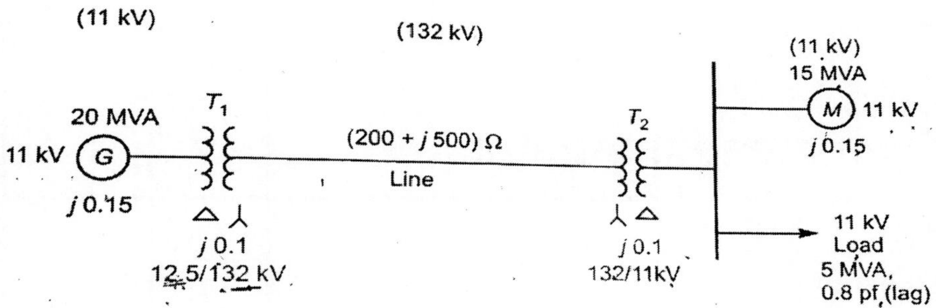

Result

previousprot =

1.

ans =

0.

the per unit reactance of the generator is: 0.12

the per unit reactance of the generator is: 0.20

the per unit reactance of the generator is: 0.08

the per unit reactance of the generator is: 0.10

previousprot =

0.

ans =

0.

the terminal generator voltage is: 14.57 kV

4

Over-Head Insulator

Introduction

The Over-head insulator for overhead lines provides insulation to the electrical conductor. It is connected to the cross arm of the supporting structure and the electrical conductor passes through the clamp (insulator). It is mainly made of either glazed porcelain (Silica 20% + Feldspar 30% + Clay 50%) or toughened glass. The porcelain must be free from any defect. It should be vitrified because the presence of pore or air will lower down its dielectric strength. Any captured air impurity will also reduce the dielectric strength of the porcelain. Therefore, it is highly desirable that the insulators should be air-free and resistant to the entrance of liquids/gases. It is desirable not to allow porcelain to come in direct contact with a hard metal screw thread. Therefore, cement is used between metal and the porcelain.

Toughened glass is also sometimes used for insulators as it has higher dielectric strength. It is very easy to detect any flaw like trapping of air as it is transparent. It has lower coefficient of thermal expansion and strain is less. The maφor drawback of glass is that moisture condenses on its surface and hence its use is limited. The design of the insulators is such that the stress due to contraction and expansion in any part of the insulator does not lead to any defect.

String efficiency in suspension insulator

String efficiency is a measure of the utilization of material in the string. It is defined as:

$$\eta = \frac{\text{Voltage across the string}}{\text{n} \times \text{Voltage across the unit near the power conductor}}$$

Method of equalizing the potential

(1) Selection of 'm', where $m = \dfrac{\text{Mutual Capacitance}}{\text{Capacitance to ground}}$

(2) Grading of insulator units

(3) Static Shielding

Determine the maximum voltage that the string of the suspension insulators in Figure 4.1 can withstand if the maximum voltage per unit is 18 kV. The ratio of *C* and *C*' is 8.

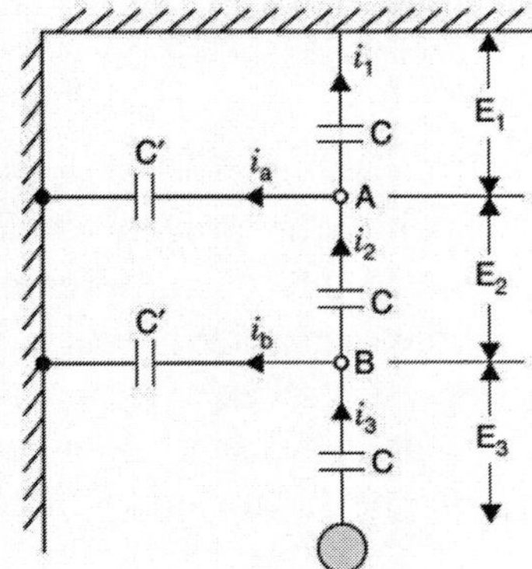

Figure 4.1: String efficiency for Insulators

MATLAB Programs

Example 1: To compute the maximum voltage that the string of the suspension insulator can withstand.

Code:

```
clear all;

clc ;

E3 =18;

E1 =64*E3/89;

E2 =9* E1/8;

7 E=E1+E2+E3;

disp('The maximum voltage that the string of the insulators can work well=%f kV', E);
```

5

Symmetrical Components

Introduction

Dr. C.L. Fortescue presented a paper in 1918 in which he proved that a system of several vectors or quantities may be resolved into equal different symmetrical groups or systems. The method of symmetrical components is a general one applicable to any poly-phase system. Because of the widespread use of 3-phase systems and the greater familiarity which electrical engineers have with them, symmetrical component equations will be developed for three phases.

The useful expression showing relationship between unsymmetrical and symmetrical component:

$$\begin{bmatrix} I_a \\ I_b \\ I_c \end{bmatrix} = \begin{bmatrix} 1 & 1 & 1 \\ 1 & \lambda^2 & \lambda \\ 1 & \lambda & \lambda^2 \end{bmatrix} \begin{bmatrix} I_a^0 \\ I_a^1 \\ I_a^2 \end{bmatrix}$$

Above equation can be written as

$$I^{abc} = AI_a^{012}$$

Where, matrix 'A' is symmetrical components transformation matrix,

$$\begin{bmatrix} 1 & 1 & 1 \\ 1 & \lambda^2 & \lambda \\ 1 & \lambda & \lambda^2 \end{bmatrix}$$

Multiplying both side of above equation by A^{-1}

$$I_a^{012} = A^{-1} I^{abc}$$

$$A-1=\frac{1}{3}\begin{bmatrix}1 & 1 & 1\\ 1 & \lambda & \lambda^2\\ 1 & \lambda^2 & \lambda\end{bmatrix}$$

$$\begin{bmatrix}I_a^0\\ I_a^1\\ I_a^2\end{bmatrix}=\frac{1}{3}\begin{bmatrix}1 & 1 & 1\\ 1 & \lambda & \lambda^2\\ 1 & \lambda^2 & \lambda\end{bmatrix}\begin{bmatrix}I_a\\ I_b\\ I_c\end{bmatrix}$$

This expression is also application for voltage phasor.

MATLAB Programs

Example 1: In a three-phase four-wire system the currents in the line a, b and c under abnormal conditions of loading were as follows:

I_a = 100(angle (30)) A, I_b = 50(angle (300)) A, I_c = 30(angle (180)) A

Calculate the zero, positive and negative phase sequence currents in line a & the return current in the neutral conductor.

Program

```
clear;
clc();
function [r,i]=d(mag,theta)
r=mag*cosd(theta);
i=mag*sind(theta);
endfunction
previousprot = funcprot(0)
funcprot(0)
mag=10;
theta=190;
[r,i]=d(mag,theta);
ia0=complex(r,i);
mag=6;
```

```
theta=20;
[r,i]=d(mag,theta);
ia1=complex(r,i);
mag=5;
theta=50;
[r,i]=d(mag,theta);
ia2=complex(r,i);
ia=(ia0+ia1+ia2);
mag=1;
theta=120;
[r,i]=d(mag,theta);
alpha=complex(r,i);
ib=(ia0 +alpha^2*ia1 +alpha*ia2);
ic=ia0+alpha*ia1+alpha^2*ia2;
mag=30;
theta=-30;
[r,i]=d(mag,theta);
va0=complex(r,i);
mag=450;
theta=0;
[r,i]=d(mag,theta);
va1=complex(r,i);
mag=225;
theta=40;
[r,i]=d(mag,theta);
va2=complex(r,i);
va=(va0+va1+va2);
mag=1;
```

```
theta=120;
[r,i]=d(mag,theta);
alpha=complex(r,i);
vb=(va0 +alpha^2*va1 +alpha*va2);
vc=va0+alpha*va1+alpha^2*va2;
s=3*va0*conφ(ia0)+ 3*va1*conφ(ia1)+ 3*va2*conφ(ia2);
printf("The complex power with symmetrical components in VA is:");
disp(s);
s1=va*conφ(ia)+vb*conφ(ib)+ vc*conφ(ic);
printf("The complex power unbalanced three phase components in VA is:");
disp(s1);
```

Result

```
previousprot=
1.
ans=
0.
The zero-sequence component of current in amperes is:
27.200847 + 2.2329099i
The positive sequence component of current in amperes is:
42.200847 + 39.760677i
The negative sequence component of current in amperes is:
17.200847 + 8.0064126i
The return current to the neutral conductor in amperes is:
81.60254 + 6.6987298i
```

Example 2 Determine the symmetrical components of currents in a three-phase system, the original phasors of which are

$I_a = 12 + \varphi 6$

Ib = 12 -φ12

Ic = - 15 + φ10

Program

```
clear;
clc;
function [r,i]=d(mag,theta)
r=mag*cosd(theta);
i=mag*sind(theta);
endfunction
previousprot = funcprot(0)
funcprot(0)
ia=complex(12,6);
ib=complex(12,-12);
ic=complex(-15,10);
mag=1;
theta=120;
[r,i]=d(mag,theta);
alpha=complex(r,i);
ia0=(1/3)*(ia+ib+ic);
printf("The zero sequence component of current in amperes is:");
disp(ia0);
ia1=(1/3)*(ia+alpha*ib+alpha^2*ic);
printf("The positive sequence component of current in amperes ia1=");
disp(ia1)
ib1=alpha^2*ia1;
printf("The positive sequence component of current in amperes ib1=");
disp(ib1);
ic1=alpha*ia1;
printf("The positive sequence component of current in amperes ic1=");
disp(ic1);
```

```
ia2=(1/3)*(ia+alpha^2*ib+alpha*ic);
printf("The negative sequence component of current in amperes ia2=");
ib2=alpha*ia2;
printf("The negative sequence component of current in amperes ib2=");
disp(ib2);
ic2=alpha^2*ia2;
printf("The negative sequence component of current in amperes ic2=");
disp(ic2);
```

Result

```
previousprot=
1.
ans=
0.
The zero-sequence component of current in amperes is:
3. + 1.3333333i
The positive sequence component of current in amperes ia1=
10.850853 + 10.127562i
The positive sequence component of current in amperes ib1=
3.3452995 - 14.460895i
The positive sequence component of current in amperes ic1=
- 14.196152 + 4.3333333i
The negative sequence component of current in amperes ia2=
The negative sequence component of current in amperes ib2=
5.6547005 + 1.127562i
The negative sequence component of current in amperes ic2=
- 3.8038476 + 4.3333333i
```

Example 3: A balanced star-connected load takes 150 A from a balanced 3-phase 4-wire supply. If the fuses in two of the lines are removed, find the symmetrical components of the line currents before and after the fuses are removed.

Program

```
clear;
clc;
function [r,i]=d(mag,theta)
r=mag*cosd(theta);
i=mag*sind(theta);
endfunction
previousprot = funcprot(0)
funcprot(0)
mag=150;
theta=0;
[r,i]=d(mag,theta);
ia=complex(r,i);
mag=150;
theta=-120;
[r,i]=d(mag,theta);
ib=complex(r,i);
mag=150;
theta=120;
[r,i]=d(mag,theta);
ic=complex(r,i);
mag=1;
theta=120;
[r,i]=d(mag,theta);
alpha=complex(r,i);
ia0=(ia+ib+ic)/3;
```

```
printf("The zero sequence component of current in amperes is:");
disp(round(ia0));
ia1=(ia+alpha*ib+ (alpha^2)*ic)/3;
printf("The positive sequence component of current in amperes ia1=");
disp(round(ia1));
ia2=(ia+alpha^2*ib+alpha*ic)/3;
printf("The negative sequence component of current in amperes ia2=");
disp(round(ia2));
//b).after fuse removal between two lines
i_b=0;
i_c=0;
i_a0=(ia+i_b+i_c)/3;
printf("The zero sequence component of current in amperes is:");
disp(i_a0);
ia1=(ia+alpha*i_b+alpha^2*i_c)/3;
printf("The positive sequence component of current in amperes ia1=");
disp(round(ia1));
ia2=(ia+alpha^2*i_b+alpha*i_c)/3;
printf("The negative sequence component of current in amperes ia2=");
disp(ia2);
```

Result

```
previousprot=
1.
ans=
0.
The zero-sequence component of current in amperes is:
0
```

The positive sequence component of current in amperes ia1=

150.

The negative sequence component of current in amperes ia2=

0

The zero-sequence component of current in amperes is:

50.

The positive sequence component of current in amperes ia1=

50.

The negative sequence component of current in amperes ia2=

50.

Example 4: The resolution of a set of three-phase unbalanced voltage into symmetrical components gave the following results:

$V_{a0} = 30\angle -30^0 \ V$

$V_{a1} = 450\angle 0^0 \ V$

$V_{a2} = 225\angle 40^0 \ V$

The component currents are

$I_{a0} = 10\angle 190^0 \ A$

$I_{a1} = 6\angle 20^0 \ A$

$I_{a2} = 5\angle 50^0 \ A$

Determine the complex power represented by these voltages and currents by (a) symmetrical components (b) unbalanced phase components

Program

```
clear;

clc;

function [r,i]=d(mag,theta)

r=mag*cosd(theta);

i=mag*sind(theta);
```

```
endfunction
previousprot = funcprot(0)
funcprot(0)
mag=10;
theta=190;
[r,i]=d(mag,theta);
ia0=complex(r,i);
mag=6;
theta=20;
[r,i]=d(mag,theta);
ia1=complex(r,i);
mag=5;
theta=50;
[r,i]=d(mag,theta);
ia2=complex(r,i);
ia=(ia0+ia1+ia2);
mag=1;
theta=120;
[r,i]=d(mag,theta);
alpha=complex(r,i);
ib=(ia0 +alpha^2*ia1 +alpha*ia2);
ic=ia0+alpha*ia1+alpha^2*ia2;
mag=30;
theta=-30;
[r,i]=d(mag,theta);
va0=complex(r,i);
mag=450;
theta=0;
```

```
[r,i]=d(mag,theta);
va1=complex(r,i);
mag=225;
theta=40;
[r,i]=d(mag,theta);
va2=complex(r,i);
va=(va0+va1+va2);
mag=1;
theta=120;
[r,i]=d(mag,theta);
alpha=complex(r,i);
vb=(va0 +alpha^2*va1 +alpha*va2);
vc=va0+alpha*va1+alpha^2*va2;
s=3*va0*conφ(ia0)+ 3*va1*conφ(ia1)+ 3*va2*conφ(ia2);
printf("The complex power with symmetrical components in VA is:");
disp(s);
s1=va*conφ(ia)+vb*conφ(ib)+ vc*conφ(ic);
printf("The complex power unbalanced three phase components in VA is:");
disp(s1);
```

Result

```
previousprot=
1.
ans=
0.
The complex power with symmetrical components in VA is:
10245.796 - 2777.9169i
The complex power unbalanced three phase components in VA is:
10245.796 - 2777.9169i
```

6

Symmetrical and Unsymmetrical Faults

Introduction

Dr. C.L. Fortescue presented a paper in 1918 in which he proved that a system of several vectors or quantities may be resolved into equal different symmetrical groups or systems.

The method of symmetrical components is a general one applicable to any polyphase system.

Because of the widespread use of 3-phase systems and the greater familiarity which electrical engineers have with them, symmetrical component equations will be developed for three phases.

In the Analysis of Unsymmetrical Faults, the following assumptions will be made:

1. The generated emf system is of positive sequence only.
2. No current flows in the network other than due to fault i.e. load currents are neglected.
3. The impedance of the fault is zero.
4. Phase R shall be taken as the reference phase.

In each case of Analysis of Unsymmetrical Faults, emf per phase are denoted by E_R, E_Y and E_B and the terminal potential difference per phase by V_R, V_Y and V_B.

Relation between unsymmetrical and symmetrical component

The useful expression showing relationship between unsymmetrical and symmetrical component:

$$\begin{bmatrix} I_a \\ I_b \\ I_c \end{bmatrix} = \begin{bmatrix} 1 & 1 & 1 \\ 1 & \lambda^2 & \lambda \\ 1 & \lambda & \lambda^2 \end{bmatrix} \begin{bmatrix} I_a^0 \\ I_a^1 \\ I_a^2 \end{bmatrix}$$

Above equation can be written as

$$I^{abc} = AI_a^{012} \tag{1}$$

Matrix 'A' is symmetrical components transformation matrix,

$$\begin{bmatrix} 1 & 1 & 1 \\ 1 & \lambda^2 & \lambda \\ 1 & \lambda & \lambda^2 \end{bmatrix}$$

Multiplying both side of equation (1) by A^{-1},

$$I_a^{012} = A^{-1} I^{abc}$$

$$A^{-1} = \frac{1}{3}\begin{bmatrix} 1 & 1 & 1 \\ 1 & \lambda & \lambda^2 \\ 1 & \lambda^2 & \lambda \end{bmatrix}$$

$$\begin{bmatrix} I_a^0 \\ I_a^1 \\ I_a^2 \end{bmatrix} = \frac{1}{3}\begin{bmatrix} 1 & 1 & 1 \\ 1 & \lambda & \lambda^2 \\ 1 & \lambda^2 & \lambda \end{bmatrix}\begin{bmatrix} I_a \\ I_b \\ I_c \end{bmatrix}$$

This expression is also application for voltage phasor.

Single-Line-to-Ground Fault

Consider a 3-phase system with an earthed neutral. Let a single line-to-ground fault occur on the red phase as shown in Figure 6.1. It is clear from this figure that:

$V_R = 0$ and $I_B = I_Y = 0$

The sequence currents in the red phase in terms of line currents shall be:

$$I_0 = \frac{1}{3}\left(I_R + I_B + I_Y\right) = \frac{1}{3} I_R \qquad I_1 = \frac{1}{3}\left(I_R + aI_Y + a^2 I_B\right) = \frac{1}{3} I_R$$

$$I_2 = \frac{1}{3}\left(I_R + a^2 I_Y + aI_B\right) = \frac{1}{3} I_R \qquad I_0 = I_1 = I_2 = \frac{1}{3} I_R$$

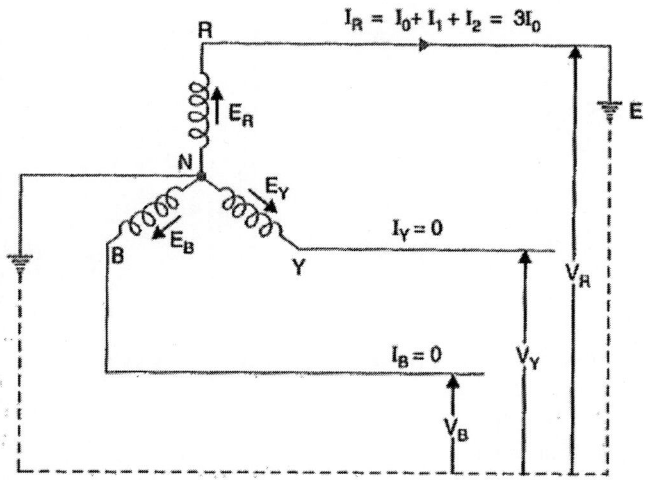

Figure 6.1: Single-line-to-ground fault

Fault current

First of all, expression for fault current I_R will be derived. Let Z_1, Z_2 and Z_0 be the positive, negative and zero sequence impedances of the generator respectively. Consider the closed loop NREN. As the sequence currents produce voltage drops due only to their respective sequence impedances, therefore, we have,

$$E_R = I_1Z_1 + I_2Z_2 + I_0Z_0 + V_R \qquad V_R = 0 \text{ and } I_0 = I_1 = I_2$$

$$E_R = I_0(Z_1 + Z_2 + Z_0) \qquad I_0 = \frac{E_R}{Z_1 + Z_2 + Z_0}$$

$$I_R = 3I_0 = \frac{3E_R}{Z_1 + Z_2 + Z_0} \tag{1}$$

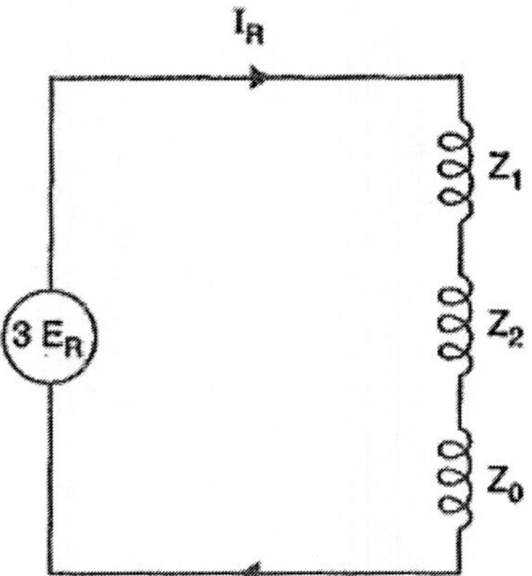

Figure 6.2: Sequence network for Single-line-to-ground fault

Examination of exp. (1) shows that the equivalent circuit from which fault current may be calculated is as given in Figure 6.2. It is clear that fault current is obtained by connecting the phase sequence impedances in series across an imaginary generator of voltage 3 E_R. This is a wonderful part of the method of symmetrical components and makes the analysis easy and interesting. In fact, this method permits to bring any Analysis of Unsymmetrical Faults into a simple circuit of interconnection of sequence impedances appropriate to the fault condition prevailing.

The assumption made in arriving at exp. (i) is that the fault impedance is zero. However, the fault impedance is Z_e, then expression for fault current becomes:

$$I_R = \frac{3E_R}{Z_1 + Z_2 + Z_0 + 3Z_e}$$

It may be added here that if the neutral is not grounded, then zero sequence impedance will be infinite and the fault current is zero. This is expected because now no path exists for the flow of fault current.

For line (R-phase)-to-ground fault:

$$I_R = \text{fault current} = \frac{3E_R}{Z_1 + Z_2 + Z_0 + 3Z_e} \qquad I_Y = 0 \quad I_B = 0$$

$$V_R = 0; \quad V_Y = V_0 + a^2V_1 + aV_2; \quad V_B = V_0 + aV_1 + a^2V_2$$

Line-to-Line Fault

Consider a line-to-line fault between the blue (B) and yellow (Y) lines as shown in Figure 6.3. The conditions created by this fault lead to

$V_Y = V_B$; $I_R = 0$ and $I_B + I_Y = 0$

Again, taking R-phase as the reference, we have,

$$I_0 = \frac{1}{3}\left(I_R + I_B + I_Y\right) = 0$$

$$V_Y = V_B$$

Expressing in terms of sequence components of red line, we have,

$$V_0 + a^2V_1 + aV_2 = V_0 + aV_1 + a^2V_2$$

$$V_1(a^2 - a) = V_2(a^2 - a)$$

$$V_1 = V_2 \tag{1}$$

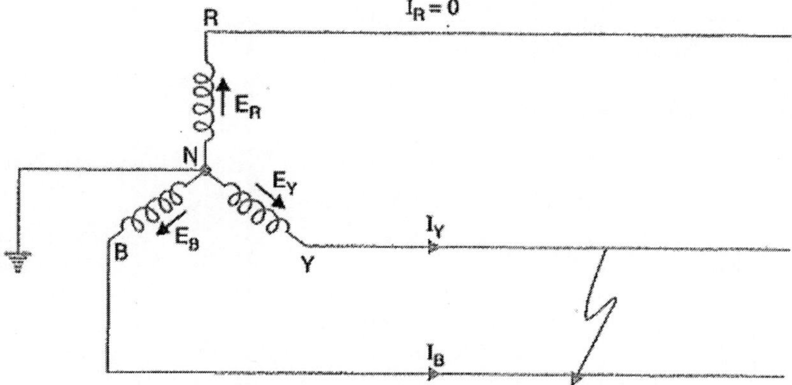

Figure 6.3: Line to Line fault

$$I_Y + I_B = 0$$

$$(I_0 + a^2I_1 + aI_2) + (I_0 + aI_1 + a^2I_2) = 0$$

$$(a^2 + a)(I_1 + I_2) + 2I_0 = 0$$

$$I_1 + I_2 = 0 \tag{2}$$

Fault current

Examination of exp. (i) and exp (ii) reveals that sequence impedances should be connected as shown in Figure 6.4. It is clear from the figure that:

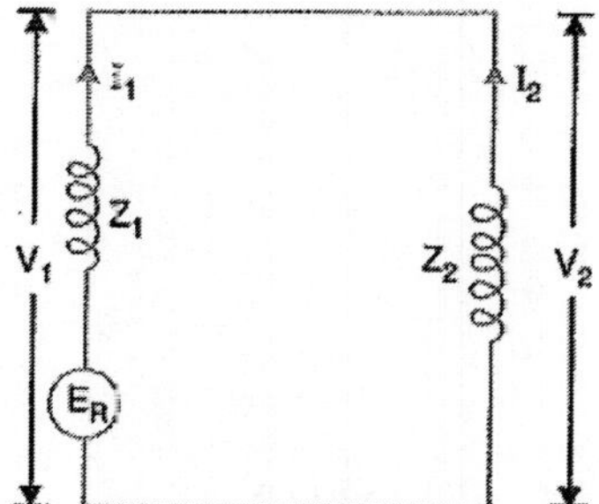

Figure 6.4: Sequence network of Line-to-Line fault

For line-to-line fault (Blue and Yellow lines):

$$I_R = 0; \;\; I_Y = -I_B = \frac{-j\sqrt{3}E_R}{Z_1 + Z_2}$$

$$V_Y = V_B = -\frac{Z_2}{Z_1 + Z_2}E_R \;\; \text{and} \;\; V_R = \frac{2Z_2}{Z_1 + Z_2}E_R$$

Double Line-to-Ground Fault

Consider the double line-to-ground fault involving Y-B lines and earth as shown in Figure 6.5. The conditions created by this fault lead to:

$V_Y = V_B = 0; I_R = 0$

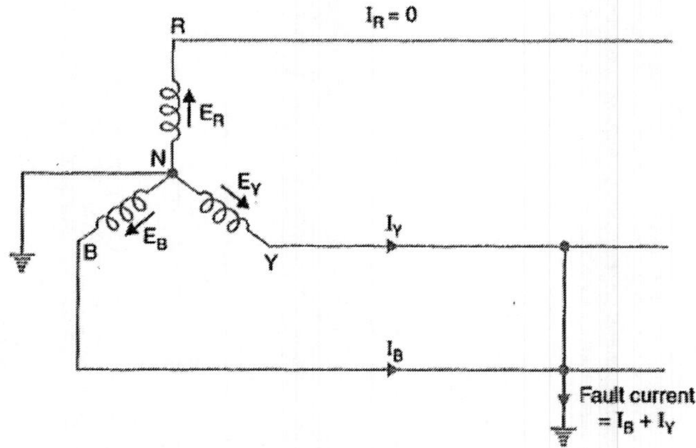

Figure 6.5: Double line to Ground fault

$V_Y = V_B = 0$

$$V_1 = V_2 = V_0 = \frac{1}{3} V_R \quad (1)$$

$$I_R = I_1 + I_2 + I_0 \quad (2)$$

Fault current

Examination of exp. (1) and exp. (2) reveals that sequence impedances should be connected as shown in Figure 6.6. It is clear that:

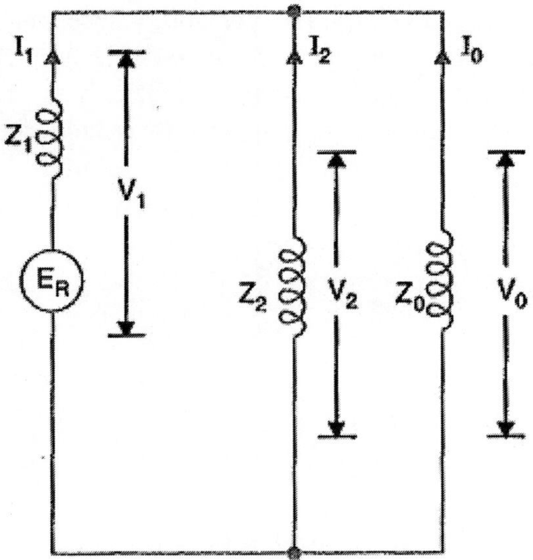

Figure 6.6: Sequence network of Double line to Ground fault

$$\text{fault current } (I_F) = I_Y + I_B = 3I_0 = 3\left(-I_1 \frac{Z_2}{Z_0 + Z_2}\right)$$

$$= -\frac{3Z_2}{Z_0 + Z_2} \times \frac{E_R}{Z_1 + \frac{Z_0 Z_2}{Z_0 + Z_2}} = -\frac{3Z_2 E_R}{Z_0 Z_1 + Z_0 Z_2 + Z_1 Z_2}$$

Phase Voltages: The sequence voltages for phase R are:

$$V_1 = E_R - I_1 Z_1; \quad V_2 = 0 - I_2 Z_2; \quad V_0 = 0 - I_0 Z_0$$

$$V_1 = V_2 = V_0 = \frac{1}{3} V_R$$

$$V_R = V_0 + V_1 + V_2 = 3V_2$$

$$V_Y = V_0 + a^2V_1 + aV_2 = (a^2 + a + 1)V_2 = 0$$

$$V_B = V_0 + a^2V_2 + aV_1 = (a^2 + a + 1)V_2 = 0$$

MATLAB Programme

Example 1: An 11 kV, 25 MVA synchronous generator has positive, negative and zero sequence reactance's of 0.12, 0.12 and 0.08 per unit, respectively. The generator neutral is grounded through a reactance of 0.03 per unit. A single line-to-ground fault occurs at the generator terminals. Determine the fault current and line-to-line voltages. Assume that the generator was unloaded before the fault.

Program

```
clear;
clc;
function [r,i]=d(mag,theta)
 r=mag*cosd(theta);
i=mag*sind(theta);
endfunction
previousprot = funcprot(0)
funcprot(0)
ea=1;
za0=(%i)*0.08;
za1=(%i)*0.12;
za2=(%i)*0.12;
sb=25*10^6;
vb=11000;
zf=(%i)*0.03;
ia1=ea/(za0+za1+za2+3*zf);
```

```
ia0=ia1;
ia2=ia1;
mag=1;
theta=120;
[r,i]=d(mag,theta);
alpha=complex(r,i);
ia=ia0+ia1+ia2;
ibas=sb/(sqrt(3)*vb);
ia=ia1*ibas;
ib=0;
ic=0;
va1=ea-za1*ia1;
va2=-za1*ia1;
va0=-(za0+3*zf)*ia1;
va=va0+va1+va2;
vb=va0+alpha^2*va1+alpha*va2;
v_c=va0+alpha*va1+alpha^2*va2;
vab1=va-vb;
vbc1=vb-v_c;
vca1=v_c-va;
vbas=11/sqrt(3);
vab=vab1*vbas;
printf("The voltage of line ab in kV is vab=");
disp(vab);
vbc=vbc1*vbas;
printf("The voltage of line bc in kV is vbc=");
disp(round(vbc));
vca=vca1*vbas;
```

```
printf("The voltage of line ca in kV is vca=");
disp(vca);
```

Result

```
previousprot =
 1.
ans =
 0.
 The voltage of line ab in kV is vab=
 3.9499207 + 5.5i
The voltage of line bc in kV is vbc=
 - 11.i
The voltage of line ca in kV is vca=
 - 3.9499207 + 5.5i
```

Example 2: Draw the sequence networks for the system shown in fig below. Determine the fault current when (a) LLG and (b) LL fault occurs at point F. the per unit reactance's all referred to the same base are as follows:

	X_0	X_1	X_2
Generator G_1	0.05	0.3	0.2
Generator G_2	0.03	0.25	0.15
Line 1	0.70	0.3	0.3
Line 2	0.70	0.3	0.3
Transformer T_1	0.12	0.12	0.12
Transformer T_2	0.10	0.1	0.1

Both the generators are generating 1 pu voltage.

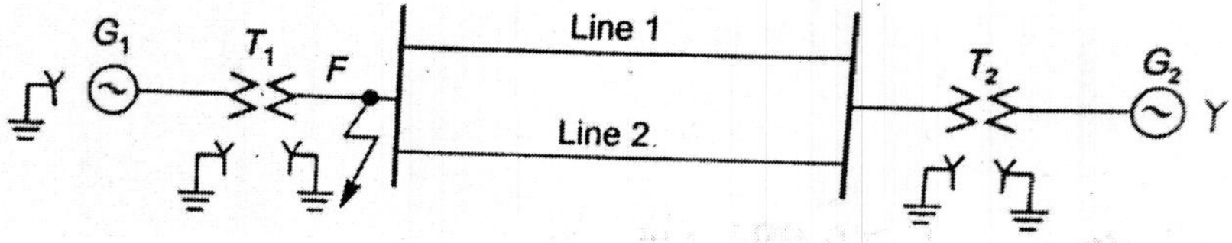

Program

```
clear;
clc;
```

```
function [r,i]=d(mag,theta)
 r=mag*cosd(theta);
i=mag*sind(theta);
endfunction
previousprot = funcprot(0)
funcprot(0)
x0g1=0.05;
x1g1=0.3;
x2g1=0.2;
x0g2=0.03;
x1g2=0.25;
x2g2=0.15;
x0l1=0.70;
x1l1=0.3;
x2l1=0.3;
x0l2=0.7;
x1l2=0.3;
x2l2=0.3;
x0t1=0.12;
x1t1=0.12;
x2t1=0.12;
x0t2=0.1;
x1t2=0.1;
x2t2=0.1;
vf=1;
ia01=0;
z1=(%i)*((x1g1+x1t1)*(x1g2 +x1t1 + x1l1*0.5))/(x1g1 +x1t1 + x1g2+
x1t1+x1l1*0.5);
```

```
z2=(%i)*((x2g1+x2t1)*(x2g2+ x2t2+ x2l2*0.5))/(x2g1 +x2t1 +x2g2 +x2t2
+x2l2*0.5);
z0=(%i)*(x0g1+x0t1);
//a).LLG fault
za0=z0;
za2=z2;
ia1=vf/(z1+(z0*z2/(za0+za2)));
printf("The positive sequence current in amperes is:");
disp(ia1);
ia0=-(3.1729*z2/(z0+z2));
printf("The zero sequence current in amperes is:");
disp(ia0);
ia2=-ia1*z0/(z0+z2);
printf("The negative sequence current in amperes is:")
disp(ia2);
//b).LL fault
ia11=vf/(z1+z2);
printf("The positive sequence current in amperes is:");
disp(ia11);
ia21=-ia11;
printf("The negative sequence current in amperes is:");
disp(ia21);
ia=ia01+ia11+ia21;
printf("phase a fault current is:")
disp(ia);
mag=1;
theta=120;
```

```
[r,i]=d(mag,theta);
alpha=complex(r,i);
ib=ia01+alpha^2*ia11+alpha*ia21;
printf("phase b fault current in pu is:")
disp(ib);
ic=ia01+alpha*ia11+alpha^2*ia21;
printf("phase c fault current in pu is:")
disp(ic);
```

Result

```
previousprot =
1.
ans =
0.
The positive sequence current in amperes is:
- 3.132426i
The zero sequence current in amperes is:
- 1.6219297
The negative sequence current in amperes is:
1.5311859i
The positive sequence current in amperes is:
- 2.4383214i
The negative sequence current in amperes is:
2.4383214i
phase a fault current is:
0
phase b fault current in pu is:
- 4.2232966 + 6.661D-16i
```

phase c fault current in pu is:

4.2232966 - 6.661D-16i

Example 3: A three phase 10 MVA, 11 kV synchronous generator with a solidly grounded neutral point supplies a feeder. The relevant impedances of the generator and feeder are as follows:

	Generator	**Feeder**
To positive sequence currents, Z_1	φ *1.2 Ω*	φ *1.0 Ω*
To negative sequence currents, Z_2	φ *0.9 Ω*	φ *1.0 Ω*
To zero sequence currents, Z_0	φ *0.4 Ω*	φ *3.0 Ω*

A fault from one phase to ground occurs at the far end of the feeder. Determine the voltage to neutral of the faulty phase at the terminal of the gender.

Program

```
clear;

//clc();

xz1g=(%i)*1.2;

xz1f=(%i)*1.0;

xz2g=(%i)*0.9;

xz2f=(%i)*1.0;

xz0g=(%i)*0.4;

xz0f=(%i)*3.0;

vf=11000/sqrt(3);

za0=xz0g +xz0f;

za1=xz1g +xz1f;

za2=xz2g +xz2f;

ia0=vf/(za0+za1+za2);

ia1=ia0;

ia2=ia0;

ifault=3*ia0;

va=vf-ia0*(xz0g+xz1g+xz2g);

printf("The voltage to neutral of the faulty phase is:%.2f V",va);
```

Result

The voltage to neutral of the faulty phase is:4233.90 V

7

Protection of Generator and Transformer

Introduction

A **generator** is subφected to electrical stresses imposed on the insulation of the machine, mechanical forces acting on the various parts of the machine, and temperature rise. These are the main factors which make protection necessary for the generator or alternator. Even when properly used, a machine in its perfect running condition does not only maintain its specified rated performance for many years, but it does also repeatedly withstand certain excess of overload.

Preventive measures must be taken against overloads and abnormal conditions of the machine so that it can serve safely. Even ensuring an efficient design, construction, operation, and preventive means of protection – the risk of a fault cannot be completely eliminated from any machine. The devices used in generator protection, ensure that when a fault arises, it is eliminated as quickly as possible.

An electrical generator can be subφected to either an internal fault or external fault or both. The generators are normally connected to an electrical power system, hence any fault occurred in the power system should also be cleared from the generator as soon as possible otherwise it may create permanent damage in the generator.

The number and variety of faults occur in a generator are huge. That is why generator or alternator is protected with several protective schemes. Generator protection is of both discriminative and non-discriminative type. Great care is to be taken in coordinating the systems used and the settings adopted to ensure that a sensitive, selective and discriminative generator protection scheme is achieved.

Any internal fault inside the stator winding is cleared by mainly differential protection scheme of the generator or alternator.

The differential protection is provided in the generator by using longitudinal differential relay.

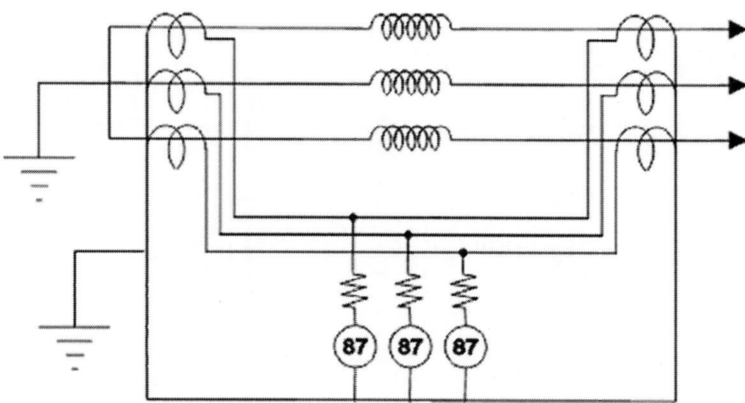

Figure 7.1: Longitudinal Differential protection scheme

Generally instantaneous attracted armature type relays are used for this purpose because all they have high speed operation and also, they are free from being affected by any AC transient of the power circuit.

There are two sets of current transformers, one CT is connected to the line side of the generator and other is connected to the neutral side of the generator in each phase. It is needless to say that the characteristics of all current transformers installed against each phase must be matched. If there is any maφor mismatched in the current transformer's characteristics of both sides of the generator, there may be high chance of malfunctioning of differential relay during the fault external to the stator winding and also may be during normal operating conditions of the generator. To ensure that the relay does not operate for the faults external to the operated zone of the protection scheme, a stabilizing resistor is fitted in series with the relay operating oil. It also ensures that if one set of CT has been saturated, there will be no possibility of malfunctioning of the differential relay.

It is always preferable to use dedicated current transformers for differential protection purpose because common current transformers may cause unequal secondary loading for other functionalities imposed on them. It is also always preferable to use all current transformers for differential protection of generators or alternators should be of same characteristics. But practically there may be some difference in characteristics of the current transformers installed at line side to those installed in neutral side of the generator.

These mismatches cause spill current to flow through the relay operating coil. To avoid the effect of spill current, percentage biasing is introduced in differential relay.

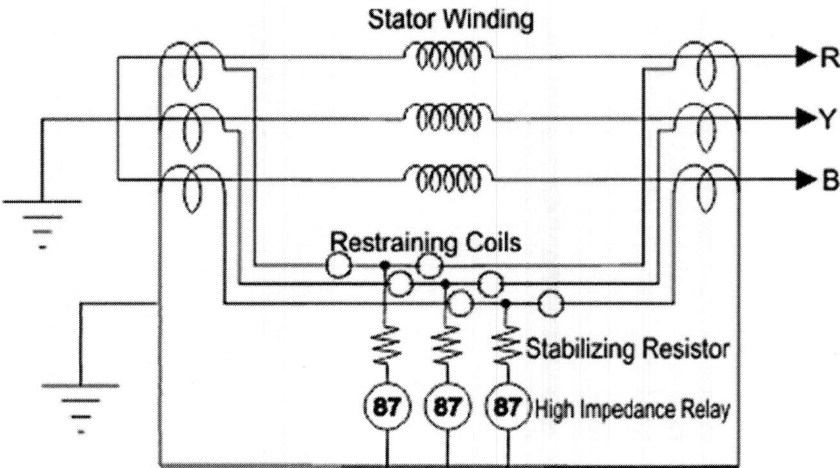

Figure 7.2: Modified Differential protection scheme

The percentage biased differential relay comprises two restraint coils and one operating coil per phase. In the relay, the torque produced by operating coil tends to close the relay contacts for instantaneous tripping of circuit breakers but at the same time the torque produced by the restraint coils prevents to close the relay contacts as restraint coils torque is directed opposite of the operating coil torque. Hence during through fault, the differential relay would not be operated because the setting of the relay is increased by restraint coils and also it prevents malfunctioning of relay due to spill current. But during internal fault in the winding of the stator, the torque produced by restraint coils is ineffective and the relay closes its contact when setting current flows through the operating coil.

Differential current pickup setting/bias setting of the relay is adopted based on the maximum percentage of allowable mismatch adding some safety margin. The spill current level for the relay is to φust operate it; is experienced as a percentage of the through fault current causing it. This percentage is defined as bias setting of the relay.

$$\text{Percentage Bias} = \frac{\text{spill current for relay operation}}{\text{through fault current causing it}} \times 100\%$$

MATLAB Programes

Example 1: A generator is provided with restricted earth-fault protection. The ratings are 11kV, 5000 kVA. The percentage of winding protected against phase to ground fault is 80%. The relay setting such that it trips for 25% out of balance. Calculate the resistance to be added in neutral to ground connection.

Program

```
clear;
clc;
v=11e3/sqrt(3);
v=round(v);
r=5e6;
per=20;
i=r/(3*v);
i=round(i);
i0=i*25/100;
R=per*v/(i0*1000);
R=round(R*100)/100;
printf("the resistance to be added=%f ohms",R);
```

Result

the resistance to be added=1.940000ohms

Example 2: The neutral point of a 10,000 V alternator is earthed through a resistance of 10 Ω, the relay is set to operate when there is an out of balance current of 1A. The CT's have a ratio of 1000/5. What percentage of the winding is protected against fault to earth and what must be minimum value of earthing resistance to give 90% protection to each phase winding?

Program

```
clear;
clc;
v=10e3/sqrt(3);
R=10;
i=1;
ct=1000/5;
```

```
ip=i*ct;

per=R*ip*100/v;

p=10;

res=p/100*v/ip;

mprintf("The percentage of unprotected winding=%fpercent\nResistance for 90percent winding protection=%fohms",100-(per),res);
```

Result

The percentage of unprotected winding=65.358984percent

Resistance for 90 percent winding protection=2.886751ohms

Example 3: A 3 phase, 2 pole, 11 kV, 10,000 kVA alternator has neutral earthed through a resistance of 7 ohms. The machine has current balance protection which operates upon out of balance current exceed 20% of full load. Determine % of winding protected against earth fault.

Program

```
clear; // Remove clear, clc from code if you want to access existing stored variable from the memory

clear;

clc;

per=.2;

r=10e6;

R=7;

v=11e3;

i=r/(sqrt(3)*v);

i=round(i);

i0=per*i;

v=v/sqrt(3);

p=R*i0/v*100;

p=round(p*10)/10;

printf("Percentage of unprotected winding for earth fault=%fpercent",p);
```

Results

Percentage of unprotected winding for earth fault=11.600000percent

Example 4: A 11 kV, 3 phase Alternator has full load rated current of 200 A. reactance of armature winding is 15%. The differential protection system is set to operate on earth fault currents of more than 200 A. find the neutral earthing resistance, which gives earth fault protection to 90% of stator winding.

Program

```
clear; // Remove clear, clc from code if you want to access existing stored variable from the memory

clear;

clc;

i=200;

c=.1;

v=11e3/sqrt(3);

per=.15;

x=per*v/(i);

ru=c*x;

vi=v*c;

y=i\vi;

r=sqrt((y^2)-(ru^2));

r=round(r*100)/100;

printf("the neutral earthing resistance=%fohms",r);
```

Result

the neutral earthing resistance=3.140000ohms

Example 5: A star-connected, 3-phase, 10-MVA, 6·6 kV alternator has a per phase reactance of 10%. It is protected by Merz-Price circulating-current principle which is set to operate for fault currents not less than 175 A. Calculate the value of earthing resistance to be provided in order to ensure that only 10% of the alternator winding remains unprotected.

Program

```
clear;
clc;
mva=10*1e6;
v=6.6*1e3;
ph_z=0.1;
i=175;
awp=0.1;
vph=v/sqrt(3);
flc=mva/sqrt(3)/v;
x=v*mva/1e6/flc/sqrt(3)/100;
zw=x*ph_z;
emfw=vph*ph_z;
r=sqrt((emfw/i)^2-zw^2);
printf("Voltage per phase = %.0f V \n\n", vph);
printf("Full load current = %.0f V \n\n", flc);
printf("Reactance per phase = %.3f ohm \n\n", x);
printf("Reactance of 10 percent winding = %.4f ohm \n\n", zw);
printf("EMF induced in 10 percent winding = %.0f V \n\n", emfw);
printf("Impedance offered by fault by 10 percent winding is Zf=sqrt(%.4f^2+r^2)
\n\n", zw);
printf("Earthing resistance to be provided = %.3f ohm \n\n", r);
```

Result

Voltage per phase = 3811 V

Full load current = 875 V

Reactance per phase = 0.436 ohm

Reactance of 10 percent winding = 0.0436 ohm

EMF induced in 10 percent winding = 381 V

Impedance offered by fault by 10 percent winding is Zf=sqrt(0.0436^2+r^2)

Earthing resistance to be provided = 2.177 ohm

Example 6: A 3-phase transformer of 220/11,000 line volts is connected in star/delta. The protective transformers on 220 V side have a current ratio of 600/5. What should be the CT ratio on 11,000 V side?

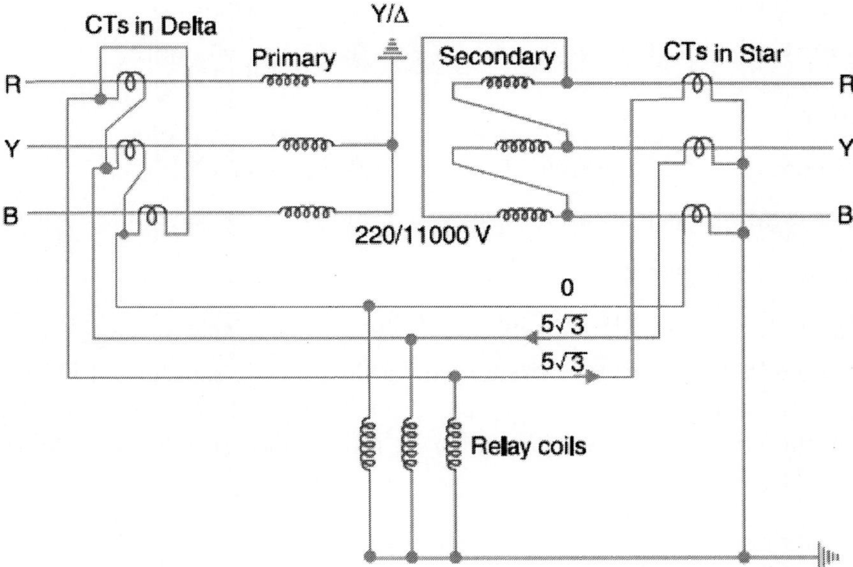

Program

```
clear;

clc;

v1=220;

v2=11000;

v=220;

i1=600;

i2=5;

lc=sqrt(3)*i2;

pap=sqrt(3)*v*i1;
```

```
i=pap/sqrt(3)/v2;

ratio=i/lc;

printf("Line current of delta connected CTs on %d V side = %.2f A \n\n", v1, lc);

printf("Phase current of star connected CTs on %d V side = %.2f A \n\n", v2, lc);

printf("The line current on the %d V side = %.2f A \n\n",v2, i);

printf("Turn ratio of CTs on %d V side = %.3f:1 \n\n", v2, ratio);
```

Result

Line current of delta connected CTs on 220 V side = 8.66 A

Phase current of star connected CTs on 11000 V side = 8.66 A

The line current on the 11000 V side = 12.00 A

Turn ratio of CTs on 11000 V side = 1.386:1

Example 7: A 3-phase transformer having line-voltage ratio of 0·4 kV/11kV is connected instar-delta and protective transformers on the 400 V side have a current ratio of 500/5. What must be the ratio of the protective transformers on the 11 kV side?

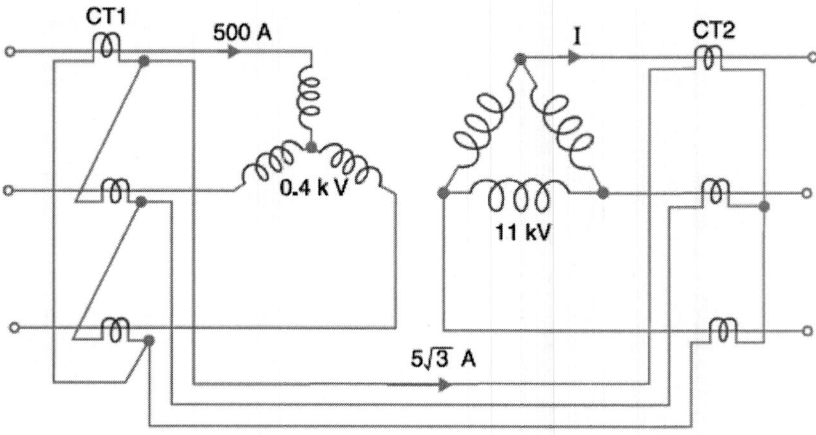

Program

```
clear;

clc;
```

```
v1=0.4;
v2=11;
v=400;
i1=500;
i2=5;
lc=sqrt(3)*i2;
pap=sqrt(3)*v*i1;
i=pap/sqrt(3)/v2/1000;
ratio=i/lc;
printf("Line current of delta connected CTs on %d V side = %.2f A \n\n", v1, lc);
printf("Phase current of star connected CTs on %d V side = %.2f A \n\n", v2, lc);
printf("The line current on the %d V side = %.2f A \n\n",v2, i);
printf("Turn ratio of CTs on %d V side = %.1f:1 \n\n", v2, ratio);
```

Result

Line current of delta connected CTs on 0 V side = 8.66 A

Phase current of star connected CTs on 11 V side = 8.66 A

The line current on the 11 V side = 18.18 A

Turn ratio of CTs on 11 V side = 2.1:1

Example 8: A 30-MVA, 34.5 kV Y/138 kV D transformer is protected by differential relays with taps. Select CT ratios, CT connections, and relay tap settings. Also determine currents in the transformer and in the CTs at rated conditions. Assume that the available relay tap settings are 5:5, 5: 5.5, 5: 6.6, 5: 7.3, 5: 8, 5: 9, and 5: 10, giving relay tap ratios of 1.00, 1.10, 1.32, 1.46, 1.60, 1.80, and 2.00.

Program

```
clc;
clear;
```

```
Srated = 30; //power rating in MVA
Vprtr = 34.5; //primary side of transformer voltage in kV
Vsectr = 138; //secondary side of transformer voltage in kV
IArated = (Srated*10^6)/(sqrt(3)*Vsectr*10^3); //Rated current on the 138-kV side of the transformer in Amperes
CTratiosec = 150/5; //CT ratio on the 138-kV side
IA = IArated/CTratiosec; //differential current in 138kV side in Amperes
Iarated = (Srated*10^6)/(sqrt(3)*Vprtr*10^3); //Rated current on the 34.5-kV side of the transformer in Amperes
CTratiopr = 500/5; //CT ratio on the 34.5-kV side
Ia = Iarated/CTratiopr; //differential current in 138kV side in Amperes
Iab = Ia*sqrt(3); //differential current in left hand re-straining winding of figure 10.37 in Amperes
crtratio = Iab/IA; //ratio of the currents in the left- to right hand restraining winding
TA = 5;
Tab = 10;
tapratio = Tab/TA; //closest relay tap ratio
%mismatch = (((Iab/Tab)-(IA/TA))/(Iab/Tab))*100; //percentage mismatch for tap setting
printf('\nRated current on the 138kV side of the transformer is %f A',IArated);
printf('\nRated current on CT ratio in 138 kV side of the transformer is %f A',IA);
printf('\nRated current on the 34.5kV side of the transformer is %f A',Iarated);
printf('\nRated current on CT ratio in 34.5kV side of the transformer is %f A',Ia);
printf('\nThe percentage mismatch for the tap setting is %f',%mismatch);
```

Result

Rated current on the 138kV side of the transformer is 125.510928 A

Rated current on CT ratio in 138 kV side of the transformer is 4.183698 A

Rated current on the 34.5kV side of the transformer is 502.043712 A

Rated current on CT ratio in 34.5kV side of the transformer is 5.020437 A

The percentage mismatch for the tap setting is 3.774955

8

Fundamentals of Fault Clearing and Switching Phenomena

Introduction

The insulating material (may be fluid or air) used in circuit breaker should serve two important functions. They are written as follows:

1. It should provide sufficient insulation between the contacts when circuit breaker opens.
2. It should extinguish the arc occurring between the contacts when circuit breaker opens.

The second point needs more explanation. To understand this point let us consider a situation if there is some fault or short circuit in the system, the relay provides desired signals to the circuit breaker so as to prevent system from ongoing fault. Now when circuit breaker opens its contacts, due to this an arc is drawn. The arc is interrupted by suitable insulator and technique.

Methods of Arc Interruption

There are two methods by which interruption is done.

1. High resistance method,
2. Low resistance method or current zero interruption method.

In high interruption method we can increase the electrical resistance many times to such a high value that it forces the current to reach to zero and thus restricting the possibility of arc being restrike. Proper steps must be taken in order to ensure that the rate at which the resistance is increased or decreased is not abnormal because it may lead to generation of harmful induced voltages in the system. The arc resistance can be increased by various methods like lengthening or cooling of the arc etc.

Limitations of High Resistance Method

Arc discharge has a resistive nature due to this most of the energy is received by circuit breaker itself hence proper care should be taken during the manufacturing of circuit breaker like mechanical strength etc. Therefore, this method is applied in DC power circuit breaker, low and medium AC power circuit breaker. Low resistance method is applicable only for ac circuit and it is possible there because of presence of natural zero of current. The arc gets extinguished at the natural zero of the AC waves and is prevented from restricting again by rapid building of dielectric strength of the contact space.

There are two theories which explain the phenomenon of arc extinction:

1. Energy balance theory,
2. Voltage race theory.

Before going in details about these theories, we should know the following terms

Restriking Voltage

It may be defined as the voltages that appears across the breaking contact at the instant of arc extinction.

Recovery Voltage

It may be defined as the voltage that appears across the breaker contact after the complete removal of transient oscillations and final extinction of arc has resulted in all the poles.

Active Recovery Voltage

It may be defined as the instantaneous recovery voltage at the instant of arc extinction.

Arc Voltage

It may be defined as the voltages that appears across the contact during the arcing period, when the current flow is maintained in the form of an arc. It assumes low value except for the point at which the voltage rises rapidly to a peak value and current reaches to zero.

Energy Balance Theory

When the contact of circuit breaker is about to open, restriking voltage is zero, hence generated heat would be zero and when the contacts are fully open there is infinite resistance this again make no production of heat. We can conclude from this that the maximum generated heat is lying between these two cases and can be approximated, now this theory is based on the fact that the rate of generation of heat between the contacts of circuit breaker is lower than the rate at which heat between the contact is dissipated. Thus, if it is possible to remove the generated heat by cooling, lengthening and splitting the arc at a high rate the generation, arc can be extinguished.

Voltage Race Theory

The arc is due to the ionization of the gap between the contacts of the circuit breaker. Thus the resistance at the initial stage is very small i.e. when the contact are closed and as the contact separates the resistance starts increasing. If we remove ions at the initial stage either by recombining them into neutral molecules or inserting insulation at a rate faster than the rate of ionization, the arc can be interrupted. The ionization at zero current depends on the voltages known as restriking voltage.

Let us define an expression for restriking voltage. For loss-less or ideal system we have

$$v = V\left[1 - \cos\frac{t}{\sqrt{LC}}\right]$$

Here,

v = restriking voltage.

V = value of voltage at the instant of interruption.

L and C are series inductor and shunt capacitance up to fault point. Thus, from above equation we can see that lower the value of product of L and C, higher the value of restriking voltage.

The variation of v versus time is plotted below:

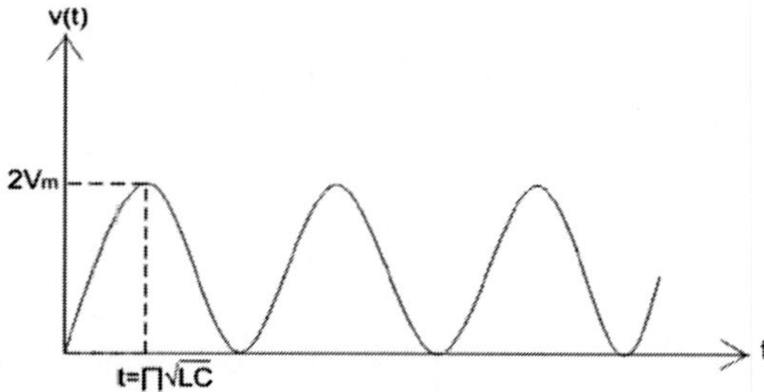

Figure 8.1: Variation of restriking voltage with time for loss less system

Now let us consider a practical system, or assume their finite loss in the system. As figure shown below, in this case the restriking voltage is damped out due to the presence of some finite resistance. Here it is assumed that the current lags behind the voltage by an angle (measured in degrees) of 90. However in practical situation angle may varies depending upon time in cycle at which the fault is occurred.

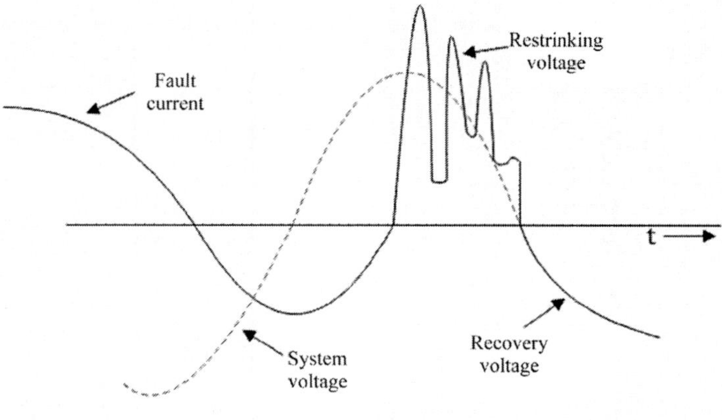

Figure 8.2: Variation of restriking voltage in practical system

Let us consider the effect of arc voltage, if arc voltage is included in the system, there is an increment in the restriking voltage. However, this is offset by another effect of an arc voltage which opposes the current flow and making change in the phase of current, thus bringing it more into phase with the applied voltages. Hence the current is not at its peak value when voltage passes through zero value.

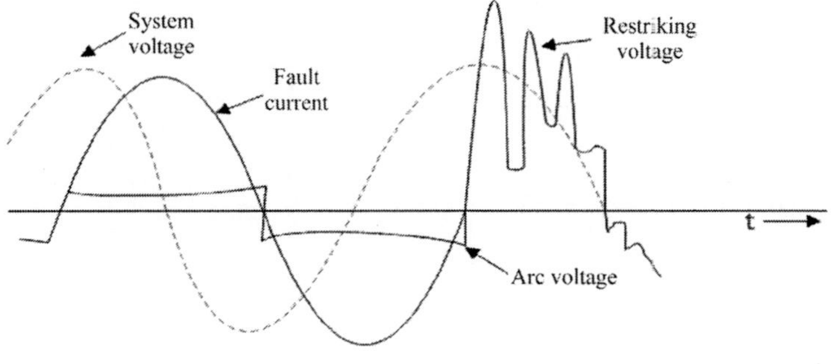

Figure 8.3: Variation of current and restriking voltage

Rate of Rise of Restriking Voltage (RRRV)

It is defined as the ratio of peak value of restriking voltage to time taken to reach to peak value. It is one of the most important parameters as if the rate at which the dielectric strength developed between the contacts is greater than RRRV, and then the arc will be extinguishes.

MATLAB Programes

Example 1: A 50 Hz sinusoidal voltage of amplitude 400 volts is applied to a series circuit of resistance 10 ohm ans inductance of 0.1 H. find the expression for the value of the current at any instant ater the voltage is applied, assuming the voltage is zero at the instant of application. Calculate the value of transient current 0.02 sec after switching on.

Program

```
clear ;
close ;
clc ;
R=10;
L=0.1;
f=50;
w=2*%pi*f;
k=sqrt((R^2)+((w*L)^2));
 angle=atan(w*L/R);
E=400
A=E*sin(angle)/k;
i=A*exp((-R)*.02/L);
i=round(i*100)/100;
mprintf("the transient current =%fA",i)
```

Result

```
E = 400.
the transient current =1.560000A
```

Example 2: A 50- cycle alternating voltage is applied to an R-L series circuit by closing a switch. The resistance is 10 ohms. Inductance is 0.1 Henry. The r.m.s. value of applied voltage is 100 volts.

a) Find the value of DC component of current upon closing the switch if instantaneous value of voltage is 50 at that time.

b) What value of instantaneous voltage will produce a maximum DC component of current upon closing the switch?

c) What is the instantaneous value of voltage which will result in the absence of any DC component upon closing the switch?

d) If the switch is closed when instantaneous voltage is zero, find the instantaneous current 0.5, 1.5, 5.5 cycles later.

Program

```
clear;
close;
clc;
R=10;
L=0.1;
f=50;
w=2*%pi*f;
k=sqrt((R^2)+((w*L)^2));
angle=atan(w*L/R);
E=100;
Em=sqrt(2)*E;
A=Em*sin(angle)/k;
i1=A;
Em=round(Em*10)/10;
i1=round(i1*10)/10;
mprintf("current in amperes for part1=%fA\n",i1);
mprintf("current in part 2& part 3= 0\n");
```

mprintf("the DC component vanishes if e=%fV",Em);//the error is due to the erroneous values in the textbook

t1=0.5*.02;

i2=A*exp((-R)*t1/L);

mprintf("\ncurrent at .5 cycles for t1=%fsec \ncurrent in the problem = %fA",t1,i2);

t2=1.5*.02;

i3=A*exp((-R)*t2/L);

mprintf("\ncurrent at 1.5 cycles for t2=%fsec \ncurrent in the problem = %fA",t2,i3);

t3=5.5*.02;

i4=A*exp((-R)*t3/L);

mprintf("\ncurrent at 5.5 cycles for t3=%fsec \ncurrent in the problem = %fA",t3,i4);

disp("the difference in result is due to erroneous value in textbook.")

Result

current in amperes for part1=4.100000A

current in part 2& part 3= 0

the DC component vanishes if e=141.400000V

current at .5 cycles for t1=0.010000sec

current in the problem = 1.503684A

current at 1.5 cycles for t2=0.030000sec

current in the problem = 0.203502A

current at 5.5 cycles for t3=0.110000sec

current in the problem = 0.000068A

the difference in result is due to erroneous value in textbook.

Example 3: A 50-cycles, 3-phase alternator with grounded neutral has inductance of 1.6 mH per phase and is connected to busbar through a circuit-breaker. The capacitance to earth between the alternator and circuit-breaker is 0.003 μF per phase. The circuit breaker opens when r.m.s. value of current is

7500 A. Determine analytically the following

a) Maximum rate of rise of restriking voltage.

b) Time for maximum rate of rise of restriking voltage.

c) Frequency of oscillation.

Neglect first-pole-to-clear factor.

Program

```
clear;
close;
clc;
C=.003e-6
L=1.6e-3
y=sqrt(L*C);
y=round(y*1e7)/1e7;
f=(2*3.14*y)^-1;
f=round(f/100)*100;
i=7500;
E=i*2*3.15*L*50;
Em=1.414*E;
Em=round(Em/10)*10
t=y*%pi/2;
t=t*1e6;
t=round(t*100)/100;
e=Em/y;
e=round((e)/1e6)*1e6;
e=fix(e/1e7)*1e7
mprintf("frequency of oscillations=%fc/s",f);
mprintf("\ntime of maximum restriking voltage=%fmicrosec",t);
mprintf("\nmaximum restriking voltage=%dV/microsecs",e/1e6);
```

Result

```
C =

   3.000D-09

L =

   0.0016

Em =

   5340.

e =

   2.420D+09

frequency of oscillations=72400.000000c/s

time of maximum restriking voltage=3.460000microsec

maximum restriking voltage=2420V/micro secs
```

Example 4: A three phase alternator has the line voltage of 11 kV. The generator is connected to a circuit-breaker. The inductive reactance upto the circuit breaker is 5 ohm per phase. The distributed capacitance upto circuit-breaker between phase and neutral is 0.01 μF. determine the following:

Neglect first pole to clear factor

a) Peak restriking voltage across the circuit-breaker.
b) Frequency of restriking voltage transient.
c) Average rate of restriking voltage upto peak restriking voltage.
d) Maximum rate of rise of restriking voltage.

Program

```
clear;

close;

clc;

R=5

f=50

L=R/(2*%pi*f);
```

```
V=11e3;
Vph=11/sqrt(3);
C=0.01d-6;
y=sqrt(L*C);
Em=sqrt(2)*Vph;
ep=2*Em;
ep=round(ep*10)/10;
y=round(y*1e7)/1e7;
t=y*%pi;
t=fix(t*1e7)/1e7
ea=ep/t;
ea=round(ea/1e3)*1e3
fn=(2*3.14*y)^-1;
Em=round(Em)
Emax=Em/y;
Emax=round(Emax/1000)*1e3;
mprintf("peak restriking voltage=%dkV",ep);
printf("\nfrequency of oscillations=%dc/s",fn);
printf("\naverage rate of restriking voltage=%fkV/microsecs",ea/1e6);
printf("\nmax restriking voltage=%dV/microsecs",Emax/1e3);
```

Results:

```
R =
   5.
f =
   50.
t =
   0.0000395
```

```
ea =

  456000.

Em =

  9.

peak restriking voltage=18kV

frequency of oscillations=12637c/s

average rate of restriking voltage=0.456000kV/micro secs

max restriking voltage=714V/micro secs
```

Example 5: In a system the r.m.s. voltage is 19.1 kV, L is 10 mH, C is 0.02 mF. Determine the average rate of rise of restriking voltage, when the circuit breaker opens.

Program

```
clear;

close;

clc;

E=19.1*1e3;

L=10*1e-3;

C=.02*1e-6;

Em=sqrt(2)*E;

y=sqrt(L*C);

t=%pi*y*1e6;

emax=2*Em;

eavg=emax/t;

eavg=round(eavg/10)*10

printf("average restriking voltage=%dV/microsecs",eavg);
```

Results

```
eavg =

  1220.

average restriking voltage=1220V/micro secs
```

Example 6: In a short-circuit test on a circuit-breaker, the following readings obtained on a frequency transient:

a) Time to reach the peak restriking voltage 70 μ-sec.

b) The peak restriking voltage 100 kV.

Calculate the average rate of rise of restriking voltage and the natural frequency of the circuit.

Program

```
clear;
clc;
Em=100e3
t=70e-6
Ea=Em/t/1e6
f=1/(2*t);
Ea=round(Ea/10)*10;
f=round(f);
printf("average voltage in volts=%dV/microsecs\n",Ea);
printf("frequency of oscillation =%dc/s",f);
```

Results

```
Em =
  100000.
t =
  0.00007
Ea =
  1428.5714
average voltage in volts=1430V/microsecs
frequency of oscillation =7143c/s
```

Example 7: In a system of 132 kV, the circuit phase to ground capacitance is 0.01 mF, the series inductance is 6 H. calculate the voltage appearing across the pole of a circuit breaker. Is a magnetizing current of 10 amp is interrupted (instantaneous). Calculate the value of resistance to be used across contact space to eliminate the restriking voltage transient.

Program

```
clear;
clc;
L=6;
C=0.01e-6;
i=10;
v=i*sqrt(L/C);
R=.5*v/i;
R=round(R/10)*10;
printf("damping resistance in ohms=%fkohms",R/1e3);
```

Results

damping resistance in ohms=12.250000kohms

9

Fundamentals of Fuse and Relays

Introduction

In normal working condition of electrical network, the current flows through the network is within the rated limit. If a fault occurs in the network mainly phase to phase short circuit fault or phase to ground fault, the network current crosses the rated limits. This high current may have a very high thermal effect which will cause permanent damage to the valuable pieces of equipment connected to the electrical network. So this high fault current should be interrupted as fast as possible. This is what an electrical fuse does. A fuse is a part of the circuit which consists of a conductor which melts easily and breaks the connection when current exceeds the predetermined value. An electrical fuse is the weakest part of an electrical circuit which breaks when more than predetermined current flows through it.

Fuse Wire

The function of fuse wire is to carry the normal current without excessive heating but more than normal current when passes through fuse wire, it rapidly heats up and melts.

Materials used for fuse wires

The materials used for fuse wires are mainly tin, lead, zinc, silver, antimony, copper, aluminum etc.

Fuse wire rating

The melting point and specific resistance of different metals used for fuse wire

Metal	Melting point	Specific Resistance
Aluminium	240°F	2.86 $\mu\,\Omega$ – cm
Copper	2000°F	1.72 $\mu\,\Omega$ – cm
Lead	624°F	21.0 $\mu\,\Omega$ – cm
Silver	1830°F	1.64 $\mu\,\Omega$ – cm
Tin	463°F	11.3 $\mu\,\Omega$ – cm
Zinc	787°F	6.1 $\mu\,\Omega$ – cm

Fuse law

It determines the current carrying capacity of a fuse wire. We can establish the law in the following way. At steady state condition that is when fuse carries normal current without increasing its temperature to the melting limit. That means at this steady state condition, heat generated due to the current through fuse wire is equal to heat dissipated from it.

Heat generated= I^2R

Where, R is the resistance of the fuse wire.

$$I^2R = I^2\left(\rho\frac{l}{a}\right)$$

Where, ρ is the resistivity, l is the length and a is the cross-sectional area of fuse wire.

$$I^2\left(\rho\frac{l}{a}\right) = I^2\left(\rho\frac{l}{\Pi d^2/4}\right)$$

Where, d is the diameter of fuse wire.

$$I^2\left(\rho\frac{l}{\Pi d^2/4}\right) = I^2.K1\left(\frac{l}{d^2}\right)$$

Where, K1 is a constant.

Heat lost ∝ surface area of fuse wire ∝ πdl

Therefore, Heat loss= K_2dl (ii)

Where, K_2 is a constant.

Now, equating above expressions, we get,

$$I^2.K1\left(\frac{l}{d^2}\right) = K_2\text{dl}$$

$I^2\ Kd^3$

where K is an another constant.

$I = Kd^{3/2}$

$$I \propto d^{3/2}$$

This is known as fuse law.

Relay

A relay is automatic device which senses an abnormal condition of electrical circuit and closes its contacts. These contacts in turns close and complete the circuit breaker trip coil circuit hence make the circuit breaker tripped for disconnecting the faulty portion of the electrical circuit from rest of the healthy circuit.

Now let's have a discussion on some terms related to protective relay. Pickup Level of Actuating Signal: The value of actuating quantity (voltage or current) which is on threshold above which the relay initiates to be operated.

If the value of actuating quantity is increased, the electromagnetic effect of the relay coil is increased, and above a certain level of actuating quantity, the moving mechanism of the relay φust starts to move.

Reset level

The value of current or voltage below which a relay opens its contacts and comes in original position.

Operating time of relay

9ust after exceeding pickup level of actuating quantity the moving mechanism (for example rotating disc) of relay starts moving and it ultimately closes the relay contacts at the end of its φourney. The time which elapses between the instant when actuating quantity exceeds the pickup value to the instant when the relay contacts close.

Reset time of relay

The time which elapses between the instant when the actuating quantity becomes less than the reset value to the instant when the relay contacts return to its normal position.

Reach of relay

A distance relay operates whenever the distance seen by the relay is less than the pre-specified impedance. The actuating impedance in the relay is the function of distance in a distance protection relay. This impedance or corresponding distance is called the reach of relay.

Power system protection relays can be categorized into different types of relays.

Types of protection relays are mainly based on their characteristic, logic, on actuating parameter and operation mechanism.

Based on operation mechanism protection relay can be categorized as electromagnetic relay, static relay and mechanical relay. Actually, a relay is nothing but a combination of one or more open or closed contacts. These all or some specific contacts the relay change their state when actuating parameters are applied to the relay. That means open contacts become closed and closed contacts become open. In an electromagnetic relay, these closing and opening of relay contacts are done by the electromagnetic action of a solenoid.

In the mechanical relay, these closing and opening of relay contacts are done by mechanical displacement of different gear level system.

In static relay it is mainly done by semiconductor switches like thyristor. In digital relay on and off state can be referred as 1 and 0 state.

Based on Characteristic the protection relay can be categorized as

1. Definite time relays
2. Inverse time relays with definite minimum time (IDMT)
3. Instantaneous relays.
4. IDMT with inst.
5. Stepped characteristic.
6. Programmed switches.
7. Voltage restraint over current relay.

Based on of logic the protection relay can be categorized as

1. Differential.
2. Unbalance.
3. Neutral displacement.
4. Directional.
5. Restricted earth fault.
6. Over fluxing.
7. Distance schemes.
8. Bus bar protection.

9. Reverse power relays.
10. Loss of excitation.
11. Negative phase sequence relays etc.

Based on actuating parameter the protection relay can be categorized as

1. Current relays.
2. Voltage relays.
3. Frequency relays.
4. Power relays etc.

Based on application the protection relay can be categorized as

1. Primary relay.
2. Backup relay.

Primary relay or primary protection relay is the first line of power system protection whereas backup relay is operated only when primary relay fails to be operated during a fault. Hence backup relay is slower in action than primary relay. Any relay may fail to be operated due to any of the following reasons,

1. The protective relay itself is defective.
2. DC Trip voltage supply to the relay is unavailable.
3. Trip lead from relay panel to the circuit breaker is disconnected.
4. The trip coil in the circuit breaker is disconnected or defective.
5. Current or voltage signals from Current Transformers (CTs) or Potential Transformers (PTs) respectively is unavailable.

As because backup relay operates only when primary relay fails, backup protection relay should not have anything common with primary protection relay.

During study of electrical protective relays, some special terms are frequently used. For proper understanding, the functions of different protective relays, the definition of such terms must be understood properly. Such terms are,

1. Pick up current.
2. Current setting.
3. Plug setting multiplier (PSM).
4. Time setting multiplier (TSM).

Pick up current of relay

In all electrical relays, the moving contacts are not free to move. All the contacts remain in their respective normal position by some force applied to them continuously. This force is called the controlling force of the relay. This controlling force may be the gravitational force, may be a spring force, or may be a magnetic force. The force applied to the relay's moving parts for changing the normal position of the contacts is called deflecting force. This deflecting force is always in opposition of controlling force and is present always in the relay. Although the deflecting force always presents in the relay directly connected to live line, as the magnitude of this force is less than controlling force in normal condition, the relay does not operate. If the actuating current in the relay coil increases gradually, the deflecting force in electromechanical relay is also increased. Once the deflecting force crosses the controlling force, the moving parts of the relay initiate to move to change the position of the contacts in the relay. The current for which the relay initiates its operation is called **pick up current of relay**.

Current setting of relay

The minimum picks up the value of the deflecting force of an electrical relay is constant. Again, the deflecting force of the coil is proportional to its number of turns and the current flowing through the coil Now, if we can change the number of active turns of any coil, the required current to reach at minimum pick value of the deflecting force, in the coil also changes. That means if active turns of the relay coil are reduced, then proportionately more current is required to produce desired relay actuating force Similarly, if active turns of the relay coil are increased, then proportionately reduced current is required to produce same desired deflecting force.

Practically same model relays may be used in different systems. As per these system requirements, the pickup current of the relay is adjusted. This is known as the current setting of the relay. This is achieved by providing the required number of tapping in the coil. These taps are brought out to a plug bridge. The number of active turns in the coil can be changed by inserting a plug-in different point in the bridge.

The **current setting of relay** is expressed in percentage ratio of relay pick up current to the rated secondary current of CT.

That means,

$$\text{Currrent Setting} = \frac{\text{Pick up current}}{\text{Rated secondary current of CT}} \times 100\%$$

For example, suppose, you want that, an over current relay should operate when the system current φust crosses 125% of rated current. If the relay is rated with 1 A, the normal pick up current of the relay is 1 A and it should be equal to secondary rated current of current transformer connected to the relay.

Then, the relay will be operated when the current of CT secondary becomes more than or equal to 1.25 A. As per definition,

$$\text{Current Setting} = \frac{1.25}{1} \times 100\% = 125\%$$

The current setting is sometimes referred as current plug setting. The current setting of overcurrent relay is generally ranged from 50 % to 200 %, in steps of 25 %. For earth fault relay it is from 10% to 70% in steps of 10%.

Plug setting multiplier of relay

Plug setting multiplier of relay is referred as ratio of fault current in the relay to its pick up current.

$$\text{PSM} = \frac{\text{Fault current in relay coil}}{\text{Pick up current}} = \frac{\text{Fault current in relay coil}}{\text{Rated secondary current of CT} \times \text{Currrent Setting}}$$

Suppose we have connected on protection CT of ratio 200/1 A and current setting is 150%. Hence, pick up current of the relay is, 1 × 150 % = 1.5 A

Now, suppose fault current in the CT primary is 1000 A. Hence, fault current in the CT secondary, i.e., in the relay coil is, 1000 × 1/200 = 5A Therefore PSM of the relay is, 5 / 1.5 =3.33

Time setting multiplier of relay

The operating time of an electrical relay mainly depends upon two factors:

1. What a long distance to be traveled by the moving parts of the relay for closing relay contacts and
2. How fast the moving parts of the relay cover this distance.

So far by adφusting relay operating time, both of the factors to be adφusted. The adφustment of traveling distance of an electromechanical relay is commonly known as time setting. This adφustment is commonly known as **time setting multiplier of relay**. The time setting dial is calibrated from 0 to 1 in steps of 0.05 sec. But by adφusting only time setting multiplier, we cannot set the actual time of operation of an electrical relay. As we already said, the time of operation also depends upon the speed of operation. The speed of moving parts of relay depends upon the force due to the current in the relay coil. Hence, it is clear that the speed of operation of an electrical relay depends upon the level of fault current. In other words, the time of operation of the relay depends upon plug setting multiplier. The relation between time of operation and plug setting multiplier is plotted on a graph paper, and this is known as time/PSM graph. From this graph one can determine the total time taken by the moving parts of an electromechanical relay, to complete its total traveling distance for different PSM. In time setting multiplier, this total traveling distance is divided and calibrated from 0 to 1 in steps of 0.05.

So, when the time setting is 0.1, the moving parts of the relay have to travel only 0.1 times of the total traveling distance, to close the contact of the relay. So, if we get the total operating time of the relay for a particular PSM from time/PSM graph and if we multiply that time with the time setting multiplier, we will get, the actual time of operation of the relay for said PSM and TSM. For getting a clear idea, let us have a practical example. Say a relay has a time setting 0.1 and you have to calculate actual time of operation for PSM 10. From the time / PSM graph of the relay as shown below, we can see the total operating time of the relay is 3 seconds. That means the moving parts of the relay take total 3 seconds to travel 100 % traveling distance. As the time setting multiplier is 0.1 here, actually the moving parts of the relay have to travel only 0.1 × 100% or 10% of the total travel distance, to close the relay contacts. Hence, actual operating time of the relay is 3 × 0.1 = 0.3 sec. i.e., 10% of 3 sec.

Time vs. PSM curve of relay

This is relation curve between operating time and plug setting multiplier of an electrical relay. The x-axis or horizontal axis of the Time/PSM graph represents PSM and Y-axis, or vertical axis represents the time of operation of the relay. The time of operation represents in this graph is that, which required to operate the relay when time setting multiplier set at 1. From the Time/PSM curve of a typical relay shown below, it is seen that, if PSM is 10, the time of operation of the relay is 3 sec. That means the relay will take 3 seconds to complete its operation, with the time setting 1. It is also seen from the curve that, for a lower value of plug setting multiplier, i.e., for a lower value of fault current, the time of operation of the relay is inversely proportional to the fault current. But when PSM becomes more than 20, the operating time of relay becomes almost constant. This feature is necessary to ensure discrimination on very heavy fault current flowing through sound feeders.

MATLAB Examples

Example 1: A circuit breaker is rated as 1500 A, 1000 MVA, 33 kV, 3-second, 3-phase oil circuit breaker. Find (i) rated normal current (ii) breaking capacity (iii) rated symmetrical breaking current (iv) rated making current (v) short-time rating (vi) rated service voltage.

MATLAB Code

```
Clear;

clc;


i=1500;

mva=1000;

v=33*1e3;

t=3;


printf("(i) Rated normal current = %.0f A \n\n", i);

printf("(ii) Breaking capacity = %.0f MVA \n\n", mva);

rsbc=mva*1e6/sqrt(3)/v;

printf("(iii) Rated symmetrical breaking capacity = %.0f A \n\n", rsbc);

rmc=2.55*rsbc;

printf("(iv) Rated making current = %.0f A \n\n", rmc);

printf("(v) Rated short time rating = %.0f A for %.0f sec \n\n", rsbc, t);

printf("(vi) Rated service voltage = %.0f kV \n\n", v/1000);
```

Result

(i) Rated normal current = 1500 A

(ii) Breaking capacity = 1000 MVA

(iii) Rated symmetrical breaking capacity = 17495 A

(iv) Rated making current = 44613 A

(v) Rated short time rating = 17495 A for 3 sec

(vi) Rated service voltage = 33 kV

Example 2: A 50 Hz, 11 kV, 3-phase alternator with earthed neutral has a reactance of 5 ohms per phase and is connected to a bus-bar through a circuit breaker. The distributed capacitance up to circuit breaker between phase and neutral in 0·01 μF. Determine

(i) peak re-striking voltage across the contacts of the breaker

(ii) frequency of oscillations

(iii) The average rate of rise of re-striking voltage up-to the first peak

MATLAB Code

```
clear;
clc;
f=50;
v=11*1e3;
x=5;
c=0.01*1e-6;
l=x/2/%pi/f;
emax=sqrt(2)*v/1000/sqrt(3);
fn=1/2/%pi/sqrt(l*c);
t=1/2/fn;
avg=2*emax/t;
printf("Inductance per phase = %.5f H \n\n", l);
printf("(i) Maximum value of recovery voltage = %.2f kV \n\n", emax);
printf(" Peak restriking voltage = %.2f kV \n\n", 2*emax);
printf("(ii) Frequency of oscillation = %.0f Hz \n\n", fn);
printf("(iii) Time t at which peak restriking voltage occurs = %.2f us \n\n", t*1e6);
printf(" Average rate of rise of recovery voltage = %.0f kV/sec \n\n", avg);
```

Result

Inductance per phase = 0.01592 H

(i) Maximum value of recovery voltage = 8.98 kV

Peak restriking voltage = 17.96 kV

(ii) Frequency of oscillation = 12616 Hz

(iii) Time t at which peak restriking voltage occurs = 39.63 us

Average rate of rise of recovery voltage = 453228 kV/sec

Example 3: In a short circuit test on a circuit breaker, the following readings were obtained on single frequency transient:

(i) time to reach the peak re-striking voltage, 50 μ sec

(ii) the peak re-striking voltage, 100 kV

Determine the average RRRV and frequency of oscillations.

MATLAB Code

```
clear;
clc;
tp=50*1e-6;
rvp=100*1e3;
rrrv=rvp/tp;
fn=1/2/tp;
printf("Average RRRV = %.0f kV/sec \n\n", rrrv/1000);
printf("Natural frequency of oscillations = %.0f Hz \n\n", fn);
```

Result

Average RRRV = 2000000 kV/sec

Natural frequency of oscillations = 10000 Hz

Example 4: An air-blast circuit breaker is designed to interrupt a transformer magnetizing current of 11 A (r.m.s.) chops the current at an instantaneous value of 7 A. If the values of L and C in the circuit are 35·2 H and 0·0023 μF, find the value of voltage that appears across the contacts of the breaker. Assume that all the inductive energy is transferred to the capacitance.

MATLAB Code

```
clear;
clc;
```

```
i=7;
l=35.2;
c=0.0023*1e-6;
e=i*sqrt(l/c);
printf("Voltage across breaker = %.0f kV \n\n", e/1000);
```

Result

Voltage across breaker = 866 kV

Example 5: A fuse wire of circular cross-section has a radius of 0·8 mm. The wire blows off at a current of 8A. Calculate the radius of the wire that will blow off at a current of 1A.

MATLAB Code

```
clear;
clc;
r1=0.8;
i1=8;
i2=1;
r2=r1*(i2/i1)^(2/3);
printf("Radius = %.1f mm \n\n", r2);
```

Result:

Radius = 0.2 mm

Example 6: Determine the time of operation of a 5-ampere, 3-second overcurrent relay having a current setting of 125% and a time setting multiplier of 0·6 connected to supply circuit through a 400/5 current transformer when the circuit carries a fault current of 4000 A. Use the curve shown in Figure below.

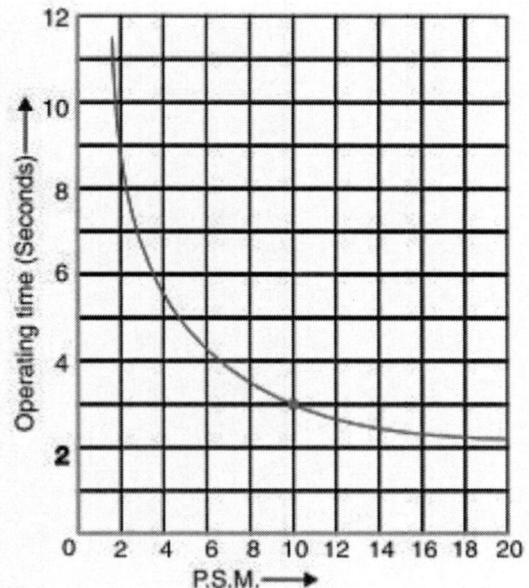

MATLAB Code

```
i=5;
cs=1.25;
ts=0.6;
tv1=400;
tv2=5;
fc=4000;
pup=i*cs;
frc=fc*tv2/tv1;
psm=frc/pup;
ot=3.5*0.6;
printf("Pick up current = %.2f A \n\n", pup);
printf("Fault current in relay coil = %.0f A \n\n", frc);
printf("Plug multiplier setting = %.0f \n\n", psm);
printf("Corresponding to the plug multiplier setting of %.0f, the time of operation is 3.5 seconds.\n \t Therefore actual relay operating time = %.2f s \n\n", psm, ot);
```

Result

Pick up current = 6.25 A

Fault current in relay coil = 50 A

Plug multiplier setting = 8

Corresponding to the plug multiplier setting of 8, the time of operation is 3.5 seconds.

Therefore, actual relay operating time = 2.10 s

10

Fundamentals of Neutral Grounding

Inroduction

The process of connecting neutral point of 3-phase system to earth (i.e. soil) either directly or through some circuit element is called neutral grounding. Neutral grounding provides protection to personal and equipment. It is because during earth fault, the current path is completed through the earthed neutral and the protective devices (e.g. a fuse etc.) operate to isolate the faulty conductor from the rest of the system. This point is illustrated in Figure 10.1.

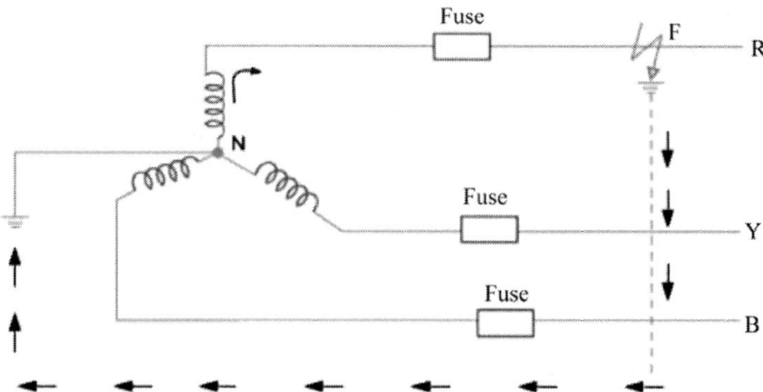

Figure 10.1: Neutral Grounding

Figure 10.1 shows a 3-phase, star-connected system with neutral earthed. Suppose a single line to ground fault occurs in line R at point F. This will cause the current to flow through ground path as shown in Figure1. Note that current flows from R phase to earth, then to neutral point N and back to R-phase. Since the impedance of the current path is low, a large current flow through this path. This large current will blow the fuse in R-phase and isolate the faulty line R. This will protect the system from the harmful effects of the fault. One important feature of grounded neutral is that the potential difference between the live conductor and ground will not exceed the phase voltage of the system i.e., it will remain nearly constant.

Advantages of neutral grounding

The following are the advantages of neutral grounding

1. Voltages of the healthy phases do not exceed line to ground voltages i.e., they remain nearly constant.
2. The high voltages due to arcing grounds are eliminated.
3. The protective relays can be used to provide protection against earth faults. In case earth fault occurs on any line, the protective relay will operate to isolate the faulty line.
4. The over voltages due to lightning are discharged to earth.
5. It provides greater safety to personnel and equipment.
6. It provides improved service reliability.
7. Operating and maintenance expenditures are reduced

Methods of neutral grounding

The methods commonly used for grounding the neutral point of a 3-phase system are:

1. Solid or effective grounding
2. Resistance grounding
3. Reactance grounding
4. Peterson-coil grounding

The choice of the method of grounding depends upon many factors including the size of the system, system voltage and the scheme of protection to be used.

MATLAB Examples

Example 1: Calculate the reactance of Peterson coil suitable for a 33 kV, 3-phase transmission line having a capacitance to earth of each conductor as 4.5 μF. Assume supply frequency to be 50 Hz.

MATLAB Code

```
clear;

clc;

f=50;
```

```
v=33*1e3;
c=4.5*1e-6;
xl=1/(3*2*%pi*f*c);
printf("Reactance of Peterson Coil = %.2f ohm \n\n", xl);
```

Result

Reactance of Peterson Coil = 235.79 ohm

Example 2: A 230 kV, 3-phase, 50 Hz, 200 km transmission line has a capacitance to earth of 0.02 μF/km per phase. Calculate the inductance and kVA rating of the Peterson coil used for earthing the above system.

MATLAB Code

```
clear
clc;
v=230*1e3;
lt=200;
c=0.02*1e-6;
f=50;
cle=c*lt;
l=1/(3*(2*%pi*f)^2*cle);
xl=2*%pi*f*l;
vph=v/sqrt(3);
fi=vph/xl;
rating=vph*fi;
printf("Capacitance of line to earth = %.6f F \n\n", cle);
printf("Required inductance of Peterson coil = %.2f F \n\n", l);
printf("Current through Peterson coil = %.0f A \n\n", fi);
printf("Rating of Peterson coil = %.0f kVA \n\n", rating/1000);
```

Result

Capacitance of line to earth = 0.000004 F

Required inductance of Peterson coil = 0.84 F

Current through Peterson coil = 501 A

Rating of Peterson coil = 66476 kVA

Example 3: A 50 Hz overhead line has line to earth capacitance of 1.2 μF. It is desired to use earth fault neutralizer (Peterson coil). Determine the reactance to neutralize the capacitance of (i) 100% of the length of the line (ii) 90% of the length of the line and (iii) 80% of the length of the line.

MATLAB Code

```
clear;

clc;

f=50;

c=1.2*1e-6;

c1=c;

c2=0.9*c;

c3=0.8*c;

xl1=1/(3*2*%pi*f*c1);

xl2=1/(3*2*%pi*f*c2);

xl3=1/(3*2*%pi*f*c3);

printf("Inductive reactance of coil to neutralize capacitance of 100%% of the
length of the line is %.2f ohm \n\n", xl1);

printf("Inductive reactance of coil to neutralize capacitance of 90%% of the
length of the line is %.2f ohm \n\n", xl2);

printf("Inductive reactance of coil to neutralize capacitance of 80%% of the
length of the line is %.2f ohm \n\n", xl3);
```

Result

Inductive reactance of coil to neutralize capacitance of 100% of the length of the line is 884.19 ohm

Inductive reactance of coil to neutralize capacitance of 90% of the length of the line is 982.44 ohm

Inductive reactance of coil to neutralize capacitance of 80% of the length of the line is 1105.24 ohm

11

MATLAB Application for Renewable Energy Resources

Solar Photovoltaic

A single solar cell cannot provide required useful output. Therefore, to increase output power level of a PV system, it is required to connect number of such **PV solar cells**. A solar module is normally series connected sufficient number of solar cells to provide required standard output voltage and power. One solar module can be rated from 3 watts to 300 watts. The solar modules or PV modules are commercially available basic building block of a solar electric power generation system.

Actually, a single solar PV cell generates very tiny amount that is around 0.1 watt to 2 watts. But it is not practical to use such low power unit as building block of a system. So required number of such cells are combined together to form a practical commercially available solar unit which is known as **solar module or PV module**.

In a solar module the solar cells are connected in same fashion as the battery cell units in a battery bank system. That means positive terminals of one cell connected to negative terminal voltage of solar module is simple sum of the voltage of individual cells connected in series in the module.

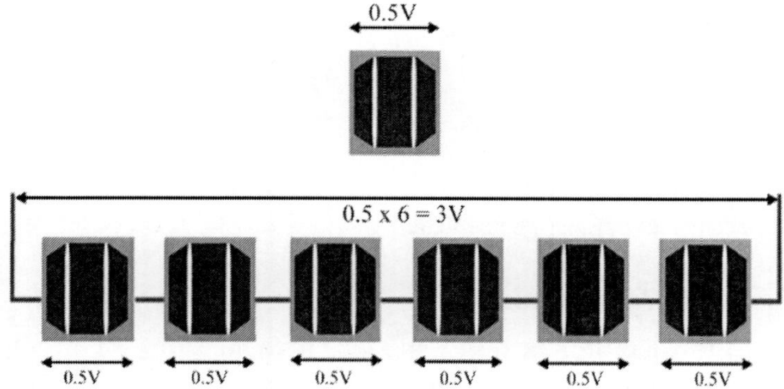

Figure 11.1: Series Connected Solar Module

The normal output voltage of a solar cell is approximately 0.5 V hence if 6 such cells are connected in series then the output voltage of the cell would be $0.5 \times 6 = 3$ Volt.

Ratings of solar module

The output from a solar module depends upon some conditions such as ambient temperature and intensity of incidence light. Hence the rating of a solar module must be specified under such conditions. It is standardized practice to express rating of PV or solar module at 25°C temperature and 1000 w/m² light radiation. The solar modules are rated with their output open circuit voltage (V_{oc}), short circuit current (I_{sc}) and peak power (W_p). That means these three parameters (V_{oc}, I_{sc} and W_p) can be delivered by a solar module safely at 25°C and 1000 w/m² solar radiations.

These conditions i.e. 25°C temperature and 1000 w/m² solar radiations are collectively called Standard Test Conditions. The Standard Test Conditions may not be available at site where the solar modules to be installed. This is because the solar radiations and temperature vary with location and time.

V-I characteristic of solar module

If we draw a graph by taking X-axis as voltage axis and Y-axis as currents of a solar module, then the graph will represent V-I characteristic of a solar module.

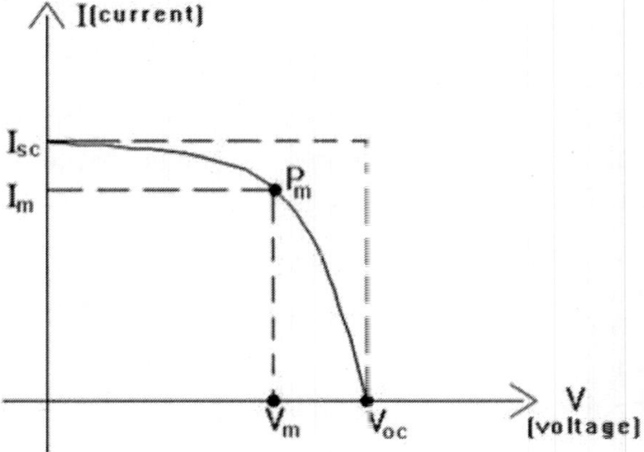

Figure 11.2: V-I Characteristucs of Solar Photovoltaic

Short circuit current of PV module

Under Standard Test Condition positive and negative terminal of a solar module are short circuited, then the current delivered by the module is short circuit current. Bigger value of this current indicates betterness of the module.

Although under standard test condition, this current also depends upon the area of the module exposed to the light. As it depends upon area, it is better to express by short circuit current per unit area.

This is denoted as ϑ_{SC}.

$$Jsc = \frac{Isc}{A}$$

Hence

Where, A is the area of the module exposed to the standard light radiation ($1000w/m^2$). Short circuit current of a PV module also depends upon solar cell manufacturing technology.

Open circuit voltage (V_{oc})

The voltage output of a solar module under standard test condition, when the terminals of the modules are not connected with any load. This rating of solar module mainly depends upon the technology used to manufacture solar cells of the module. More V_{OC} indicates betterness of the solar module. This open circuit voltage of a solar module also depends upon operating temperature.

Maximum power point

This is the maximum amount of power which can deliver by the module Under Standard Test Conditions. For a fixed dimension of a module higher the maximum power better the module. Maximum power also called peak power and this is denoted as W_m or W_p.

A solar module can be operated in any voltage and current combination upto V_{oc} and I_{sc}. But for a particular current and voltage combination under standard conditions the output power is maximum. If we proceed through y-axis of the V-I characteristic of a solar module, we will find the power output increases nearly linearly with current but after a certain current power output will fall down as it approaches to the short circuit current as at short circuit condition the voltage is considered to be ideally zero across the terminals of the solar module. So it is clear that maximum output power of a solar module does not occur at maximum current i.e. short circuit current instead it occurs at certain current which is less than short circuit current (I_{sc}). This current at which maximum output power occurs is denoted as I_m.

Similarly maximum power of a solar cell does not occur at open circuit voltage as it is open circuit condition and current through the cell is considered to be ideally zero, at this condition. But similarly, as previous case, maximum power in a solar module occurs at a voltage lower than open circuit voltage (V_{oc}). The voltage at which maximum power output occurs is denoted as V_m. The maximum power of a solar module is given as

$$Pm \text{ or } Wm = Vm * Im$$

The current and voltage at which maximum power occurs are referred as, current and voltage at maximum power point respectively.

Fill factor of a solar module

Fill factor of a solar module is defined as the ratio of maximum power ($P_m = V_m \times I_m$) to product of open circuit voltage (V_{oc}) and short circuit current (I_{sc}).

$$Fill\ Factor\ (FF) = \frac{Pm}{Voc * Isc}$$

Higher the Fill Factor (FF), better is the solar module.

Efficiency of solar module

Efficiency of solar module is defined as the ratio of maximum power at standard test condition, to the input power. Input power of a solar module is solar radiation which is considered as 1000 w/m^2. So, actual input power to the cell is 1000A W. Where, A is the exposed area of the solar module.

Therefore, efficiency,

$$N = \frac{Pm}{1000 * A} * 100$$

The V_{cc} of a solar cell depends upon mainly its manufacturing techniques. The table below shows the open-circuit voltage of different solar cells at standard test conditions.

Table 11.1: Open Circuit Voltages of Different Solar Cell

S. No.	Types of Solar Cell	Open circuit voltage at STC
1	Mono Crystalline Silicon Solar Cell	0.55 to 0.68 V
2	Poly Crystalline Silicon Solar Cell	0.55 to 0.65 V
3	Amorphous Silicon Solar Cell	0.7 to 1.1 V
4	Cadmium Telluride Solar Cell	0.8 to 1.0 V
5	Copper Indium Gallium Selenide Solar Cell	0.5 to 0.7 V
6	Gallium Indium Phosphide/ Gallium Arsenide/ Gallium Solar Cell	1 to 2.5 V

MATLAB Examples

Example 1: Write a MATLAB code to draw V-I and P-V curve for solar photovoltaic.

MATLAB Code

```
%PVcharacteristics.m
%This code plots the P-V and I-V curves of the photovoltaic array
 clc
clear
Tcell = 25; %Ambient cell temperature (degrees celcius)
G = 1000;%Solar Irradiation (W/m2)
Voc = 43.2; %Open circuit voltage
%Uses photovoltaic model to calculate values
Va = 0:01:Voc;
Ipv = PVcharacteristics_func(Va,G,Tcell);
Ppv = Va.*Ipv; %Calculating PV output power
Pmax = max (Ppv); %Finding the maximum power of the array
%Plots P-V characteristic
% plot(Va,Ppv,'r','Linewidth', 3)
% title('P-V Characteristic (Irradiance 1kW/m^2)');
% xlabel('Voltage (V)');
% ylabel('Power (W) ');
% grid on
%Plots I-V characteristic
figure
plot(Va,Ipv,'b','Linewidth', 3)
figure
plot(Va,Ppv,'b','Linewidth', 3)
title('I-V Characteristic(Irradiance 1kW/m^2)');
xlabel('Voltage (V)');
```

```
ylabel('Current (A)');
grid on
```

Results

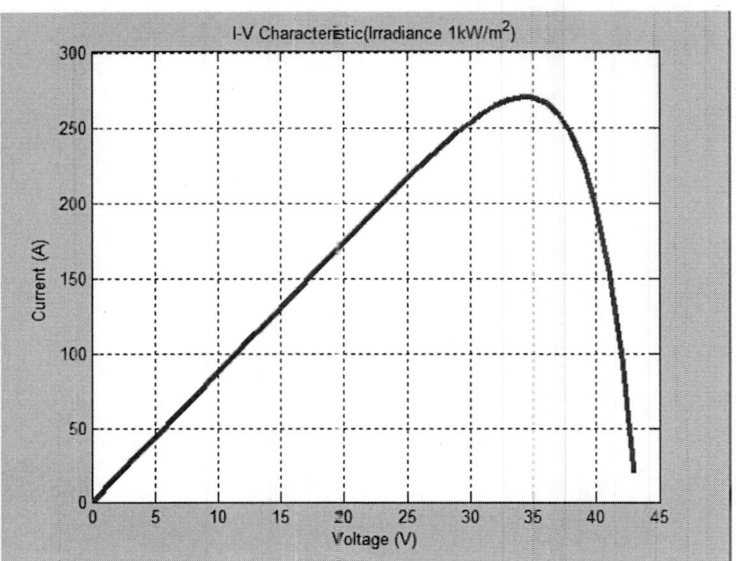

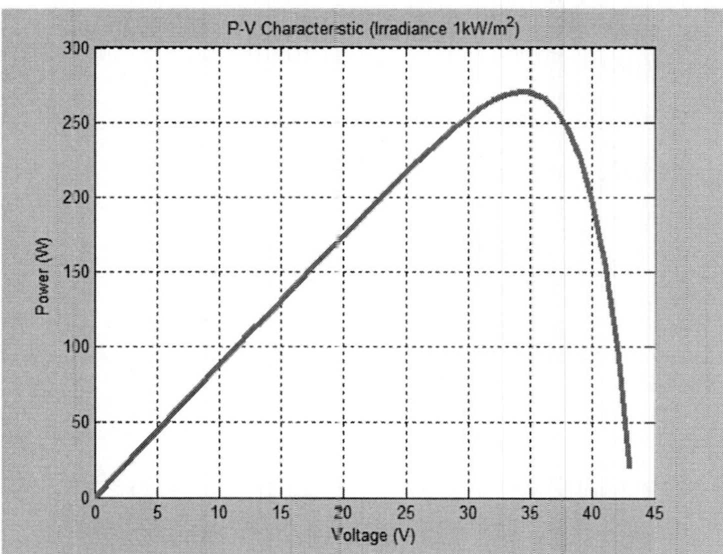

Example 2: Write a MATLAB program to calculate number of batteries required to store solar energy.

MATLAB Code

```
close all; clear all; clc
I_min=4.57; %Minimum insolation in kWh
U_battery=12; %Battery voltage
Eta_solarcell=0.12;   %Efficiency of the solar cell
Eta_battery=0.8;      %Efficiency of the solar cell
Eta_CR=0.95;             %Efficiency of the solar cell
Eta_invertor=0.92;       %Efficiency of the solar cell
Eta_cable=0.99; %     Efficiency of the solar cell
A_module=0.6312;      %Area of the chosen solar cell module
E_day_total=2.7167;      %Total energy need per day in kWh
n_day=3;                 %Number of days the battery will be designed for
B_battery=190;         % Storage capacity in Ah for the chosen battery
B_day_produced=I_min*Eta_solarcell*1000/U_battery;
                      %Ah produced per day and square meter of solar cells
B_day_consumed=E_day_total*1000/U_battery;
                      %Total number of ampere hours consumed per day
B_day_required=B_day_consumed/(Eta_battery*Eta_CR*Eta_invertor*Eta_
cable);
                      %Total ampere hours needed including system losses
A_required=B_day_required/B_day_produced;
                      %The required solar cell area
n_modules=ceil(A_required/A_module);
                      %The number of solar cell modules that will be needed
B_total=B_day_consumed*n_day;    %Total back-up storage needed in Ah
n_battery=ceil(B_total/B_battery);     %Number of batteries needed to supply
the needed storage capacity
disp([' Maximum daily energy output per m^2 is ', num2str(B_day_produced),'
Ah'])
```

disp([' The daily required energy conversion is ', num2str(B_day_required),' Ah'])

disp([' The area needed is ', num2str(A_required), ' m^2, and for that',num2str(n_modules), ' solar cell modules will be used'])

disp([' The total needed battery storage is ', num2str(B_total), ' Ah, and for that ',num2str(n_battery), ' batteries will be used.'])

Results

Maximum daily energy output per m^2 is 45.7 Ah

The daily required energy conversion is 327.0573 Ah

The area needed is 7.1566 m^2, and for that12 solar cell modules will be used

The total needed battery storage is 679.175 Ah, and for that 4 batteries will be used.

Example 3: Write a MATLAB program to calculate altitude angle.

MATLAB Code

lati=29.4; % latitude of the place (degrees)

9ulian_date=100;

t=0.94; % transmittance (unitless)

S=1367; %solar constant (w/m^2)

p=100; %air pressure (KPa)

% Define the time at which we want to calculate the solar radiation;

hours=[7,8,9,10,11,12,13,14,15,16,17];

% Calculate the hour angle;

hangle=(12.0-hours)*15.0*pi/180.0; % Note pi/180.0 is the factor to convert angles in degrees in radiances.

% Calculate the sun declination angle;

declangle=23.45*sin(2.0*pi*(284.0+9ulian_date)/365.0)*pi/180.0;

% Calculate the solar zenith angle;

cosz=sin(lati*pi/180.0)*sin(declangle)+cos(lati*pi/180.0)*cos (declangle)*cos(hangle);

```
% Calculate optical air mass, m.
m=p./(101.3*cosz);
plot(hours,90-cosz*180/pi,'-');
myD=Julian_date;
dom=day(myD);
[n,m]=month(myD);
myday=[num2str(n,2) '/' num2str(dom,2)];
til=['Altitude Angle: at ',num2str(lati),' deg. Latitute on', myday];
title(til);
grid on
xlabel('Hours of Day');
ylabel('Altitude Angle (Degree)');
```

Result

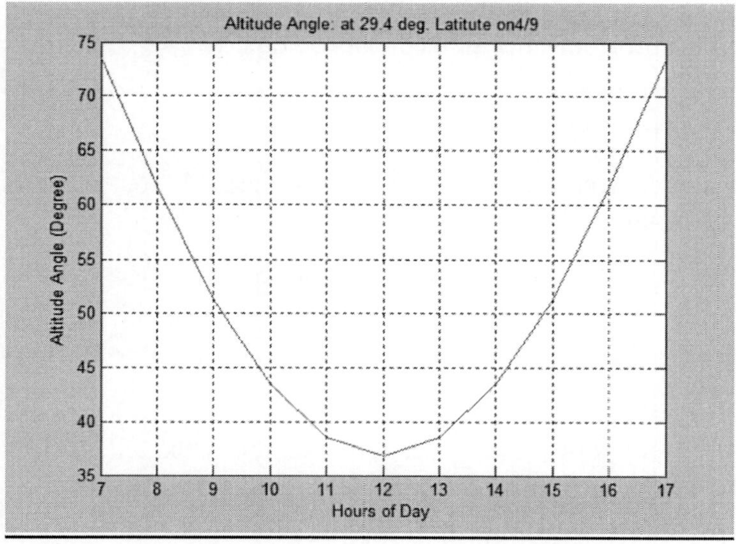

Example 4: Write a MATLAB program to forecast the Weather conditions.

MATLAB Code

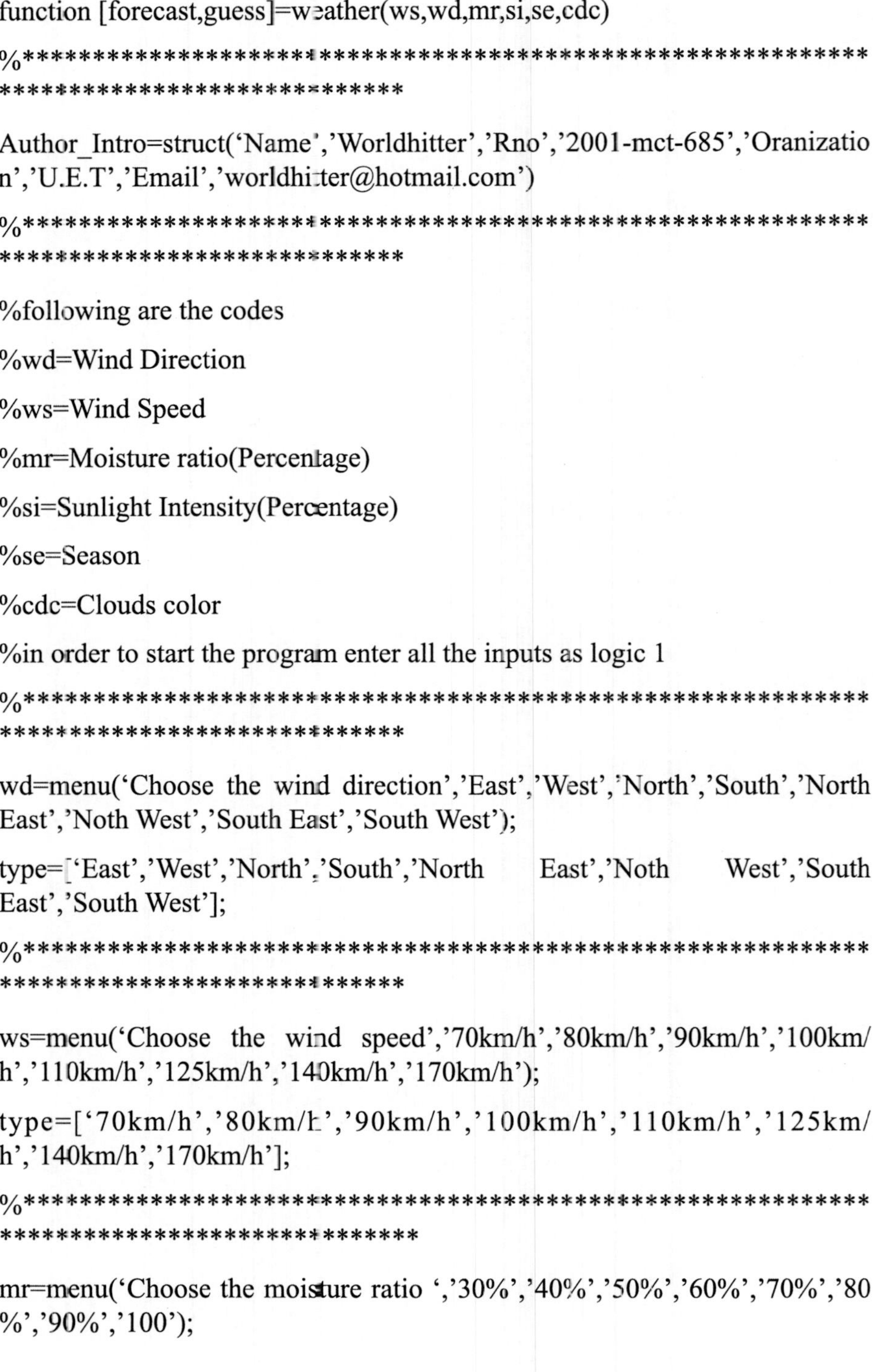

```
function [forecast,guess]=weather(ws,wd,mr,si,se,cdc)

%***************************************************
************************

Author_Intro=struct('Name','Worldhitter','Rno','2001-mct-685','Oranizatio
n','U.E.T','Email','worldhitter@hotmail.com')

%***************************************************
************************

%following are the codes

%wd=Wind Direction

%ws=Wind Speed

%mr=Moisture ratio(Percentage)

%si=Sunlight Intensity(Percentage)

%se=Season

%cdc=Clouds color

%in order to start the program enter all the inputs as logic 1

%***************************************************
************************

wd=menu('Choose the wind direction','East','West','North','South','North
East','Noth West','South East','South West');

type=['East','West','North','South','North East','Noth West','South
East','South West'];

%***************************************************
************************

ws=menu('Choose the wind speed','70km/h','80km/h','90km/h','100km/
h','110km/h','125km/h','140km/h','170km/h');

type=['70km/h','80km/h','90km/h','100km/h','110km/h','125km/
h','140km/h','170km/h'];

%***************************************************
*************************

mr=menu('Choose the moisture ratio ','30%','40%','50%','60%','70%','80
%','90%','100');
```

```
type=['30%','40%','50%','60%','70%','80%','90%','100'];
%******************************************************
***************************
si=menu('Choose sunlight intensity)','0%','20%','40%','50%','60%','70%','80%','100%');
type=['0%','20%','40%','50%','60%','70%','80%','100%'];
%******************************************************
**************************
se=menu('Choose the season','Winter','Summer','Spring','Autumn');
type=['Winter','Summer','Spring','Autumn'];
%******************************************************
**************************
cdc=menu('Choose the color of clouds','Grey','White','Black','Greenish grey','Clear Sky');
type=['Grey','White','Black','Greenish grey','Clear Sky'];
%******************************************************
**************************
if( (wd == 1) | (ws==1) | (mr == 1) | (si == 2) | (se == 1) | (cdc == 2))
forecast='The weather will remain pleasent for next 2 days sky will be clear afterein'
guess='The temperature ranges from 15 to 20 degrees centigrade'
else if((wd == 1) | (ws == 2) | (mr == 1) | (si == 4) | (se == 1) | (cdc == 2))
forecast='The weather will pleasent throughout the day'
guess='The temperature ranges from 13 to 17 degrees celciuus'
else if((wd == 1) | (ws == 2) | (mr == 1) | (si == 4) | (se == 1) | (cdc == 2))
forecast='The weather will pleasent throughout the day'
guess='The temperature ranges from 13 to 17 degrees celciuus'
else if((wd == 1) | (ws == 3) | (mr == 2) | (si == 4) | (se == 1) | (cdc == 2))
forecast='There might be light shower after noon '
guess='The temperature ranges from 7 to 14 degrees celciuus'
```

```
else if((wd == 1) | (ws == 2) | (mr == 1) | (si == 4) | (se == 1) | (cdc == 2))
forecast='The weather will pleasent throughout the day'
guess='The temperature ranges from 13 to 17 degrees celciuus'
elseif((wd == 8) | (ws == 5) | (mr == 5) | (si ==1) | (se == 2) | (cdc == 3))
forecast='Thunder storm and heavy showers are expected'
elseif((wd == 3) | (ws == 2) | (mr == 2) | (si == 2) | (se == 3) | (cdc == 4))
forecast='This is an indication of svere storm'
guess='The temperature ranges from 20 to 30 degrees celcius'
elseif((wd == 4) | (ws == 1)  (mr == 3) | (si == 6) | (se == 2) | (cdc == 5))
forecast='The weather will remain hot for a week'
guess='The temperature ranges from 38 to 49 degrees celcius'
elseif((wd == 4) | (ws == 4)  (mr == 7) | (si == 4) | (se == 2) | (cdc == 2))
forecast='These are clear indication of rain after one day'
guess='The temperature ranges from 30 to 41 degrees celcius'
elseif((wd == 7) | (ws == 7)  (mr == 8) | (si == 1) | (se == 2) | (cdc == 4))
forecast='Cruel clouds may distroy the crops, they can be activated at any time'
guess='The temperature ranges from 15 to 34 degrees celcius'
elseif((wd == 3) | (ws == 5) | (mr == 6) | (si == 4) | (se == 3) | (cdc == 5))
forecast='This is beautiful weather it will be maintained for 2 to 3 days'
guess='The temperature ranges from 23 to 28 degrees celcius'
elseif((wd == 6) | (ws == 3) | (mr == 7) | (si == 5) | (se == 2) | (cdc == 2))
forecast='Hot moisty air will blow all the day'
guess='The temperature ranges from 31 to 44 degrees celcius'
elseif((wd == 5) | (ws == 6) | (mr == 5) | (si == 8) | (se == 2) | (cdc == 5))
forecast='Hot air will blow for two days'
guess='The temperature ranges from 35 to 54 degrees celcius'
elseif((wd == 8) | (ws == 8) | (mr == 4) | (si == 5) | (se == 2) | (cdc == 1))
```

```
forecast='Hot moisty air will blow all the day'
guess='The temperature ranges from 31 to 44 degrees celcius'
elseif((wd == 2) | (ws == 3) | (mr == 4) | (si == 6) | (se == 4) | (cdc == 3))
forecast='Smooth weather will be through out the day'
guess='The temperature ranges from 26 to 37 degrees celcius'
else forecast='Weather will be all right soon '
guess='Temperature will be in the normal range today'
end

%*************************************************************
******************************
if (wd == 8)
 forecast='Be careful! there may be a rain session for a day after 30 hours'
 guess='The temperature will be in the range of 12 to 34 degrees celcius'
end
if ((si == 2) | (se == 1))
 forecast='OOps! too cold ,save your self by wearing ,of something wool'
 guess ='The temperature ranges from 1 to 12 degrees celcius'
end
```

Experiments

Date

Experiment No.-1

Aim: Study the characteristics between voltage and time of over voltage relay.

Apparatus Required

S. No.	Apparatus	Type	Range
1	Voltmeter	MI	0-300 V
2	Auto-transformer	-	0-250 V
3	Indicating lights	-	-
4	Over voltage relay	-	-
5	Timer with starter stop facility	-	-
6	Push button for timer START of STOP	-	-
7	Rotary switch	-	-
8	MCB (Isolator)	-	-
9	Insulating terminals	-	-

Theory

There are several instruments in power system which may be exposed to over voltage like motor, transformer, capacitor etc. To protect instruments from over voltage, relay is necessary. The several type of over voltage relay is studying here. It is electromechanical type and very rigid in use. Its failure is almost negligible.

Here the relay has an operating coil which core developed a rotating flush in disk due to this disk rotates at certain voltage. Before the certain voltage a vehicle spring makes force in opposite directions. So, the disk is not is the condition to rotate. A socket is given to different setting of tripping voltage. Time setting multiplier is given to setting direction so the disk is not in the condition to rotate.

We apply tripping voltage to pin no. 9 and 10 to this relay; the pin no. 9 will extend to a transformer, disk operating coil and through moving this contact to a bridge. Pin no. 10 will extend to the common pole of plug setting socket all the primary tapings, are connected and secondary will fed to operating disk coil of bridge rectifier. The output of bridge is filleted by capacitor C, and resistor R_1. The filtered DC voltage is fed to flag voltage capacity of flag relay resistance R_q and it is meant for increasing the operating voltage capacity of disk operating coil.

PSM = applied voltage plug setting.

Diagram

Observation Table

S. No.	Voltage (V)	Time (S)
1.	131	26.66 Sec.
2.	141	16 Sec
3.	176	7.48 Sec

Result

We have successfully studied the over voltage relay.

Merits of Relay

1. Normally no false tripping.
2. The relay is electromechanical type can tolerate rough tough use.
3. High accuracy in respect of thermal relay.

Demerits of Relay

1. The relay is electromechanical type, so it should be used in dust for environment.
2. The frequency variations can change parameters.

Precaution

1. The relay should be used carefully and test beyond its rating.
2. Suppose the 1A relay can withstand up to 1A continuously and it, more than 1A relay should trip the system within few second.

Date

Experiment No.-2

Aim: To Study the characteristics between voltage and time for Inverse Definite minimum time over current relay.

Apparatus Required:

S. No.	Apparatus	Type	Range
1	Voltmeter	MI	0-300 V
2	Auto-transformer	-	0-250 V
3	Indicating lights	-	-
4	Over voltage relay	-	-
5	Timer with starter stop facility	-	-
6	Push button for timer START of STOP	-	-
7	Rotary switch	-	-
8	MCB (Isolator)	-	-
9	Insulating terminals	-	-

Theory

It is defined as the relay, which operates only when the value of the current is greater than the relay setting time. It protects the equipment's of the power system from the fault current. In an over-current relay, the alerting quantity is only current. This is only one current operated element in the relay. No voltage coil is required to construct this protection relay.

Working Principle

These would be essentially a current coil-when normal current flows through this coil, the magnetic affect generated by the coil is not sufficient to move the moving element of the relay, as in this condition the restraining force is greater than deflecting force. But when the current through the coil increase, the magnetic effect increase, and after a certain level of current, the deflecting force generated by the magnetic effect of the coil, crosses the restraining force. As a result, the moving element starts moving to change the contact position in the relay. Although there are different types of over current relays but basic working type of over current relays but basic working principle of overcorrect relay is less same.

Diagram

Observation

Deoperating current (Io) = 1.14

Operating current (I) = 1.4

S. No.	Current (Ic) A	Tripping Time (Sec)
1.	1.5	25-52
2.	2	0.100
3.	2.5	5.314
4.	3	3.722

Result

We have successfully studied the over current relay.

Merits of Relay

1. Normally no false tripping.
2. The relay is electromechanical type can tolerate rough tough use.
3. High accuracy in respect of thermal relay.

Demerits of Relay

1. The relay is electromechanical type, so it should be used in dust for environment.
2. The frequency variations can change parameters.

Precaution

1. The relay should be used carefully and test beyond its rating.
2. Suppose the 1A relay can withstand up to 1A continuously and it, more than 1A relay should trip the system within few second.

Date

Experiment No.-3

Aim: Study about the voltage v/s tripping time characteristic of under voltage relay.

Apparatus Required

S. No.	Apparatus	Type	Range
1	Voltmeter	MI	0-300 V
2	Auto-transformer	-	0-250 V
3	Indicating Lights	-	-
4	under voltage relay	-	-
5	Timer with starter stop facility	-	-
6	Push button for timer START of STOP	-	-
7	Rotary switch	-	-
8	MCB (Isolator)	-	-
9	Insulating terminals	-	-

Theory

The principal of relay is based to generate a rotary flex in an aluminum disc. Due to this disc gets a torque of rotates clockwise direction. A vehicle spring changed in opposite direction of the rotation torque. The position of stopper sets the tripping time of relay. As the stopper was most clockwise direction. So, the disc has to travel more at the time of tripping so, it takes more time.

Steps to perform

1. Set the voltage at the relay by plug setting (PS) as desired.
2. Set the voltage adφust / relay test selector switch at voltage adφust.
3. Switch On the MCB.
4. Start the system by pushing the on button.
5. Keep the voltage -10% of the plug set voltage by the voltage source.
6. New Change the voltage adφust / relay test selector switch at relay test.

Diagram

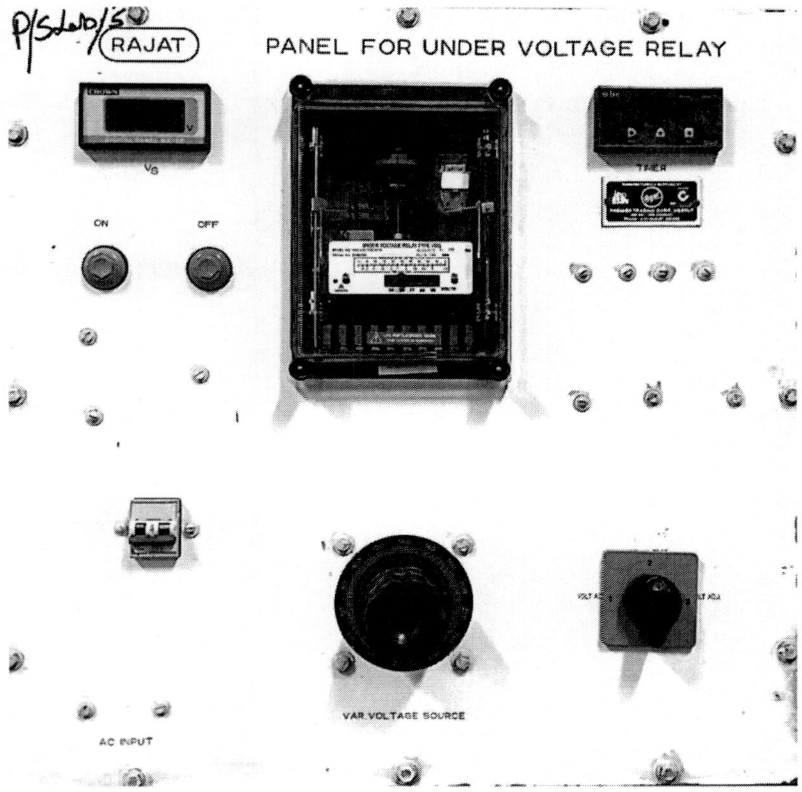

Observation Table

S. No.	Plug Setting (PS)	Tripping Time
1.	53V	13.61S
2.	45V	8.07S
3.	33V	5.412S
4.	25V	4.56S

Result

We have successfully studied the under-voltage relay.

Merits of the Relay

1. The relay is electromechanical type can tolerance rough-tough one.
2. High accuracy in respect of thermal relays.
3. Normally no false tripping.

Demerits of the relay

1. The relay is electromechanical type so it should be used in dust free environment.
2. The frequency variations can change the parameters.

Precautions

1. The relay should be used very carefully and fast, beyond its rating.
2. Suppose the no volts relay can withstands up to 110V continuously and if more than 110V the relay is not safe.

Date

Experiment No.-4

Aim: To study about directional over current relay.

Apparatus Required:

S. No.	Apparatus	Type	Range
1	Voltmeter	MI	0-300 V
2	Ammeter	MI	0-10 A
3	Auto-transformer	-	0-250 V
4	Transformer	-	110 V
5	Indicating height	-	-
6	Directional over voltage relay	-	-
7	Timer with starter stop facility	-	-
8	Push button for timer START of STOP	-	-
9	Rotary switch	-	-
10	MCB (Isolator)	-	-
11	Insulating terminals	-	-

Theory

There are several over current protections such as fuse, thermal, IDMT, directional relay etc. Directional over current relay is a high accuracy over current relay. The beauty of this relay over IDMT that it will operate when the fault is towards this. If the fault is at but or reverse side of line, it will block the tripping. A special arrangement is given at the top of this relay. An armature and cup will move and make a contact when the fault reverse side of the line, the armature / cup will move reverse direction and restrict to make the contact. It is widely used to prevent over current on transmission lines, power transformer etc. because the error and tripping time of the relay is tolerable by the lines and transformer.

The time multiplier setting for an inverse time relay is defined as:

$$\text{TMS} = \frac{T}{T_m}$$

Where,

T - The required time of operation.

T_m - The time obtained from the relay characteristic curve at

TMS = 1.0 using PSM equivalent to most fault current.

Current setting is adφusted by means of a tapped plug bridge hence Known as PSM,

$$\text{PSM}=\frac{\text{Fault Current}}{\text{Plug Setting}}$$

Selection Time Setting

For selective operation, when there is a no. of relay connected in series, the relay farthest from the source should be set to operate. In the minimum possible time. For succeeding relays toward source, a time delay step is given. It will automatically have a higher selectivity at the minimum fault current as the curve is more inverse on lower current region.

Selection of current setting

It is necessary to calculate the maximum fault current which can occur at each relay position. On a redial system the lowest setting must be at the farthest and the setting being increased for the subsequent relay towards the sources. As per Indian standard the operating value should not exceed 130% of setting.

$$1 \text{ Setting} = \frac{\text{Minimum short circuit current}}{1.3}$$

Diagram

Observations

	Fault Current	Tripping Time
Plug Setting	0.80	33.78
0.5A	1.00	13.63
	1.30	9.498
	1.62	7.77

Result

Thus, we have studied the directional over current relay.

Precautions

1. The relay should be used carefully, should not be touched.
2. Ratings of the relay should not be ignored

Date

Experiment No.-5

Aim: To study Earth fault Relay.

Apparatus Required:

S. No.	Apparatus Required	Type	Range
1	Voltmeter	MI	0-300 V
2	Auto-transformer	-	0-270 V
3	Indicating lamp and Push button	-	-
4	Earth Fault relay	-	-
5	Timer with starter stop facility	-	-
6	Rotary switch	-	-

Theory

Earth fault relay is used to protect feeder against faults involving ground. Typically earth fault are single line to ground and line to ground fault. For the purpose of setting and coordinating only single line to ground faults are considered. There can be significant variation in the earth fault current values. They can even be below the load current due to large impedance to ground. Hence, to provide sensitive protection, earth fault relay uses zero sequence current rather than phase current for fault diction.

Diagram

Observation Table

S. No.	Current (A)	Time
1.	0.22A	7.47
2.	0.29A	4.202
3.	0.35A	3.22 sec
4.	0.44A	2.73 Sec
5.	0.52 A	2.41 Sec
6.	0.65A	2.026 Sec

Result

Thus, we have studied the relay by examining it and performing by varying current and noting time. We conclude that with increase in current, time decreases.

Precautions

3. The relay should be used carefully, should not be touched.
4. Ratings of the relay should not be ignored

Viva Questions based on Laboratory

1. What is the main objective behind conducting experiments in laboratories?

Ans. It is to verify the theoretical aspects. This ultimately leads to better understanding of principles of science and engineering. In turn a skill is developed for application of basic principles and laws.

2. What should be done if a person gets an electric shock?

Ans. He should be immediately removed from the contact of the live conductor using any one method explained in this chapter. Then, he should be given artificial respiration. Next, his burns should be given first aid. Finally, the doctor should be called or he should be taken to a hospital.

3. Name the different methods of artificial respiration.

Ans. The different methods of artificial respiration are

(i) Mouth to mouth artificial respiration.

(ii) Schafer's method for artificial respiration.

(iii) Silvestre's method for artificial respiration.

4. What do you understand by Earthing?

Ans. Earthing means all the metal parts of an appliance or an equipment which are supposed not to carry any current are effectively connected to solid earth.

5. Name the different factors on which the soil resistivity depends.

Ans. The soil resistivity depends on, moisture contents of the soil, its chemical composition and temperature.

6. What artificial agents are used to reduce the resistivity of the soil?

Ans. The artificial agents used to reduce the resistivity of the soil are sodium chloride, calcium chloride, sodium carbonate and copper sulphate.

7. What materials would you like to recommend for earth electrodes?

Ans. Copper, copper clad or zinc coated iron electrodes are recommended materials for earth electrodes.

8. What are the various purposes of Earthing?

Ans. The various purposes of earthing are (i) Safety of the persons working with electricity. (ii) Safety of the equipment. (iii) To provide proper and reliable operation of protective equipment.

9. What are the recommended maximum values of the earth resistances during dry season?

Ans. The recommended values of earth resistances during dry season are the following: 1. High rated power plants = 0.5 2. 2. Moderate and low rated plants = 192. 3. Large substations = 22. 4. Small substations = 2.5 2. 5. All other installations = 5 2.

10. What are the various methods of earthing commonly used?

Ans. The common methods of earthing are (a) Pipe/rod earthing. (b) Plate earthing.

11. What are the limitations of battery as a source of DC supply?

Ans. The batteries are costly, they require frequent replacement. Arrangement of recharging a battery has to be made. It is not maintenance free. The battery gets consumed as well as discharged even if it is not under use.

12. Can the lamps in a house-lighting circuit be connected in series?

Ans. No, the lamps in a house-lighting circuit shouldn't be connected in series. The reason being that if a break occurs in any part of the circuit, the entire circuit will be inoperative. The supply voltage would be divided among the lamps connected in series. This would require a higher supply voltage.

13. What electric power supply system should be available in an electrical engineering laboratory?

Ans. The electric power supply systems which should be available in an electric engineering laboratory are 220 V d.c.; 230 V a.c. 50 Hz, and 3-phase 440 V, 50 Hz. Sometimes 110 V, d.c. and 110 V a.c. 50 Hz systems are also made available.

14. In what different ways a rheostat is used in practical circuit?

Ans. A rheostat can be used as a variable resistance or as a potential divider.

15. What is the frequency of AC supply available in our houses?

Ans. The frequency of AC supply available in our houses is 50 Hz.

16. What is the difference between direct current and alternating current?

Ans. The instantaneous value of direct current is positive at all instants of time, whereas the instantaneous value of alternating current changes from +ve to -ve and vice versa 50 times in a second i.e., equal to the supply frequency.

17. How can you obtain a variable AC supply in the laboratory?

Ans. A variable AC supply is obtained by means of a Variac. It is basically an autotransformer. It has two pairs of terminals. One pair is meant for input, which are connected to a constant voltage a.c. source. Another pair is meant for output, at which a variable AC voltage is available. A variable 3-phase supply can also be obtained using 3-phase Variac.

18. How an ammeter is connected in an electrical circuit?

Ans. Ammeter is always connected in series of an electrical circuit, because it is used to measure current through a branch of the circuit.

19. How a voltmeter is connected in an electric circuit?

Ans. Voltmeter is always connected in parallel because it is used to measure voltage across two points in a circuit.

20. How a wattmeter is connected in an electric circuit?

Ans. There are four terminals in a wattmeter. There are two coils in a wattmeter, one being the current coil (low resistance) and another being the pressure coil (high resistance). The current coil is always connected in series and pressure coil is always connected in parallel. This is the case for a single-phase wattmeter. In case of a 3-phase wattmeter, there are six terminals. Three terminals are connected as input and three terminals are connected as output.

21. What will happen when a permanent magnet moving coil instrument be connected in an AC circuit?

Ans. A permanent magnet moving coil instrument indicates the average value of the quantity under measurement. It can be safely used for measurement of d.c. quantities. But, since the average value of an a.c. quantity is zero, if a permanent magnet moving coil instrument be connected in an a.c. circuit, it would indicate a zero value.

22. How do you provide loading to an electric motor in electrical engineering laboratory?

Ans. There are two ways of loading an electric motor. First, is by using a belt and pulley. It provides the mechanical loading. Another way is to couple the motor with a d.c. generator, and d.c. generator in turn is loaded electrically.

23. How, do you measure speed in laboratory?

Ans. We can measure speed by tachometer. For accurate measurement of speed, synchroscope is also used.

24. How various lamps are connected in a lamp bank load?

Ans. These are connected in parallel.

25. What do you mean by a series connection?

Ans. When two or more components are connected such that the same (not merely equal) current passes through them, they are said to be connected in series.

26. What do you mean by a parallel connection?

Ans. When two or more components are connected such that the same (not merely equal) potential difference occurs across them, they are said to be connected in parallel.

27. What are the general requirements of an indicating instrument from construction and operation points of views?

Ans. Following are the general requirements of an indicating instrument from construction and operation points of views.

1. Pointer and scale.
2. Deflection system and deflecting torque.
3. Controling system and controlling torque.
4. Damping system and damping torque.
5. Supporting system for the moving parts.
6. Casing.

28. What is the purpose of deflecting torque in indicating type of measuring instruments?

Ans. The purpose of deflecting torque is to produce a deflection in the moving system of the instrument which should be proportional to the quantity under measurement.

29. What is purpose of the controlling torque in indicating type of measuring instruments?

Ans. In the absence of controlling torque, the moving system of a measuring instrument would continue to rotate. The controlling torque enables the moving system to indicate a constant reading. This controlling torque is equal and opposite to the deflecting torque under steady condition.

30. What is the purpose of damping in indicating type of measuring instruments?

Ans. In the absence of damping torque, the moving system of the measuring instrument would vibrate under the combined influence of deflecting torque and the controlling torque for an inordinately long time. The damping provided in these instruments damps the vibrations of the moving system in a short period of time.

31. How the controlling torque can be provided in measuring instruments?

Ans. The controlling torque can be provided in measuring instruments by any one of the following methods.

(i) By a pair of hair springs attached to the moving system. (ii) By gravity control. (iii) By a suspension wire made of phosphor bronze.

32. What different methods of damping are employed in indicating type of measuring instruments?

Ans. Following are the different methods of damping employed in indicating type of measuring instruments.

(i) Air friction. (ii) Fluid friction. (iii) Eddy currents.

33. What do you mean by the supporting system in reference to the indicating type measuring instruments?

Ans. Supporting system is meant for supporting the moving system of the instruments.

34. What are the different types of supporting systems used in the indicating type instruments?

Ans. The different types of supporting systems used in the indicating type of measuring instruments are

(i) Pivot method. (ii) Suspension method.

35. What is the advantage of using steel castings for casing for meters?

Ans. By using the steel casting for meters, the effect of extraneous magnetic fields on the moving system of the instrument is eliminated.

36. In a moving coil galvanometer why the magnets and the core of the moving coil are made cylindrical?

Ans. It serves the following two purposes. (i) The length of air gap is reduced. This improves the sensitivity of the instrument. (ii) The magnetic field becomes radial. Due to this the deflection becomes directly proportional to the current though the coil and thus a uniform scale can be used.

37. What will happen if an AC current is passed through a moving coil instrument?

Ans. Meter will not show any deflection.

38. What are the various types of ohmmeters? What are their applications?

Ans. There are three types of ohmmeters. (i) Shunt type ohmmeters. These are used to measure low resistances. (ii) Series-type ohmmeters. These are used to measure medium resistances. (iii) Meggar type ohmmeters. These are used to measure high resistances.

39. In a shunt-type ohmmeter, if the test terminals are short circuited, what reading would it show?

Ans. The shunt-type ohmmeter would show a zero reading and zero deflection.

40. In a shunt-type ohmmeter, if the test terminals are open circuited, what reading would it show?

Ans. A shunt-type ohmmeter would show a reading of infinity and full-scale deflection.

41. In a series-type ohmmeter, if the test terminals are short circuited, what reading would it show?

Ans. The series-type ohmmeter would show a zero reading but a full-scale deflection because it has an inverted scale.

42. In a series-type ohmmeter, if the test terminals are left open, what reading would it show?

Ans. The series-type ohmmeter would show a reading of infinity but a deflection of zero as its scale is inverted.

43. What precaution should be taken while using an ohmmeter?

Ans. It should never be connected to an energized circuit. If it is done so, meter may be damaged or else its readings would be meaningless.

Appendices

Appendix A1.1

Power System Questions

1. What is the voltage that has been selected for HVDC transmission?

Ans. 500 kV

2. What is an electric power supply system?

Ans. The system of generation, transmission and distribution of electrical power is called the electrical power supply system.

3. What is the usable voltage for secondary distribution?

Ans. 415/240 V (415 volts for 3-phase loads and 240 V for single phase loads).

4. What is the legal requirement to which a distributor is subφected?

Ans. A distributor is subφect to the legal requirement that the voltage at the consumer's terminals should be maintained within ± 6% of the declared (or rated) voltage.

5. State the empirical formula for determining the system voltage of transmission line.

Ans. Approximate economical voltage for transmission in

$$kv = 5.5\sqrt{\frac{\text{km}}{1.6} + \frac{\text{load in kva}}{150}}$$

6. What is modified Kelvin's law?

Ans. Modified Kelvin's law states that the most economical cross- sectional area of transmission line conductor makes the variable part of annual cost on account of interest and depreciation on the capital outlay equal to the annual cost of electrical energy wasted in the conductor.

7. What are the various components of an overhead line?

Ans. The main components of an overhead line are supports/towers. cross-arms and clamps, insulators, conductors, guys and stays, lightning arresters, earth wire, V-guards, guard wires, phase guards, vibration damper etc.

8. What is the function of line support?

Ans. The function of line support is to support the phase wires and to keep them at the safe distance from the ground.

9. What are four types of line supports?

Ans. The four types of line supports are wooden poles, RCC or PCC poles, rail poles and steel structures or towers.

10. What are the drawbacks of wooden poles?

Ans. The main drawbacks of wooden poles are comparatively smaller life, poor mechanical strength, tendency to rot below the ground level and requirement of periodical inspection.

11. What are the main requirements of good conductor material used for overhead lines?

Ans. Main requirements of good conductor material used for overhead lines are high electrical conductivity, high tensile strength, low specific gravity, low cost, easy availability. free from brittleness and the material should not react with atmospheric condition.

12. What for ACSR conductor stands?

Ans. ACSR conductor stands for aluminum conductor steel reinforced conductor.

13. Why are conductors for transmission line stranded?

Ans. All conductors used for overhead transmission lines are stranded so as to increase the flexibility. Solid wires, except of smaller sizes, are difficult to handle when used for long spans and tend to crystalize at the points of supports because of swinging in winds.

14. Why is galvanized steel wire not suitable for EHT lines for the purpose of transmitting large amounts of power over a long distance?

Ans. The galvanized steel wire is not suitable for EHT lines for the purpose of transmitting large amounts of power over a long distance because of poor conductivity, high internal reactance and eddy current and hysteresis losses.

15. What is the function of cross arm?

Ans. The function of cross arm is to keep the conductors separate form each other's and form pole.

16. What is sag?

Ans. The difference in level between the points of line supports and the lowest point on the conductor is known as sag.

17. What are the factors affecting the sag in a transmission line?

Ans. Factors affecting the sag in a transmission line are weight of conductor, span length (distance between the poles), working tensile strength, ice coating, wind pressure and temperature.

18. What is stringing chart?

Ans. Stringing chart gives the data for sag and tension to be allowed at a particular temperature. It is very useful while erecting the transmission line conductors for adjusting the sag and tension properly.

19. What are the important properties that an overhead line insulator must possess?

Ans. The most important properties that an overhead line insulator must possess are: High mechanical strength, high relative permittivity, high insulation resistance, high ratio of rupturing strength to flash-over voltage and capability of withstanding temperature variations.

20. What types of insulators are used for overhead transmission and distribution lines?

Ans. The insulators used for overhead line are pin type, suspension type, strain type and shackle insulators.

21. What are the causes of failure of insulators?

Ans. The insulator failure may occur due to cracking of porcelain, porosity, puncture, mechanical stresses, flash-over etc.

22. porosity, puncture, mechanical stresses, flash-over etc. How do the insulators fail?

Ans. Electric failure of insulators may occur either by puncture or flash-over. In the case of puncture, the arc passes through the insulator body while flash-over is caused by an arc discharge between the line conductor and earth through air surrounding the insulator.

23. How are insulators tested to determine flash-over voltages?

Ans. For determination of flash-over voltage of a given insulator high voltage supply terminals are connected to the pin and insulator and the test voltage is gradually raised to a specified value in approximate 10 seconds and is maintained for one minute. The voltage is then gradually raised till a flash-over takes place. The flash-over voltage depends upon barometric pressure, temperature, humidity etc.

24. What is meant by birdage?

Ans. Birdage means the short-circuiting of the line conductor to earth through the large birds or similar obφects.

25. What is meant by efficiency of an insulator string?

Ans. The efficiency of an insulator string, called the string efficiency, is the ratio of voltage across the whole string and the product of the number of units and voltage across the unit nearest to the line conductor.

26. How are voltage distribution and the string efficiency affected by rain?

Ans. More uniform voltage distribution is obtained and the string efficiency is improved because of reduction in value of K [when insulators are wet, mutual capacitance C increases while the shunt capacitance C□ (except for the unit nearest the cross-arm) remains constant).

27. What is corona?

Ans. When the voltage between conductors of an overhead line exceeds the disruptive critical voltage value, a hissing noise accompanied by a violet glow appears. This phenomenon is called the corona.

28. What are the various factors which affect corona?

Ans. Factors affecting corona are (i) number of ions, size and charge per ion, mean free path depending on atmospheric conditions, (ii) line voltage, (iii) ratio of conductor spacing to conductor radius, (iv) contour of the conductor surface, and (v) state of the conductor surface.

29. What are the adverse effects of corona?

Ans. Adverse effects of corona are (i) power loss but not very important except under abnormal weather conditions, (ii) corrosion due to production of ozone gas, and (iii) interference with neighboring communication circuits.

30. What do you mean by dielectric strength of air?

Ans. The value of potential gradient at which complete disruption of air occurs, is called the dielectric strength of air.

31. What is meant by disruptive critical voltage?

Ans. The disruptive critical voltage is defined as the minimum phase to neutral voltage at which corona occurs.

32. What is visual critical voltage?

Ans. Visual critical voltage is defined as the minimum phase to neutral voltage at which glow appears all along the line conductors.

33. What are the various factors which affect corona loss?

Ans. The factors affecting corona loss are system frequency. system voltage, air conductivity, air density, conductor diameter, conductor surface condition, atmospheric conditions and load current.

34. How can the corona effect be minimized?

Ans. Corona effect can be minimized in a transmission line by using large diameter conductors which may be accomplished by using hollow conductors with a hemp core, ACSR conductors or bundled conductors. Corona effect can also be reduced but to a limited extent, by increasing the spacing between the conductors.

35. What is skin effect?

Ans. The phenomenon of concentration of an electric current near the surface of the conductor is known as skin effect.

36. On what factors does the skin effect depend?

Ans. The skin effect depends upon the type of material, frequency of current, diameter of conductor and shape of conductor. It increases with the increase of x-section, permeability and supply frequency.

37. What is meant by GMR of a conductor?

Ans. The product re-1/4 is known as GMR of a conductor and is equal to 0.7788 times the conductor radius.

38. What is self GMD of a conductor consisting of seven identical strands, each having a radius?

Ans. 2.176r.

39. What is bundling of conductors?

Ans. The use of more than one conductor per phase is called the bundling of conductors.

40. What is meant by transposition of line conductors?

Ans. Transposition of line conductors means changing the positions of the three phases on the line supports twice over the total length of the line. The line conductors in practice, are so transposed that each of the three possible arrangements of conductors exist for one-third of the total length of the line.

41. What are the factors which govern the performance of a transmission line?

Ans. The factors governing the performance of a transmission line are series resistance R and inductance L, shunt capacitance C and conductance G. However, the effect of conductance is negligible.

42. How do you classify transmission lines?

Ans. Transmission lines are classified depending upon the manner in which capacitance is taken into account.

43. What are the causes of voltage drop and line loss in a transmission line?

Ans. The voltage drop in a transmission line is due to its resistance and inductance whereas the line loss is due to its resistance.

44. What is Ferranti effect?

Ans. The phenomenon of rise in voltage at the receiving end of the open-circuited or lightly loaded line is called the Ferranti effect.

45. How does rise in voltage at the receiving end of an open- circuited line depend upon the length of the line and system operating voltage?

Ans. Rise in voltage at the receiving end of an open-circuited line is proportional to the square of the length of the line and varies directly in proportion to the system operating voltage.

46. How do the terms 'impedance drop', 'voltage drop' and 'voltage regulation', in connection with transmission lines differ?

Ans. The impedance drop in a line is the voltage drop in the line impedance and is equal to the phase difference between the sending-ending voltage and the receiving-end voltage. The voltage drop in a line is the arithmetical difference between the sending-end and receiving-end voltages. The voltage regulation is defined as the change in voltage at the receiving-end when full-load is thrown off, the sending- end voltage being held constant.

47. Why in medium and long transmission lines, regulation is greater than voltage drop?

Ans. The regulation is greater than voltage drop in a medium/long transmission line due to rise in voltage owing to Ferranti effect.

48. What are the units of generalised constants of a transmission line?

Ans. Generalised constants A and D are dimensionless whereas the units of generalised constants B and C are ohms and siemens (mho) respectively.

49. What is the range of surge impedance for an overhead transmission line?

Ans. 400-600 Ω.

50. What is the surge impedance loading of a line?

Ans. The unity power factor load, that can be delivered by a line of negligible resistance, is called the surge impedance loading.

51. How does synchronous phase modifier differ from a synchronous motor?

Ans. Synchronous phase modifier differs from an ordinary synchronous motor in as much as they are built for the highest economical speeds, and provided with smaller shafts and bearings and special attention is paid for securing a high overall efficiency.

52. How does the reactive power of a transmission line vary for given values of Vs. VR and X?

Ans. For fixed values of Vs. Va and X, the reactive power is proportional to the magnitude of voltage drop in the line (i.e. Vs-VR).

53. Why high and extra voltages are adopted for transmission of bulk power over long distances?

Ans. This is because the maximum power transferred over a given line increases with the increase in Vs and Vp.

54. What is meant by receiving-end power circle diagram?

Ans. The circle diagram drawn with receiving-end true and reactive power components as the horizontal and vertical ordinates is called the receiving-end power circle diagram.

55. What is meant by sending-end power circle diagram?

Ans. The circle diagram drawn with sending-end true and reactive power components as the horizontal and vertical ordinates is called the sending-end power circle diagram.

56. For what purpose metallic sheath is provided in underground cables?

Ans. The metallic sheath is provided around the insulation to protect it against ingress of moisture, gas or other damaging liquids (acid or alkalis) from the soil and atmosphere.

57. Why are copper conductors not used in power cables?

Ans. Copper conductors are not used in power cables because of scarcity of copper.

58. What are the main requirements of insulating materials used for underground cables?

Ans. The main requirements of the insulating materials used for underground cables are high insulation resistivity, high dielectric strength, good mechanical properties, non- hygroscopic, non-inflammable, immune to attacks by acids and alkalies.

59. Why single core cables are usually not provided with armouring?

Ans. Single core cables are usually not provided with armouring in order to avoid excessive loss in the armour.

60. On what factors does the insulation resistance of a cable depend?

Ans. The installation resistance of a cable depends upon core radius, insulation thickness, length of cable and resistivity of insulating material.

61. How does the insulation resistance of a cable vary with its length?

Ans. Insulation resistance of a cable varies inversely as the length of the cable.

62. On what factors does the capacitance of a single core depend?

Ans. The capacitance of a single core depends upon its length, relative permittivity of its dielectric and the ratio of core diameter to sheath inner diameter.

63. How does electric stress vary in the dielectric of a cable

Ans. The electric stress in the dielectric of a cable varies inversely as the distance from the centre of the cable and hence minimum at the sheath and maximum at the conductor surface.

64. For what value of R/r the minimum value of stress is obtained in a cable?

Ans. For value of R/r=e i.e. 2.71828 the minimum value of stress is obtained in a cable.

65. What is the modern practice adopted to avoid grading of cables?

Ans. The modern practice adopted to avoid grading of cables is to employ oil-filled or gas-pressure cables.

66. Why loss angle of a cable should be very small?

Ans. As the dielectric losses occurring in a cable are very small, the value of loss angle & is small.

67. How does dielectric loss vary with the change in voltage, frequency of supply and capacitance of cable?

Ans. The dielectric loss in a cable varies in proportion with the capacitance of a cable and frequency of supply and varies as the square of the voltage.

68. What are the factors, which are to be considered while selecting a cable for a particular service?

Ans. The factors to be considered in selection of a cable are continuous current rating, voltage grade, permissible voltage drop and short-circuit current rating (in case of cables only) and operating cost.

69. On what factors does the primary distribution voltage depend?

Ans. The voltage employed for primary distribution system depends upon the amount of power to be supplied and distances of the substations required to be supplied.

70. What is/are the primary distribution voltage/voltages adopted in India?

Ans. The usual value of primary distribution voltage in India is 11 kV bur in case of high load density area it may be higher (33 or 66 kV) even.

71. What is the declared consumer voltage as per IE rules?

Ans. The declared consumer voltage, according to IER is 415/240 V (i.e. 415 V between phases and 240 V between phase and neutral) with a permissible voltage variation of ±6%.

72. Why is it necessary to keep voltage variation at the consumer's terminals as low as possible?

Ans. All the appliances and motors are designed to be operated at a particular or declared value of voltage. Large voltage variation means either a too high or a too low voltage. Low voltage will cause not only in burning of motors/appliances/ equipment and insufficient lighting but also loss of revenue to the supplier. On the other hand, high voltage will cause the lamps to burn out permanently and also may cause failure of other appliances. So, the primary requirement of a distribution system is that voltage variations at the consumer's terminals should be the lowest possible (not more than ±6% of the rated value).

73. What is a feeder?

Ans. Feeder is a line which supplies a point of distribution network without being tapped at any intermediate point. In simple terms, the feeder may be defined as the line carrying current from the secondary substation to the distribution substations or as a primary distribution line.

74. What is a distributor?

Ans. Distributor is a line from which tapping's are taken along its length for providing supply to the consumers.

75. Why de 2-wire system is never employed for transmission purpose?

Ans. Because of poor efficiency.

76. Why is radial system of distribution is not generally used?

Ans. Because of poorest service reliability.

77. What are the advantages of ring main distributor?

Ans. Better flexibility, greater reliability, better voltage regulation and less power losses.

78. Why is neutral wire taken of half cross-section of either outer in a 3-wire dc distribution system?

Ans. Since only out of balance current flows through the middle or neutral wire, so generally neutral wire is taken of half x-section of either outer.

79. What is a booster?

Ans. Booster is a low-voltage, high-current series wound de generator inserted into a de circuit to add or inφect a certain voltage proportional to circuit current.

80. For what purpose a booster is employed?

Ans. A booster (low-voltage, high-current series wound de generator) is used for compensating voltage drop in the feeder.

81. What are the faults that are likely to occur in cables?

Ans. The faults which are likely to occur in cables are open- circuit, short-circuit and ground faults.

82. What is meant by 'ground fault' in cables?

Ans. When the current starts flowing from core to earth or to cable sheath because of damage to the cable insulation, the fault is known as ground or earth fault.

83. What is meant by 'short-circuit fault in cables?

Ans. When the conductor of a cable is broken or point is pulled out, the fault is called the open-circuit fault.

84. What principle do Murray loop and Varley loop tests employ?

Ans. Murray loop and Varley loop tests employ the principle of Wheatstone bridge.

85. What is the essential difference between Murray and Varley loop tests?

Ans. The difference between Murray and Varley loop tests is that in Varley loop test provision is made for the measurement of total loop resistance instead of obtaining it from the relation

$$R = \rho \frac{l}{a}$$

86. In what way Varley loop test is advantageous over Murray loop test?

Ans. In Murray loop test sound cable resistance must be known whereas in Varley loop test it is not so.

87. What is the essential requirement for performing capacity tests on cables?

Ans. The essential requirement for performing capacity tests on cables is that the core to be tested has a good insulation resistance.

88. What are the transmission voltages that are used in India?

Ans. 33, 66, 110, 132, 220 and 400 kV. 765/800 kV is also in used in India.

89. What are the advantages of adopting extra high voltage/ ultra-high voltage for transmission of ac electric power.

Ans. Advantages of adopting of EHV/UHV for transmission of ac electric power are reduced line losses, high transmission efficiency, improved voltage regulation, reduced conductor material requirement, flexibility for future system growth, increase in transmission capacity of the line, increase of SIL and reduction in ROW.

90. Why does surge impedance loading (SIL) increase with the increase in voltage level?

Ans. SIL varies as the square of the operating voltage, so SIL increases with the increase in voltage level.

91. What are the problems associated with an EHV transmission?

Ans. The problems associated with EHV transmission are corona loss and radio interference, requirements of heavy supporting structures and erection difficulties and insulation requirements.

92. On what basis the conductor size is determined for transmission lines up to 220 kV?

Ans. For overhead transmission lines operating at voltages up to 220 kV, the conductor size is determined on the basis of its current carrying capacity.

93. On what basis the conductor size of EHV line is determined?

Ans. The conductor size of an EHV line is decided on the basis of corona.

94. What are the sources of audible noise generation in EHV transmission systems?

Ans. The sources of noise generation in EHV transmission systems are corona, humming of transformers, cooling systems and mechanical and electrical auxiliaries.

95. What material is universally employed for overhead transmission lines?

Ans. ACSR conductors.

96. What is economical value of span for 400 kV transmission line?

Ans. 350-400 m.

97. How does the use of bundle conductors reduce corona loss in an EHV line?

Ans. The bundle conductor acts, as far as the electric field is concerned, like a conductor of diameter much larger than that of the equivalent single conductor. This reduces the voltage gradient and so the corona.

98. What for series and shunt compensation provided in EHV lines?

Ans. Series compensation is provided in EHV lines to artificially reduce the series reactance of line so as to improve stability, voltage regulation and transmission efficiency whereas shunt compensation is provided to artificially reduce the line susceptance so as to increase the voltage regulation under light load condition.

99. Why is de used for HV transmission of electric power?

Ans. There is a limit to distance over which power can be transmitted by ac using overhead lines. This is because of surge impedance limitation producing high voltage at the receiving end. That is why de is used for HV transmission of electric power as an alternative.

100. Why is voltage regulation better in case of de transmission?

Ans. Because of absence of inductance in de systems.

101. Why there is no limit on the length of the cable for de system while it is in case of ac system?

Ans. Because of absence of charging currents in DC system.

102. Why HVDC line do not require any reactive power compensation?

Ans. Because of absence of charging currents and unity power factor operation.

103. Why de systems invariably use ground or sea return?

Ans. Because of economy and reliability.

104. Why D transmission is preferable over ac transmission for long distance transmission?

Ans. Because with larger distances, the saving in cost of de overhead lines becomes greater than the additional expenditure on terminal equipment.

105. What is normal value of inductance of the DC reactor?

Ans. The value of inductance of the de reactor should be such that a resonance of the de circuit does not occur at power frequency and is in the range of 0.4-1.0 H

106. What is the equipment's that supply reactive power in HVDC converter stations?

Ans. AC filters, static shunt capacitors, synchronous condensers, static VAR compensators etc.

107. Can corona loss be ignored in the design of HVDC transmission lines?

Ans. No

Appendix A1.2

Power System Protection

1. What is meant by short-circuit?

Ans. When conductors of different phases come into contact with each other in a power line, power transformer, or any other circuit element, part of the impedance of the system is shunted out of the circuit, a large current flow in one or more phases, a short-circuit is said to have occurred.

2. What is difference between a short-circuit and an overload?

Ans. On occurrence of a short-circuit, the voltage at the fault point is reduced to zero and current of abnormally high magnitude flows through the network to the fault point. On the other hand, an overload means that load greater than the designed value have been imposed on the system. In case of an overload, the voltage at the overload point may become low, but cannot be zero. The undervoltage conditions may extend to some distance beyond the overload point in the remainder of the system. In case of an overload, the currents are high but are substantially lower than in the case of a short-circuit.

3. The probability of the failure or occurrence of abnormal conditions is more on the power line. Why?

Ans. The probability of the failure or occurrence of abnormal conditions is more on the power line. This can be explained by the fact that the power lines are widely branched, have greater length, operate under variable weather conditions and are subject to the action of atmospheric disturbances of electrical nature.

4. Name the different faults likely to occur in an overhead transmission line.

Ans. The different faults likely to occur in an overhead transmission line are (1) Single phase to ground, (2) Phase to phase, (3) Two phases to ground, (4) Phase to phase and third phase to ground. (5) All the three phases shorted and (6) All the three phases to ground.

5. Which type of fault occurs more frequently in overhead lines?

Ans. Single phase to ground (L-G) fault occurs more frequently in overhead lines.

6. Which type of faults are taken into account while determining the rating of circuit breaker?

Ans. The balanced three-phase fault, though its occurrence in quite rare (only about 5 percent of the total), imposes most severe duty on circuit breakers and is, therefore, used in determination of the circuit-breaker rating.

7. What are the sources of short-circuit power in a power system?

Ans. Mainly there are three sources viz. synchronous generators, synchronous motors and synchronous condensers and induction motors, which can feed into the short-circuit.

8. What are the basic requirements of a protection system?

Ans. The basic requirements of a protection system are essentially reliability and simplicity, selectivity and discrimination, sensitivity, speed of operation and economy.

9. Why current transformers are required in a protective relaying scheme?

Ans. The primary circuit currents which are of high magnitudes are to be reduced to values suitable for relay operation and current transformers are used for this purpose. In addition to this CTs provide insulation against the high voltage of the power circuit mid protect the apparatus and the operating personnels from contact with the high voltages of the primary circuits.

10. Why potential transformers are required in a protective relaying scheme?

Ans. It is not possible to connect the potential coils of the protective devices directly into the system in case of high voltage systems. It is, therefore, necessary to stepdown the voltage and also to insulate the protective equipment from the primary circuit (high voltage power circuit). This is achieved by employing potential (or voltage) transformers (PTS).

11. What is meant by percentage reactance?

Ans. Percentage reactance can be defined as the voltage drop across the given reactance expressed as percentage of normal voltage when carrying full-load current pertaining to normal rating.

12. Which of the electrical quantities are chosen as base values?

Ans. For common representation, base KVA (or MVA) and base voltage (kV) are chosen for base value.

13. What is meant by per unit value of any quantity?

Ans. Per-unit value of any quantity is defined as the ratio of the actual value of the quantity in any unit to the base or reference value in the same unit.

14. What is the relationship between base KVA, base XV and percentage reactance?

Ans. Percentage reactance = Actual reactance x (base KVA) /((base kV)^2 x 1,000)) x 100

15. Why the per-unit method is considered superior to per cent method for short-circuit calculation?

Ans. Per cent value is not convenient for use as the factor of 100 has to be carried in computations.

16. Define base impedance of a power system element and write an expression for it in terms of base MVA and kV.

Ans. Base impedance is defined as the ratio of base voltage in volt: and base current in amperes and it is expressed in terms of base MVA and kV as follows:

Base impedance in ohms = (Base kV)^2 / Base MVA

17. What is meant by base quantities?

Ans. For common representation, the quantities (voltage, current, power or impedance) chosen as common or reference quantities are called the base quantities.

18. Define PU impedance.

Ans. Per unit impedance is defined as the ratio of voltage drop across the given impedance to the normal voltage when carrying full- lead current pertaining to normal rating.

19. What is relation between percentage impedance and PU impedance?

Ans. Percentage impedance = PU impedance × 100

20. What is meant by fault leve!?

Ans. When fault occurs at a point in a power system, the corresponding MVA is referred to as the fault level at that point.

21. What are current limiting reactors?

Ans. Current limiting reactors are inductive coils having large inductive reactances in comparison to their resistances and are commonly installed on large power systems to limit the magnitude of short- circuit currents and reduce the resulting voltage disturbances on the rest of the system.

22. Why it is not necessary to insert current limiting reactor between the generator and the generator bus?

Ans. Since modern generators have sufficient inherent transient reactance to protect themselves against dead 3-phase short-circuit at their terminals, insertion of current limiting reactors between the generators and their bus is not necessary.

23. What are symmetrical components?

Ans. According to the theory of symmetrical components, any unbalanced system of phasors (voltages, currents or impedances etc.) can be resolved into balanced systems of phasors equal in number to the number of phases. These balanced systems of phasors are called the symmetrical components of the original unbalanced system.

24. What are uses of symmetrical components in power systems?

Ans. Symmetrical component method provides a useful analytical tool for the protection engineer and is useful in system fault calculation and system stability problems etc.

25. Differentiate between the positive-sequence system and zero- sequence system.

Ans. The positive-sequence system is a set of three symmetrical voltages (ie., all numerically equal and all displaced from each other by 120") having the same phase sequence as the original set of three unbalanced voltages i.e., A-B-C whereas the zero- sequence system is a set of three voltages, all equal in magnitude and in phase with each other.

26. Define negative-sequence impedance.

Ans. The impedance of the network offered to the flow of negative- sequence currents is called the negative-sequence impedance of the network.

27. What are sequence networks?

Ans. The single-phase equivalent circuit composed of impedances to current of any one sequence only is called the sequence network for that particular sequence. Hence corresponding to positive, negative-and zero-sequence currents we have positive, negative and zero-sequence works.

28. What are the different types of faults which occur in a power

Ans. The most common and dangerous faults that occur in a power system are the short-circuit or shunt faults such as L-G, L-L, L LG, L-L-L and L-L-L-G faults. The other type of faults that occur in a power system are series type such as one conductor open fault and two conductor open fault.

29. What are unsymmetrical faults?

Ans. The faults giving rise to unsymmetrical currents ie, currents differing in magnitude and phases in the three phases of the power system are known as unsymmetrical faults. Unsymmetrical faults may also be defined as the faults involving one or two phases (not involving 3 phases) such as L-G, L-L, L-L-G faults.

30. For a fault at a given location rank the various faults in the order of severity.

Ans. The faults in increasing order of severity are L-G, L-L, L-L-G, and symmetrical three-phase faults (L-L-L or L-L-L-G).

31. Name the faults which do not have zero-sequence currents flowing.

Ans. In line-to-line faults and symmetrical faults (L-L-L or L-L-L-G faults) no zero-sequence current flows.

32. Name the fault in which all the three sequence currents are present and are equal to each other?

Ans. In L-G fault all the three sequence currents are present and are equal to each other.

33. Which of the faults would cause a greater electromagnetic interference between the power line and a nearby communication line?

Ans. Single phase-to-ground (L-G) fault, due to heavy unbalancing of line currents, causes greater electromagnetic interference between the power line and a nearby communication line.

34. What is meant by a fuse?

Ans. Fuse is a device used in an electrical circuit for protecting electrical equipment's aga overloads and/or short-circuits.

35. On what principle does the fuse operate?

Ans. The action of a fuse is based upon the heating effect of the electric current. During overloads or short-circuits the fuse gets heated up to melting point due to excessive current and break the circuit.

36. What kind of protection is offered by a fuse?

Ans. The fuse provides protection against short-circuits and overloads.

37. What function does the fuse perform?

Ans. The function of a fuse wire is (f) to carry the normal working current safely without heating and (if) to break the circuit when current exceeds the limiting current.

38. Give the names of five materials with their melting points which are used for fuse element.

Ans. The materials commonly used for fuse elements are silver, tin, lead, copper and aluminium. Their melting points are 980°C, 240°C, 328°C, 1,090°C and 665°C respectively.

39) What material is mostly used for an ordinary fuse wire?

Ans. Copper or lead-tin alloy.

40. What is fuse link?

Ans. Fuse link is that part of the fuse which needs replacement when the fuse blows out.

41. On what factors does the current carrying capacity of a fuse depend?

Ans. The current carrying capacity of a fuse depends upon the permissible temperature rise of the contacts of the fuse holder, size of fuse wire, fuse material and the surrounding of the fuse.

42. On what factors does the fusing current depend?

Ans. The fusing current depends upon (f) diameter of wire (ii) length of wire (iii) fuse material (iv) x-sectional area (round or rectangular) (v) size and location of terminals (vi) type of enclosure used (vii) type of surface etc.

43. What are the purposes of using quartz sand as filler material in the construction of HRC fuse?

Ans. The purposes of using quartz sand as filler material in the construction of HRC fuse are to absorb heat at a very high rate, to produce a substance of high resistance and avoid the oxidation of silver element.

44. What is meant by discrimination?

Ans. Discrimination means the breaking of the faulty section of the line to an extent restricted to minimum; healthy sections of the circuit remaining in operation.

45. Can the HRC fuse be used for both ac and de circuits. If so under what condition?

Ans. -In general an HRC fuse meant for ac circuit may not be used for de circuits; but however, a fuse satisfying the requirement of arc voltage can be used on both ac and de circuits.

46. What do you understand by switchgear?

Ans. The apparatus employed for switching, controlling and protecting the electrical circuits and equipment is known as switchgear.

47. What is a circuit breaker?

Ans. Circuit breaker is a mechanical device designed to close or open contact members, thus closing or opening an electrical circuit under normal or abnormal conditions.

48. How does a circuit breaker differ from a switch ?

Ans. A circuit breaker is a device that interrupts the abnormal fault currents and in addition performs the function of a switch

49. What for isolating switches are provided along with circuit breakers?

Ans. Isolating switches are provided for separating the disconnected element from the live portion of the system for the purposes of safety.

50. Which of the gases are employed in commercial gas-blast circuit breakers?

Ans. SF, and air are used in commercial gas-blast circuit breakers.

51. What is meant by current chopping?

Ans. Current chopping is a phenomenon of current interruption before the natural current zero is reached.

52. Why is current chopping considered a serious drawback in a circuit breaker?

Ans. The current chopping is considered a serious drawback because it sets up high voltage transient across the breaker contacts.

53. How is the effect of overvoltage developed due to current chopping overcome?

Ans. By using resistance switching de., by connecting a resistance across the circuit breaker contacts.

54. Why current chopping is not common in oil circuit breakers?

Ans. Current chopping is not common in oil circuit breakers because in most of the oil circuit breakers the arc extinguishing power is proportional to the magnitude of current to be interrupted.

55. What is meant by kilometric fault?

Ans. The fault occurring between a distance of a few km to few tens km from the circuit breaker arc called kilometric faults.

56. Name the materials used for contacts of vacuum circuit breakers?

Ans. Copper-bismuth, copper-lead, copper-tellurium, copper-thallium, silver-bismuth, silver-lead and silver-tellurium are some of alloys employed as contact materials in vacuum circuit breakers.

57. What is meant by breaking or rupturing capacity of a circuit breaker?

Ans. Breaking capacity of a circuit breaker is the highest rims value of current that it is capable of breaking under specified conditions of transient recovery voltage and power frequency voltage. It is expressed in k Arms at contact separation.

58. Why asymmetrical breaking current is higher than symmetrical breaking current in a circuit breaker?

Ans. metrical breaking current is the rms value of the combined (of ac and de components of short-circuit current at the instant of separation of breaker contacts while symmetrical braking current is the rms value of the ac component alone. Thus, asymmetrical breaking current is higher than symmetrical breaking current.

59. What is meant by making capacity of a circuit breaker?

Ans. The making capacity of a circuit breaker, when closed on a short circuit, is the peak value of the maximum current wave (including de component) in the first cycle of the current after the circuit is closed by the circuit breaker.

Rated making capacity = 2.55 x symmetrical rated breaking capacity

60. What is meant by rated voltage of a circuit breaker?

Ans. The rated maximum voltage of a circuit breaker is the highest rms voltage, above nominal system voltage, for which the circuit breaker is designed and is the upper limit for operation.

61. Name the three problems encountered in HVDC circuit breakers?

Ans. The three problems encountered in HVDC circuit breakers are:

1. How to create artificial current zero?
2. How to prevent restriking of arc?
3. How to dissipate the stored energy?

62. Why a decaying dc offset current wave appears when ac circuit breaker is suddenly switched to an inductive circuit at the instant when voltage wave is passing through zero?

Ans. In an inductive circuit, current lags behind the voltage by a large angle i.e. when voltage is passing through zero, current is around maximum value. Also current in an inductive circuit cannot change instantaneously. So, when ac circuit breaker is suddenly switched on to an inductive circuit, the current may be around maximum value which cannot change instantaneously, to compensate this de off current appears which die, out as time passes and a steady- state condition is reached.

63. What is a protective relay?

Ans. Protective relay is an electrical device designed to initiate isolation of a part of an electrical installation, or to operate an alarm signal, in the event of an abnormal condition or a fault,

64. What are the functions of a protective relay?

Ans. The main functions of protective relays are to detect the presence of faults, their locations and initiate the action for quick removal from service of the faulty section of the system. Relays operate to trip the power circuit breakers to disconnect the faulty section.

65. What are the essential elements of a protective relay?

Ans. The essential elements of a protective relay are (i) sensing element (ii) comparison element, and (iii) control element.

66. What are the fundamental requirements of protective relaying?

Ans. The fundamental requirements of protective relaying are speed, selectivity, sensitivity, reliability, simplicity, and economy.

67. Define the terms 'sensitivity' and 'selectivity' of a relay.

Ans. Sensitivity of a relay means its capacity to operate reliably under the actual conditions that produce the least operating tendency while the selectivity means the ability to determine the point at which the fault occurs and select the nearest circuit breaker in the system to trip so that the fault is cleared with minimum or no damage to the system.

68. What is meant by "relay setting"?

Ans. Relay setting means actual value of the energizing or characteristic quantity at which the relay is designed to operate under given conditions.

69. What is meant by operating time of a protective relay?

Ans. The operating time of a protective relay means the time which elapses between the instant when the actuating quantity exceeds the pick-up value to the instant when the relay contacts close.

70. What is meant by 'relay time' ?

Ans. Time interval between occurrence of fault and closure of relay contacts is called the relay time.

71. How are relay time, breaker time and fault clearing time related?

Ans. Relay time plus breaker time equals the fault clearing time.

72. What are the shortcomings in electromagnetic relays?

Ans. The vibration of relay armature at doubles the supply frequency, due to pulsating nature of electromagnetic force, causes the relay to hum and produce noise and also is a source of damage to the relay contacts. This leads to sparking and unreliable operation of the relay operative circuit contacts due to make and break of the circuit.

73. How definite time lag is achieved in attraction-armature relays?

Ans. The instantaneous type attraction armature relays can be made a definite time lag or inverse-time lag by using an oil dash pot, an air-escapement chamber, a clockwork mechanism or by placing a fuse in parallel with it.

74. How induction cup type construction is superior to induction disc type?

Ans. Induction cop type construction develops more efficient torque than either shaded pole or watt-hour meter constructions. Such relays are very fast in operation and may have an operating time of less than 0.01 second.

75. What is Buchholz relay? Which equipment is protected by it?

Ans. Buchholz relay is a gas actuated relay and is used for the protection of power transformers of ratings exceeding 500 kVA.

76. For what type of fault Buchholz relay is employed?

Ans. Buchholz relay provides protection only against transformer internal faults.

77. Define PSM?

Ans. PSM stands for plug-setting multiplier and is defined as the ratio of fault current in relay coil to the pick-up value.

78. What are the main features of directional relays?

Ans. Directional relays must have the following features: (1) High speed operation, (ii) high sensitivity (iii) adequate short-time thermal rating, (iv) ability to operate with low values of voltage (v) burden must not be excessive and (v) there should be no voltage and current creep.

79. Where is directional relay used?

Ans. Directional relays are used when graded-time overload protection is applied to ring mains and interconnected networks.

80. What is distance protection?

Ans. Distance protection is the name given to the protection, whose action depends upon the distance of the feeding point to the fault point. It is non-unit type protection and the protection zone is not exact. It can be used as primary as well as backup protection.

81. In what way a distance relay is superior to overcurrent protection for the protection of transmission lines?

Ans. Distance relay is superior to overcurrent protection for the protection of transmission lines. The reasons are faster protection, simpler coordination, simpler application, permanent setting without need for readφustments, less effect of amount of generation and fault levels, fault current magnitude, permits the high line loading.

82. Why directional feature provided for impedance relay cannot be used for reactance relay?

Ans. The directional feature provided for impedance relay cannot be used for reactance relay because the reactance relay will trip even under normal operating conditions at or near unity power factor. The reactance type distance relay needs a directional unit that is inoperative under normal load conditions.

83. Why reactance type relay is very suitable as a ground relay for ground fault?

Ans. Reactance type relay is very suitable as a ground relay for ground fault because its reach is not affected by fault impedance (ie., arcing resistance).

84. For what type of protection will you recommend (i) impedance relay (ii) reactance relay and (iii) mho relay?

Ans. The impedance type relay is suitable for phase fault relaying for lines of moderate length; reactance type relays are preferred for ground fault relaying while mho type relays are best suited for long lines and particularly where a severe synchronizing power surge may occur.

85. Why a 'Mho relay is normally used for the protection of a long-distance transmission line?

Ans. Mho type relays are best suited for phase fault relaying for long lines and particularly where a severe synchronizing power surge may occur. They are most sensitive of all the distance relays. The mho relay is reliable because it combines both the directional and distance measuring functions in one unit.

86. What is meant by a differential relay?

Ans. A differential relay is defined as the relay that operates when phasor difference of two or more similar electrical quantities exceeds a predetermined amount.

87. Where differential protection principle is employed?

Ans. Differential protection principle is employed for the protection of generators, generator-transformer units, transformers, feeders, large motors and bus-bars.

88. Why is percentage-differential relay named so?

Ans. Percentage-differential relay is known by this name because the operating current required to trip can be expressed as a percentage of load current

89. Why a biased-differential relay is preferred over a simple differential relay?

Ans. Biased-differential relay is preferred over a simple differential relay because its operation is not affected by the trouble arising out of difference in CTs ratios for high values of external short- circuit currents.

90. Where is negative phase sequence relay employed?

Ans. Negative phase sequence relay is essentially employed f the protection of generators and motors against unbalanced loading that may arise due to phase-to-phase faults.

91. What are the common types of generator faults?

Ans. The common types of faults occurring in synchronous generators are enumerated below.

1. Stator winding faults - phase-to-phase faults, phase-to- earth faults and inter-turn faults.
2. Rotor winding faults conductor-to-earth faults, open- circuit faults, inter-turn faults.

The other abnormal operating conditions are (1) failure of prime mover (if) failure of field (iii) unbalanced loading (iv) overloading (v) overvoltage at generator terminals (vi) over-speed (vii) ventilation failure (viii) current leakage in the generator body.

92. The occurrence of short-circuiting between the stator windings is quite rare. Why so?

Ans. The occurrence of short-circuiting between the stator windings is quite rare because the insulation in a slot between coils of different phases is at least twice as thick as the insulation between one coil and the iron core.

93. What is differential protection?

Ans. The differential protection is that which responds to the phasor difference between two or more electrical quantities.

94. Which type of relays are used in Merz-Price protection system for alternator?

Ans. The relays used in Merz-Price protection system of alternator are instantaneous electromagnetic type (attracted-armature type).

95. How does the inclusion of resistance in neutral grounding circuit of an alternator affect the performance of the differential protection of a 3-phase synchronous generator?

Ans. With the inclusion of resistance in neutral grounding circuit of an alternator, it will not be possible to protect whole of the stator windings of a star-connected generator against earth faults. The percentage of winding protected depends on the value of neutral grounding resistor and earth-fault relay setting.

96. Does the Merz-Price system of protection applied to alternators provide any protection for turn-to-turn faults on the same phase? If not why?

Ans. Merz-Price system of protection applied to alternators does not provide any protecting against turn-to-turn faults on the same phase because the current produced by such a fault flows in a local circuit between the turns involved and does not create the difference between the currents entering and leaving the windings at its two ends where the high voltage CTs are mounted.

97. Why it is not necessary to provide protection for turn-to-turn fault in turbo-generators?

Ans. The coils of modern large turbo-generators are usually single turn and, therefore, it is not necessary to provide protection for turn-to-turn faults.

98. What type of protective device is used for the protection of an alternator against overheating of its stator?

Ans. Resistance temperature detector.

99. Why is overcurrent protection not necessary for modern alternators?

Ans. Overcurrent protection is not considered necessary for modern alternators because these are capable of withstanding a complete short-circuit at their terminals for sufficient time without much overheating and damage.

100. Why is it not necessary to provide overvoltage protection in turbo-generators?

Ans. The overvoltage in a generator occurs when the prime mover speed increases due to sudden loss of the load on the generator and the speed control governors in case of turbo-generators are very sensitive to speed variations and, therefore, generated overvoltage of significant duration or magnitude does not generally occur. That is why it is not necessary to provide over- voltage protection in turbo- generators.

101. What types of protective device are used for the protection of an alternator against overheating of its rotor.

Ans. DC relay and resistance measuring circuit.

102. Why is it necessary to suppress the field immediately after disconnection of faulty alternator from the system?

Ans. In the event of a fault on a generator winding even though the generator circuit breaker is tripped, the fault continues to be fed as long as the excitation will exist because emf is induced in the generator itself. Hence it is necessary to suppress the field immediately after disconnection of faulty generator from the system.

103. What is field suppression?

Ans. The opening of the field circuit breaker and discharge of stored energy in the field winding through a resistor is called the field suppression.

104. What type of relays are required for backup protection of alternators?

Ans. High set, definite minimum time, induction type, inverse over- current, earth-fault relays are required for backup protection of alternators?

105. What are the types of faults that are likely to occur in operation of a 3-phase induction motor?

Ans. The faults which are likely in occur in operation of a 3-phase induction motor are: Stator faults (phase-to-phase, phase-to-ground and inter-turn faults), rotor faults (earth faults and inter-turn faults), prolonged overloading, stalling, unbalanced supply voltages, single phasing, undervoltage, reverse phase,

106. What are the faults that are likely to occur in a power transformer?

Ans. The faults likely to occur in a power transformer are open-circuit faults (an open-circuit in one phase of a 3-phase transformer), earth faults, phase-to-phase faults, inter-turn faults and overheating.

107. What are incipient faults?

Ans. Incipient faults are initially minor faults causing gradual damage. These faults include loose connections in conducting path, sparking, small arcing etc., and grow into maφor faults.

108. What is through fault?

Ans. A through fault in one which is beyond the protected zone of the transformer but fed through it.

109. What are the various protective schemes available for power transformers?

Ans. Different protective schemes available for the protection of power transformers are Buchholz protection, percentage differential protection, earth-fault protection, overcurrent protection, combined leakage and overload protection, restricted earth-fault protection, over fluxing protection etc.

110. On what factors does the choice of a protective gear for transformer depend?

Ans. The choice of a protective gear for a transformer depends upon several factors such as (1) type of transformer (distribution transformer or power transformer), (if) size of the transformer, (ii) type of cooling used (iv) systems where used Le., its electrical location in the network and (v) the importance of the service for which it is required

111. In what respect core balance leakage protection is inferior to combined leakage and overload protection?

Ans. The core-balance leakage protection suffers from the drawback that it does not provide protection against phase-to-phase faults.

112. Does circulating current protection system provide protection against inter-turn fault in case of power transformers?

Ans. Yes.

113. Name the two basic requirements that are to be fulfilled by differential relay.

Ans. The two basic requirements to be satisfied by differential relay on the protection of power transformers are that the differential relay should not operate on overload or external faults and it must operate on severe internal faults.

114. If T is the turn-ratio of power transformer, what will be the turn ratio of CTs on the hv side ?

Ans. If T is the turn-ratio of power transformer, then turn-ratio of CTs on the lv side will be T times that of CTs on the hv side?

115. What are the shortcomings of differential protection scheme as applied to power transformer?

Ans. Simple differential protection, as applied to a power transformer, has the following drawbacks.

(i) Unmatched characteristic of CTs (ii) Change in ratio due to tap-changing and (iii) magnetizing inrush current

116. What is magnetizing inrush current?

Ans. When the transformer is energized initially there is no induced emf, the condition is similar to switching of an inductive circuit. The resistance being low, a transient rush of current takes place. This transient rush of current is known as magnetizing inrush current and it may be equal to 8 times the rated current.

117. What is the reason behind providing a second harmonic restraint relay?

Ans. Second harmonic restraint relay is provided to avoid maloperation of the relay's due to magnetizing inrush current.

118. What are the advantages of differential magnetic balance protection system, as applied to power transformers?

Ans. The main advantages of differential magnetic balance protection system in its application to power transformers are increased stability and sensitivity and its application to transformers provided with load tap-changers.

119. How is frame leakage protection provided in power transformers?

Ans. The transformer tank is connected to earth through CT to which an instantaneous earth-fault relay is connected.

120. What is unit type protection?

Ans. Unit system of protection is one in which the protection responds to faults in the protected zone alone and it does not respond to through faults (faults beyond the protected zone). Non-unit systems do not have exact zone boundary.

121. Give the examples of unit/and non-unit systems of protection.

Ans. Unit type protection includes pilot-wire differential protection. carrier-current protection based on phase comparison etc. while non-unit type protection includes time-graded overcurrent protection, current-graded overcurrent protection, and distance protection.

122. What is meant by time-graded protection?

Ans. Time-graded protection is a scheme of overcurrent protection, in which discrimination is incorporated ie, the time setting of relays is so graded that in the event of fault, the smallest possible part of the system is disconnected.

123. What is the purpose of time grading of protection system and where is it employed?

Ans. The purpose of time grading of protection system is to incorporate time discrimination. It is employed for the protection of radial feeders, parallel feeders and interconnected systems.

124. What type of relay is suitable for radial feeders?

Ans. Induction type inverse definite minimum time (IDMT) relays are most suitable for radial feeders.

125. What is the minimum time interval which can be permitted for the two adφacent circuit breakers to operate in radial feeders?

Ans. 0.4 second.

126. Why induction type IDMT relays are most suitable for protection of radial feeders?

Ans. Induction type IDMT relays are most suitable for protection of radial feeders because their time-current characteristics are similar in shape and in no case, they cross each other at any point.

127. On what fact is the current-graded protection based?

Ans. Current-graded protection is based on the fact that the short- circuit current along the length of the protected circuit decreases with the increase in distance between the supply end and the fault point.

128. What type of relay suitable for current-graded protection?

Ans. High-speed high-set overcurrent relays.

129. Can Merz-Price system be employed for protection of ring mains and duplicate feeders?

Ans. Yes.

130. Does Merz-Price system provide backup protection or overload protection?

Ans. No.

131. Why Merz-Price protection is not suitable beyond 33 kV?

Ans. Difficulties are experienced in balancing the secondaries of the two CTs and that is why Merz-Price protection system cannot be used beyond 33 kV.

132. Is the operation of relays affected by capacitive currents in Translay system of protection of feeders?

Ans. No.

133. Why is use of wire-pilot relaying scheme restricted to protection of short transmission lines?

Ans. In case of medium and long transmission lines the pilot-wire protection scheme becomes expensive because of cost of pilot- wires.

134. What are the main elements of current-carrier protection?

Ans. The main elements of the current-carrier protection scheme are: (1) transmitter (ii) receiver (iii) coupling equipment and (iv) line trap.

135. What purposes are served by line trap and coupling capacitors in carrier channels?

Ans. Line trap is an LC network inserted between the busbar and connection of coupling capacitor to the line and tuned to resonance at the high frequency and are used to confine the carrier currents to the protected section so as to avoid interference with or from adφacent carrier-current channels and also to avoid loss of the carrier-current signal in adφoining power circuits for any reason whatsoever.

The coupling capacitor is used to connect the high frequency (carrier) equipment to one of the line conductors and simultaneously serves to isolate the carrier equipment from the high-power line voltage

136. What is meant by voltage surge?

Ans. By voltage surge is meant any sudden excessive rise in voltage that may be dangerous to the electrical equipment of an installation.

137. What are the internal causes of overvoltage?

Ans. Internal overvoltage arise during sudden changes in the operating conditions of an installation such as a switch-in, switch-out or sudden change in load, the opening of circuits with large inductance or capacitance, the interruption of a short-circuit current, the appearance of an arc fault between one phase and ground, as well as other circuit disturbances.

138. What is meant by lightning?

Ans. Lightning is a huge spark, which is due to the electrical discharge taking place between the clouds, within the same cloud and between the clouds and ground.

139. What is difference between A stroke and B stroke?

Ans. A stroke is characterized by the comparatively long time to produce discharge, while the B stroke is characterized by its rapidity. A stroke strikes the highest and most sharply pointed obφect in the neighborhood while B stroke ignores the tall obφects and reaches the ground in a random manner, stroke is most dangerous.

140. What is earthing screen? What type of protection is provided by it? What is it application?

Ans. Earthing screen consists of a network of copper conductors, earthed at least on two points, over all the electrical equipment in the substation. An electrostatic shield against external field is presented by such a screen and so protects the system. Earthing screen protects the system from direct lightning strokes, reduces the voltages induced electrostatically or electromagnetically but does not provide any protection against voltage waves which may still reach the terminal equipment.

Substations, interconnectors and powerhouses are protected from direct strokes by earthing screen.

141. What is an earth wire?

Ans. An earth wire is a form of lightning protection using a conductor or conductors, well-grounded at regular intervals, preferably at each support, and attached from support to support above the transmission line.

142. Why earth wire is provided in overhead transmission lines?

Ans. Earth wire is provided above overhead transmission lines for protection from lightning strokes. The ground wire shields the phase or line conductors by attracting itself the lightning strokes which, in its absence would strike the phase conductors. Besides it, the ground wire reduces the voltage electrostatically or electromagnetically induced in the conductors by the discharge of a neighbouring cloud. It also provides additional protective effect by causing attenuation of travelling waves set in lines by acting as a short-circuited secondary of the line conductors.

143. What is meant by protective angle? Give its value for reliable protection.

Ans. Protective angle is the angle between the vertical line through the ground wire axis and the line passing from the ground wire axis to the outermost phase conductor. Its value for reliable protection is taken equal to 20-30 degrees.

144. What is meant by coupling factor?

Ans. The ratio of the induced voltage in the line conductors to the potential of ground wire is called the coupling factor.

145. What is meant by a lightning arrester or surge diverter?

Ans. A surge diverter is a device that is connected between line and earth, i.e., in parallel with the equipment under protection at the substation. It limits the duration and amplitude of the follow Current.

146. What is meant by impulse ratio of any lightning arrester?

Ans. Impulse ratio of any lightning arrester is the ratio of breakdown voltage of a wave of special duration to the breakdown voltage of a 50 Hz wave.

147.What is the application of an electrolytic arrester?

Ans. Electrolytic arrester is very delicate, needs daily supervision, and the film is required to be reformed whenever destroyed. It is used in conφunction with an impulse gap.

148. What is the main advantage of oxide film arrester?

Ans. The main advantage of an oxide film arrester is that it does not require daily replacement and it may thus be installed at points in transmission systems where daily attendance is difficult or expensive to provide.

149. How metal oxide lightning arrester is superior to thyrite arrester?

Ans. Metal oxide lightning arrester is simple in construction and has superior energy dissipation capacity, better surge protection, more stable protective characteristics and substantial reduction in overvoltage across the equipment as compared to thyrite arrester.

150. What is meant by "follow current"?

Ans. Follow current is the current that flows from connected supply source through lightning arrester following the passage of discharge current.

151. What is meant by power angle curve?

Ans. Power angle curve is a curve of power transfer P versus power angle 8. It is a sine curve.

152. What is meant by a swing curve?

Ans. Swing curve is a graph of power angle & versus time t.

153. On what basis do you conclude that a given synchronous machine has lost stability?

Ans. In case the power angle & reduces after attaining maxima d8 corresponding to $\frac{d\delta}{dt} = 0$ condition indicates stability. On contrary if the power angle exceeds critical angle of load, the system becomes unstable.

154. Define infinite bus in a power system.

Ans. Infinite bus has constant voltage, constant electrical angle of the voltage, constant frequency, infinite fault level and, therefore, the quantities of infinite bus do not get affected by connecting or disconnecting individual machines or transmission lines.

155. How can steady state limit be improved?

Ans. Steady state limit can be improved either by increasing the excitation of the machine/machines or by reducing transfer

156. Write swing equation for a single machine system.

Ans. The swing equation for a single machine system can be written as

$$M\frac{d^2\delta}{dt^2} = P_S - P_E = P_S - P_{Max}\sin\delta$$

where M is angular momentum, Ps is shaft power (mechanical power), Pg is electromagnetic power, δ is load angle between E and V and P is the maximum steady state power. max

157. Define inertia constants M and H.

Ans. Mis the angular momentum of the machine and is expressed in φoule-seconds per radian or in mega-φoule-seconds per electrical degree. M, the inertia constant of the machine, depends on the type and size of the machine.

Another constant H having narrow range of values for each class of machine regardless of its capacity is defined as the ratio of the kinetic energy stored at rated speed to the rated power of the machine. H is expressed in M9/MVA.

158. Give typical values of inertia constant H for turbo-generators and slow-speed way wheel generators.

Ans. The average value of turbo-generators is between 3 and 9 and for water-wheel generators it is between 2 and 4.

159. Write down the units of inertia constants M and H and their interrelationships.

Ans. M is expressed in M9-seconds/electrical degree or M9-seconds/ electrical radian and H is expressed in M9/MVA The interrelationship between M and His given as GH 180/ M= GH

160. What is equal area criterion?

Ans. Equal area criterion is a simple graphical method of predicting the transient stability of two machine systems or a single machine against infinite bus. This criterion does not require solution of swing equation for determination of stability conditions. The stability conditions are determined by equating the areas of segments on the power angle diagram between the P-curve and the new power transfer line for the given condition.

161. For the swing equation $M\frac{d^2\delta}{dt^2} = P_A$ what will be the value of P_A during steady state operation?

Ans. During steady state operation the value of accelerating power P_A will be zero.

162. If the two machines with inertias M_1 and M_2 are swinging together, what will be the inertia of the equivalent machine?

Ans. Equivalent machine inertia is given as $M_{Eq.} = \frac{M_1 M_2}{M_1 + M_2}$

163. What is the purpose served by power flow analysis?

Ans. Power flow analysis provides a systematic mathematical approach for determination of various bus voltages, their phase angles, activated reactive power flows through different branches, gene tors and loads under steady state conditions.

164. Name the three categories of buses.

Ans. Depending upon which quantities have been specified, the buses are classified into following three categories;

1. Generation or voltage-controlled or P-V buses,
2. Load or P-Q buses and
3. Slack or swing or reference bus.

165. What are quantities specified at a generator bus?

Ans. At a generator bus voltage magnitude and true or active power are specified.

166. Name the three categories of buses.

Ans. Depending upon which quantities have been specified, the buses are classified into following three categories;

1. Generation or voltage-controlled or P-V buses,
2. Load or P-Q buses and
3. Slack or swing or reference bus.

167. What are quantities specified at a generator bus?

Ans. At a generator bus voltage magnitude and true or active power are specified.

168. Mention the items specified and not specified at a reference bus

Ans. At a reference bus, the magnitude and phase angle of the voltage are specified. The unknown quantities are active and reactive power flows.

169. Why is one of the buses taken as slack bus in load flow studies.

Ans. The need to designate one of the buses as a slack bus is obviously due to the fact that the system power losses are not known initially and, therefore, the net power flow into the system cannot be fixed in advance. At his bus, the magnitude and phase angle of the voltage specified. The phase angle of the voltage is usually set equal to zero. The active and reactive power at this bus is to be determined through the solution of equations.

170. How is the swing bus selected in a load flow study?

Ans. One of the generation buses is selected as slack bus and the generators at this bus supply the difference between the specified real power inφected into the system at other buses and the total output of the system plus system power losses.

171. Define the bus admittance matrix.

Ans. General equation for n-bus network based on Kirchhoff's' current law and admittance form is [I] = [Y bas]* [V] where [I] is the n- bus current matrix, [V] is the n-bus voltage matrix and [Ybus] is called the bus-admittance matrix and above equation is written as I = Ybus* V.

172. Define tree, cotree and cut-set.

Ans. Tree is a subgraph containing all the nodes of the original graph but no closed path. The elements of the tree are known as branches.

Co-tree is a complement of a tree. The elements of the original graph not included in the tree, form a subgraph which may not necessarily be connected, is known as cotree. Cut-set is the minimal set of the elements in the connected graph which when removed, divide the graph in only two parts, i.e., two connected subgraphs.

173. Give any two operating constraints imposed in load flow studies.

Ans. Two operating constraints may be voltage and reactive power.

174. Write general expression for the complex power inφected by the source into the ith bus of a power system.

Ans. General expression for the complex power (conφugate) inφected into i^th^ bus is given as

$$S_i^* = P_i - jQ_i = V_i^* \sum_{k=1}^{n} Y_{ik} V_k \qquad i=1, 2, 3\ldots\ldots n$$

175. Name any two iterative methods used for solution of load flow problems.

Ans. The most commonly used iterative methods for solution of load flow problems are 1. Gauss-Seidel method and Newton-Raphson method.

176. Which method out Gauss-Seidel method and Newton- Raphson method is used for large power system.

Ans. Newton-Raphson methos as it needs less iteration to reach convergence, takes less computer time, convergence is certain, computer cost is less and it is more accurate and is insensitive to factors like slack bus selection, regulating transformer etc.

177. Why is Gauss-Seidel method considered superior to Gauss- iterative method?

Ans. The Gauss-Seidel method uses upgraded iterates as soon as they are available and thus has an advantage over the Gauss-iterative method which takes longer time. Gauss-Seidel method converges much faster than the Gauss-iterative method, i.e., the number of iterations required to obtain the solution is much less in Gauss- Seidel method compared to the Gauss-iterative method.

178. What is meant by acceleration factor in load flow solution and its best value?

Ans. Acceleration factor is used in Gauss-Seidel method of load flow solution to increase the rate of convergence. Acceleration factor = 1.6 is considered to be a good value for power flow studies.

179. State Gauss-Seidel load flow formula.

Ans. Gauss-Seidel load flow formula is given as

$$V_i = \frac{1}{Y_{ii}}\left[\frac{P_i - jQ_i}{V_i} - \sum_{k=1}^{n} Y_{ik}V_k\right]$$

i=2, 3,.,n Bus 1 is a slack bus with specified V.

180. What is meant by surge absorber?

Ans. The device, which reduces the steepness of the wavefront of a particular surge and thus minimizes the danger due to over- voltage is called the surge absorber.

181. What is the difference between a lightning conductor and lightning arrester?

Ans. Lightning conductor is installed at the top of tail structures and is connected to the general mass of earth through a system of connecting conductors and the earth electrodes so that the structure under protection is not subφected to the dangerously high potential of the cloud discharges. While lighting arrester is a device that is connected in parallel with the equipment, under protection from lightning strokes, at the substation.

182. What is the difference between a lightning arrester and surge absorber?

Ans. Lightning arrester limits the duration amplitude of the follow current while a surge absorber reduces the steepness of the wave front of a particular surge.

183. What are the values of natural impedances for overhead lines and cables?

Ans. The value of natural impedance for an overhead line is between 400 Ω and 600 Ω while for a cable it is between 40 Ω and 60 Ω.

184. Why velocity of propagation over all overhead lines is same?

Ans. Velocity of propagation over all overhead lines is same because product of L and C is same for all overhead lines.

185. What is meant by crest of a wave?

Ans. Crest of a wave is the maximum amplitude and is usually expressed in kV or kA.

186. What is meant by front of a wave?

Ans. Front of a wave is the portion of the wave before crest and is expressed in time from the beginning of the wave to the crest value in ms or us.

187. A transmission line of surge impedance Z_C is terminated through a resistance R. Give the coefficients of refraction and reflection.

Ans. Coefficients of refraction $= \frac{2R}{R+Z_C}$ and coefficient of reflection$= \frac{R-Z_C}{R+Z_C}$

188. While connecting station equipment to an overhead line through a cable it is important that length of the cable should not be very small. Why?

Ans. In connecting station equipment to an overhead line through a cable it is important that the length of cable should not be shorter than the expected length of the wave otherwise successive reflections at the φunction may cause piling up of voltage and the voltage at the φunction may attain the value of incident voltage.

189. What is the effect of shunt capacitance at the terminal of a transmission line?

Ans. The effect of shunt capacitance at the terminal of a transmission line is to cause the voltage at the terminal to rise to full-value gradually instead of abruptly i.e., to cause flattening of the wave front, which reduces the stress on the line-end windings of a transformer connected to the line.

190. What is the application of Bewley's lattice diagram?

Ans. With the use of Bewley's lattice diagram one can know at a glance the position and direction of motion of every incident, reflected and transmitted wave on the system at every instant of time.

191. What is difference between system earthing and equipment earthing

Ans. System earthing deals with earthing of the system neutral to ensure system security and protection while equipment earthing deals with earthing of non-current carrying parts of the equipment to ensure safety of personnel and protection against lightning.

192. What is average resistance of a human body?

Ans. About 1,000 Ω.

193. What is the current that is considered safe for human body?

Ans. About 9 mA.

194. What is meant by an earth electrode?

Ans. Any wire, rod, pipe, plate and an array of conductors embedded in ground horizontally or vertically is called the earth electrode.

195. Give the maximum permissible values of ground resistance for (a) large power stations (b) small substations.

Ans. (a) 0.5 Ω; (b) 2.0 Ω

196. What is meant by 'let go current'?

Ans. The maximum current at which the subφect is capable of releasing an electrical conductor by using muscles directly stimulated by the current is called 'let go current'.

197. On what factor does the resistivity of soil depend?

Ans. Soil resistivity depends upon moisture content, soil temperature, composition and amount of soluble salts in the soil, magnitude of current flowing through the soil and depth of the layer.

198. What is meant by earthing mat?

Ans. A number of rods when φoined together through copper conductors constitute an earthing mat. This is also called the earthing grid.

199. What is the chief argument that favours use of isolated neutral system?

Ans. The chief argument that goes in favour of isolated neutral system is the possibility of maintaining continuity of supply in case one phase gets grounded until it is convenient to disconnect and repair the faulty line.

200. What is harmonic suppressor?

Ans. Harmonic suppressor is an iron-cored reactor and is used in the neutral ground connections of a generator so that third harmonic currents are reduced to safe value

Appendix-A2

Line and Load Data for Various Standard Test Systems

Data for 12-Bus (Indian) and 85-Bus Test Systems

The line resistance (*R*), line reactance (*X*), real power load demand (P_l) and reactive power load demand (Q_l) are mentioned for the 12-bus and 85-bus systems in Tables 1 and Table 2, respectively.

12-Bus (Indian) System

The 12-bus system data is taken from literature. For study, a power base of 0.01 MVA and voltage base of 11 kV have been taken. The single line diagram of 12-bus system is shown in Figure A.1. The line parameters and bus load data are given in Table 1.

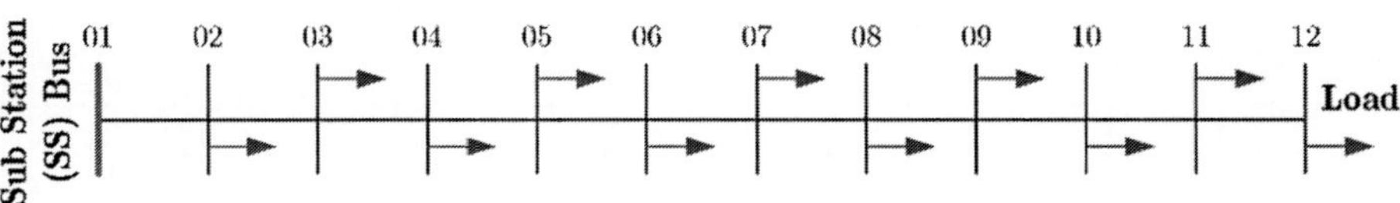

Figure 1: 12-bus RDS

Table 1: Line and load data for 12-bus system

Line Bus No.		Resistance	Reactance	Bus No.	Active Load	Reactive Load
Start	End	*R* (p.u.)	*X* (p.u.)		P_l (MW)	Q_l (MVAr)
1	2	0.00009	0.000038	2	0.06	0.06
2	3	0.000098	0.000041	3	0.04	0.03
3	4	0.000173	0.000072	4	0.055	0.055
4	5	0.000263	0.00011	5	0.03	0.03
5	6	0.00009	0.000038	6	0.02	0.015
6	7	0.000083	0.000034	7	0.055	0.055
7	8	0.000364	0.0001	8	0.045	0.045
8	9	0.000466	0.000132	9	0.04	0.04
9	10	0.000239	0.000068	10	0.035	0.03
10	11	0.000125	0.000035	11	0.04	0.03
11	12	0.000102	0.000029	12	0.015	0.015

16-Bus System

The 16-bus system data is taken from literature. For study, a power base of 100 MVA and voltage base of 23 kV have been taken. The single line diagram of 16-bus system is shown in Figure 2. The line parameters and bus load data are given in Table 2.

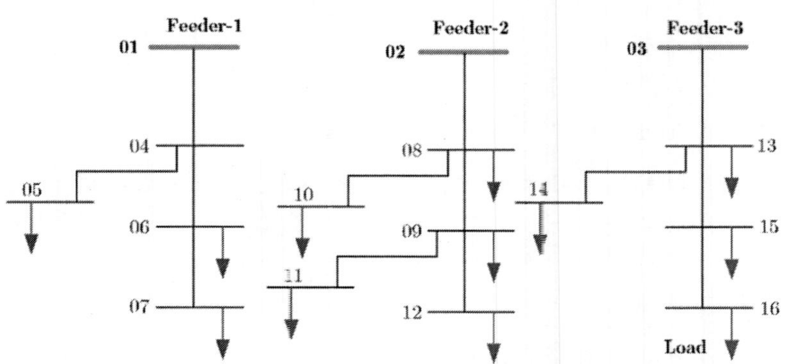

Figure 2: 16-bus RDS (Tie switches are not shown)

Table 2: Line and load data for 16-bus system

Line	From	To bus	R	X	Receiving	
No	bus i	φ	P.U.	P.U.	bus φ	
					P	Q
1	1	4	0.075	0.10	2.0	1.6
3	4	5	0.08	0.11	3.0	0.4
2	4	6	0.09	0.18	2.0	-0.4
5	6	7	0.04	0.04	1.5	1.2
7	2	8	0.11	0.11	4.0	2.7
8	8	9	0.08	0.11	5.0	1.8
9	8	10	0.11	0.11	1.0	0.9
6	9	11	0.11	0.11	0.6	-0.5
10	9	12	0.08	0.11	4.5	-1.7
15	3	13	0.11	0.11	1.0	0.9
14	13	14	0.09	0.12	1.0	-1.1
16	13	15	0.08	0.11	1.0	0.9
12	15	16	0.04	0.04	2.1	-0.8
4	5	11	0.04	0.04		
13	10	14	0.04	0.04		
11	7	16	0.12	0.12		

33-Bus System

The 33-bus system data is taken from literature. For study, a power base of 100 MVA and voltage base of 12.66 kV have been taken. The single line diagram of 33-bus system is shown in Figure 3. The line parameters and bus load data are given in Table 3.

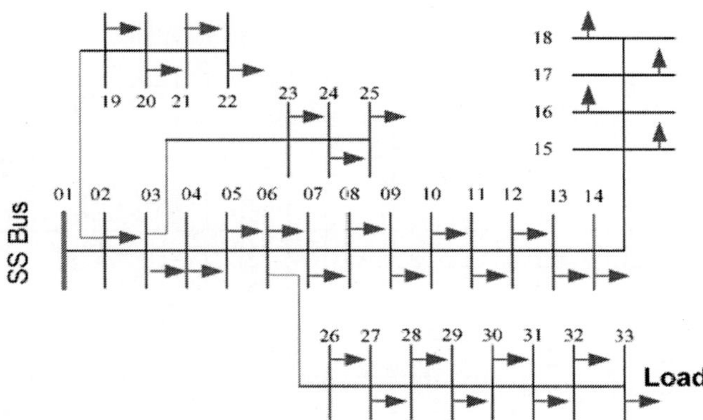

Figure 3: 33-bus Radial Distribution System

Table 3: Line and load data for 33-bus system

Line bus no		Resistance	Reactance	Load bus	Active load	Reactive load
Start	**end**					
1	2	0.0922	0.0470	2	100	60
2	3	0.4930	0.2512	3	90	40
3	4	0.3661	0.1864	4	120	80
4	5	0.3811	0.1947	5	60	30
5	6	0.8190	0.7070	6	60	20
6	7	0.1872	0.6188	7	200	100
7	8	0.7115	0.2351	8	200	100
8	9	1.0299	0.7400	9	60	20
9	10	1.0440	0.7400	10	60	20
10	11	0.1967	0.0651	11	45	30
11	12	0.3744	0.1298	12	60	35
12	13	1.4680	1.1549	13	60	35
13	14	0.5416	0.7129	14	120	80
14	15	0.5909	0.5260	15	60	10
15	16	0.7462	0.5449	16	60	20
16	17	1.2889	1.7210	17	60	20
17	18	0.7320	0.5739	18	90	40
2	19	0.1640	0.1565	19	90	40
19	20	1.5042	1.3555	20	90	40
20	21	0.4095	0.4784	21	90	40
21	22	0.7089	0.9373	22	90	40
3	23	0.4512	0.3084	23	90	50
23	24	0.8980	0.7091	24	420	200
24	25	0.8959	0.7071	25	420	200
6	26	0.2031	0.1034	26	60	25
26	27	0.2842	0.1447	27	60	25

Line bus no		Resistance	Reactance	Load bus	Active load	Reactive load
Start	end					
27	28	1.0589	0.9338	28	60	20
28	29	0.8043	0.7006	29	120	70
29	30	0.5074	0.2585	30	200	100
30	31	0.9745	0.9629	31	150	70
31	32	0.3105	0.3619	32	210	100
32	33	0.3411	0.5302	33	60	40
8	21	2.000	2.000			
9	15	2.000	2.000			
12	22	2.000	2.000			
18	33	0.5	0.5			
25	29	0.5	0.5			

69-Bus System

The 69-bus system data is taken from literature. For study, a power base of 100 MVA and voltage base of 12.66 kV have been taken. The single line diagram of 69-bus system is shown in Figure 4. The line parameters and bus load data are given in Table 4.

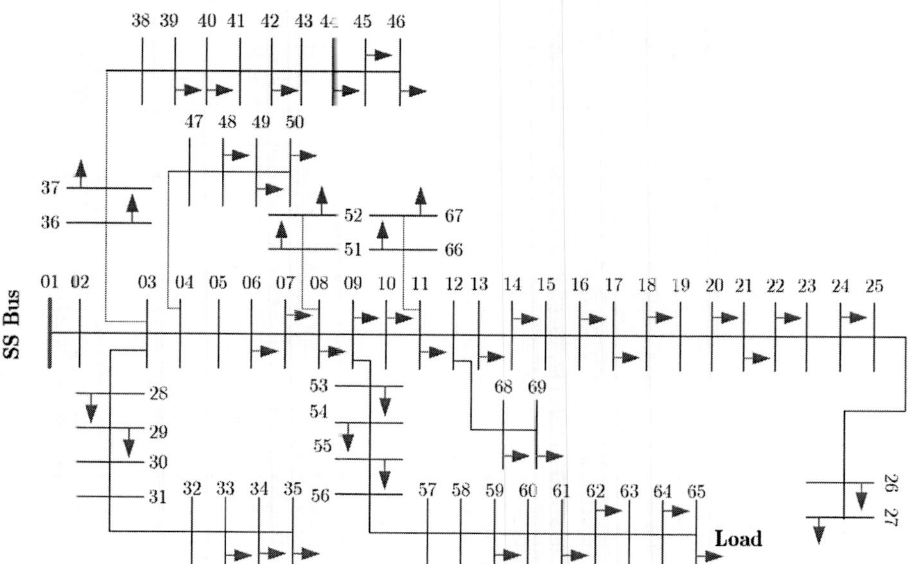

Figure 4: 69-bus RDS

Table 4: Line and load data for 69-bus system

Line number	Sending bus no.	Receiving bus no.	Resistance (Ω)	Reactance (Ω)	Load at receiving end bus	
					Real power (kW)	Reactive power (kVAr)
1	1	2	0.0005	0.0012	0	0
2	2	3	0.0005	0.0012	0	0
3	3	4	0.0015	0.0036	0	0
4	4	5	0.0251	0.0294	0	0
5	5	6	0.366	0.1864	2.6	2.2
6	6	7	0.3811	0.1941	40.4	30
7	7	8	0.0922	0.047	75	54
8	8	9	0.0493	0.0251	30	22
9	9	10	0.819	0.2707	28	19
10	10	11	0.1872	0.0691	145	104
11	11	12	0.7114	0.2351	145	104
12	12	13	1.03	0.34	8	5.5
13	13	14	1.044	0.345	8	5.5
14	14	15	1.058	0.3496	0	0
15	15	16	0.1966	0.065	45.5	30
16	16	17	0.3744	0.1238	60	35
17	17	18	0.0047	0.0016	60	35
18	18	19	0.3276	0.1083	0	0
19	19	20	0.2106	0.069	1	0.6
20	20	21	0.3416	0.1129	114	81
21	21	22	0.014	0.0046	5.3	3.5
22	22	23	0.1591	0.0526	0	0
23	23	24	0.3463	0.1145	28	20
24	24	25	0.7488	0.2745	0	0
25	25	26	0.3089	0.1021	14	10
26	26	27	0.1732	0.0572	14	10
27	3	28	0.0044	0.0108	26	18.6
28	28	29	0.064	0.1565	26	18.6
29	29	30	0.3978	0.1315	0	0
30	30	31	0.0702	0.0232	0	0
31	31	32	0.351	0.116	0	0
32	32	33	0.839	0.2816	14	10
33	33	34	1.708	0.5646	19.5	14
34	34	35	1.474	0.4673	6	4

Line number	Sending bus no.	Receiving bus no.	Resistance (Ω)	Reactance (Ω)	Load at receiving end bus	
					Real power (kW)	Reactive power (kVAr)
35	3	36	0.0044	0.0108	26	18.55
36	36	37	0.064	0.1565	26	18.55
37	37	38	0.1053	0.123	0	0
38	38	39	0.0304	0.0355	24	17
39	39	40	0.0018	0.0021	24	17
40	40	41	0.7283	0.8509	1.2	1
41	41	42	0.31	0.3623	0	0
42	42	43	0.041	0.0478	6	4.3
43	43	44	0.0092	0.0116	0	0
44	44	45	0.1089	0.1373	39.22	26.3
45	45	46	0.0009	0.0012	39.22	26.3
46	4	47	0.0034	0.0084	0	0
47	47	48	0.0851	0.2083	79	56.4
48	48	49	0.2898	0.7091	384.7	274.5
49	49	50	0.0822	0.2011	384	274.5
50	8	51	0.0928	0.0473	40.5	28.3
51	51	52	0.3319	0.1114	3.6	2.7
52	9	53	0.174	0.0886	4.35	3.5
53	53	54	0.203	0.1034	26.4	19
54	54	55	0.2842	0.1447	24	17.2
55	55	56	0.2813	0.1433	0	0
56	56	57	1.59	0.5337	0	0
57	57	58	0.7837	0.263	0	0
58	58	59	0.3042	0.1006	100	72
59	59	60	0.3861	0.1172	0	0
60	60	61	0.5075	0.2585	1244	888
61	61	62	0.0974	0.0496	32	23
62	62	63	0.145	0.0738	0	0
63	63	64	0.7105	0.3619	227	162
64	64	65	1.041	0.5302	59	42
65	11	66	0.2012	0.0611	18	13
66	66	67	0.0047	0.0014	18	13
67	12	68	0.7394	0.2444	28	20
68	68	69	0.0047	0.0016	28	20

85-Bus System

The 85-bus system data is taken from literature. For study, a power base of 100 MVA and voltage base of 11 kV have been taken. The single line diagram of 85-bus system is shown in Figure 6. the line parameters and bus load data are given in Table 6.

Table 5: Line and load data for 35-bus system

Line Bus No.		Resistance	Reactance	Load Bus	Active Load	Reactive Load
Start	**End**	***R* (p.u.)**	***X* (p.u.)**		**P_l (MW)**	**Q_l (MVAr)**
1	2	0.089256	0.061983	2	0	0
2	3	0.134711	0.092562	3	0	0
3	4	0.179339	0.12314	4	0.056	0.05713
4	5	0.089256	0.061157	5	0	0
5	6	0.359504	0.246281	6	0.03528	0.03599
6	7	0.224793	0.153719	7	0	0
7	8	0.989256	0.677686	8	0.03528	0.03599
8	9	0.089256	0.061157	9	0	0
9	10	0.494215	0.338843	10	0	0
10	11	0.449587	0.308264	11	0.056	0.05713
11	12	0.449587	0.308264	12	0	0
12	13	0.494215	0.338843	13	0	0
13	14	0.224793	0.153719	14	0.03528	0.03599
14	15	0.269421	0.184298	15	0.03528	0.03599
2	16	0.601653	0.249587	16	0.03528	0.03599
3	17	0.376033	0.156198	17	0.112	0.11426
5	18	0.677686	0.280992	18	0.056	0.05713
18	19	0.526446	0.218182	19	0.056	0.05713
19	20	0.376033	0.156198	20	0.03528	0.03599
20	21	0.67686	0.280992	21	0.03528	0.03599
21	22	1.279339	0.530579	22	0.03528	0.03599
19	23	0.150413	0.061983	23	0.056	0.05713
7	24	0.752066	0.312397	24	0.03528	0.03599
8	25	0.376033	0.156198	25	0.03528	0.03599
25	26	0.300826	0.124793	26	0.056	0.05713
26	27	0.45124	0.186777	27	0	0
27	28	0.22562	0.093388	28	0.056	0.05713
28	29	0.45124	0.186777	29	0	0
29	30	0.45124	0.186777	30	0.03528	0.03599
30	31	0.22562	0.093388	31	0.03528	0.03599
31	32	0.150413	0.061983	32	0	0
32	33	0.150413	0.061983	33	0.014	0.014283
33	34	0.67686	0.280992	34	0	0
34	35	0.526446	0.218182	35	0	0
35	36	0.150413	0.061983	36	0.03528	0.03599
26	37	0.300826	0.124793	37	0.056	0.05713
27	38	0.828099	0.343802	38	0.056	0.05713
29	39	0.45124	0.186777	39	0.056	0.05713
32	40	0.376033	0.156198	40	0.03528	0.03599
40	41	0.828099	0.343802	41	0	0

Line Bus No.		Resistance	Reactance	Load Bus	Active Load	Reactive Load
Start	End	R (p.u.)	X (p.u.)		P_l (MW)	Q_l (MVAr)
41	42	0.22562	0.093388	42	0.03528	0.03599
41	43	0.376033	0.156198	43	0.03528	0.03599
34	44	0.828099	0.343802	44	0.03528	0.03599
44	45	0.752893	0.312397	45	0.03528	0.03599
45	46	0.752893	0.312397	46	0.03528	0.03599
46	47	0.45124	0.186777	47	0.014	0.014283
35	48	0.526446	0.218182	48	0	0
48	49	0.150413	0.061983	49	0	0
49	50	0.300826	0.124793	50	0.03628	0.03599
50	51	0.376033	0.156198	51	0.056	0.05713
48	52	1.128926	0.468595	52	0	0
52	53	0.376033	0.156198	53	0.03528	0.03599
53	54	0.45124	0.186777	54	0.056	0.05713
52	55	0.45124	0.186777	55	0.056	0.05713
49	56	0.45124	0.186777	56	0.014	0.014283
9	57	0.22562	0.093388	57	0.056	0
57	58	0.67686	0.280992	58	0	0
58	59	0.150413	0.061983	59	0.056	0.05713
58	60	0.45124	0.186777	60	0	0
60	61	0.601653	0.249587	61	0.056	0.05713
61	62	0.828099	0.342975	62	0.056	0.05713
60	63	0.150413	0.061983	63	0.014	0.014283
63	64	0.601653	0.249587	64	0	0
64	65	0.150413	0.061983	65	0	0
65	66	0.150413	0.061983	66	0.056	0.05713
64	67	0.376033	0.156198	67	0	0
67	68	0.752066	0.312397	68	0	0
68	69	0.902479	0.37438	69	0.056	0.05713
69	70	0.376033	0.156198	70	0	0
70	71	0.45124	0.186777	71	0.03528	0.03599
67	72	0.150413	0.061983	72	0.056	0.05713
68	73	0.978512	0.405785	73	0	0
73	74	0.22562	0.093388	74	0.056	0.05713
73	75	0.828099	0.343802	75	0.03528	0.03599
70	76	0.45124	0.186777	76	0.056	0.05713
65	77	0.075207	0.030579	77	0.014	0.014283
10	78	0.526446	0.218182	78	0.056	0.05713
67	79	0.45124	0.186777	79	0.03528	0.03599
12	80	0.601653	0.249587	80	0.056	0.05713
80	81	0.300826	0.124793	81	0	0
81	82	0.075207	0.030579	82	0.056	0.05713

Line Bus No.		Resistance	Reactance	Load Bus	Active Load	Reactive Load
Start	End	R (p.u.)	X (p.u.)		P_l (MW)	Q_l (MVAr)
81	83	0.902479	0.37438	83	0.03528	0.03599
83	84	0.828099	0.343802	84	0.014	0.014283
13	85	0.67686	0.280992	85	0.03528	0.03599

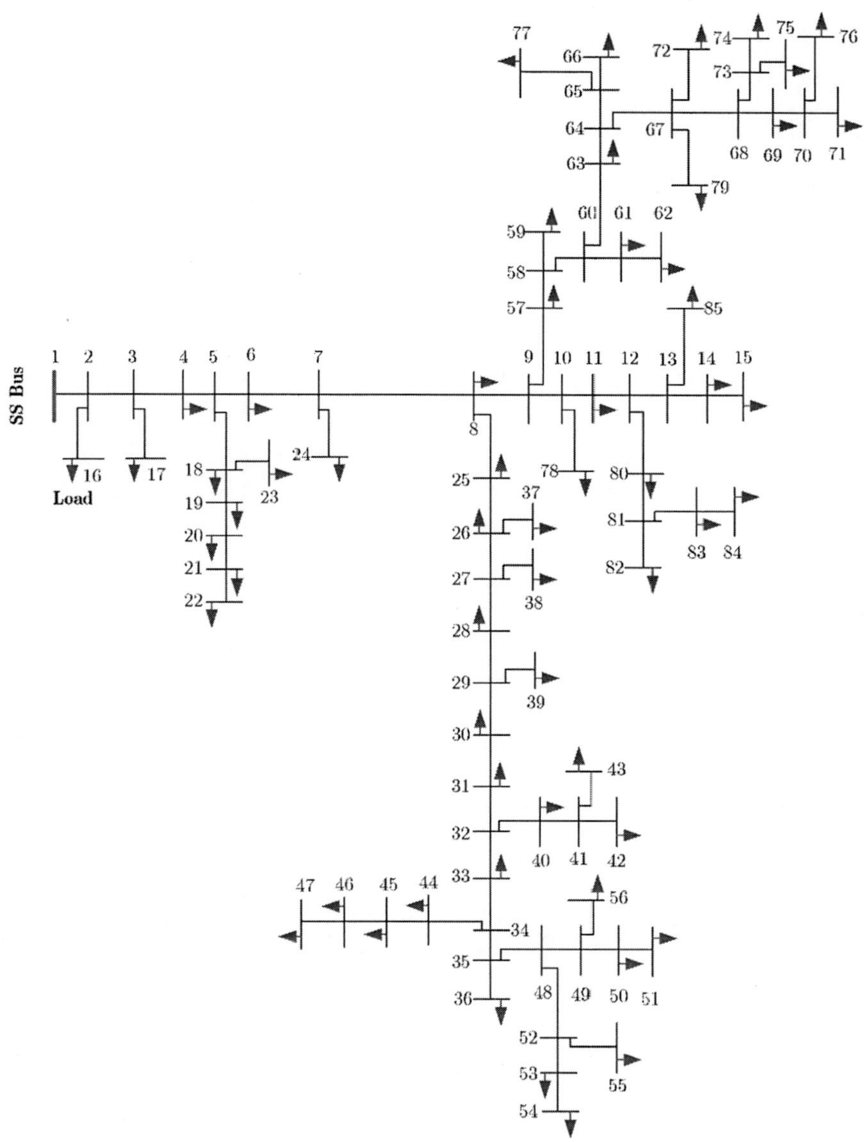

Figure 5: 85-bus RDS

मैं आंबेडकर बोल रहा हूँ

मैं आंबेडकर बोल रहा हूँ

सं. दिनकर कुमार

प्रतिभा प्रतिष्ठान, नई दिल्ली

प्रकाशक : प्रतिभा प्रतिष्ठान,
694-बी, (निकट अजय मार्केट) चावड़ी बाजार, दिल्ली-110006
सर्वाधिकार : सुरक्षित / संस्करण : 2022 / मूल्य : चार सौ पचास रुपए
मुद्रक : जयलक्ष्मी प्रिंटर्स, दिल्ली ISBN 978-93-83111-30-5

MAIN AMBEDKAR BOL RAHA HOON

Ed. Dinkar Kumar ₹ 450.00

Published by Pratibha Pratishthan, 694-B, (Near Ajay Market)
Chawri Bazar, Delhi-110006

अपनी बात

"जो कुछ मैं कर सका, वह जीवन भर मुसीबतें सहन करके विरोधियों से टक्कर लेने के बाद कर पाया हूँ। जिस कारवाँ को आप यहाँ देख रहे हैं, उसे मैं अनेक कठिनाइयों से यहाँ ला पाया हूँ। अनेक अवरोध, जो इसके मार्ग में आ सकते हैं, के बावजूद इस कारवाँ को बढ़ते रहना है। अगर मेरे अनुयायी इसे आगे ले जाने में असमर्थ रहे तो उन्हें इसे यहीं पर छोड़ देना चाहिए, जहाँ पर यह अब है; पर किन्हीं भी परिस्थितियों में इसे पीछे नहीं हटने देना है। मेरी जनता के लिए मेरा यही संदेश है।"

—डॉ. भीमराव आंबेडकर

बाबा साहब भीमराव आंबेडकर भारतीय दलित आंदोलन एवं सामाजिक न्याय अभियान के इतिहास में एक ऐसा नाम है, जो दूर से ही चमकता है।

अनेक कष्ट और परेशानियों को सहते हुए उन्होंने उच्च शिक्षा प्राप्त की तथा दलित व पिछड़े समाज को अपने साथ लेकर उनके अधिकारों की लड़ाई लड़ी। अपने जीवनकाल में ही उन्होंने पिछड़ों को उनके अधिकार दिलाए। अपनी योग्यता के बल पर भारत का संविधान लिखा और सामाजिक न्याय के लिए संघर्ष किया, दलित व पिछड़े समाज को शिक्षा के लिए प्रेरित किया। उसे विशेष सुविधाएँ दिलाईं तथा राजनीतिक शक्ति का हथियार थमाकर उसे इतना मजबूत बना दिया कि वह अपना सदियों का खोया मनोबल फिर से प्राप्त कर राष्ट्र की मुख्य धारा में सीना तानकर खड़ा हो गया है।

जन्म और माता-पिता

भीमराव का जन्म 14 अप्रैल, 1891 को मध्य प्रदेश के रत्नागिरि जिले में महू नामक स्थान पर हुआ था। महू इंदौर नगर के निकट है। उनके पिता का नाम रामजी राव तथा माता का नाम भीमाबाई था। पिता सेना में सूबेदार थे। आंबेडकर अपने माता-पिता की चौदहवीं संतान थे। उनसे पहले जनमे भाई-बहनों में से केवल दो भाई और दो बहनें जीवित थीं, अन्य की मृत्यु हो चुकी थी। उनके एक भाई का नाम आनंदराव था, दूसरे का बसंतराव; एक बहन का नाम मंजुला था और दूसरी का तुलसी।

कहने के लिए तो पिता सूबेदार थे, परंतु उन्हें वेतन केवल 50 रुपए मासिक मिलता था। उन दिनों सेना में नौकरी की अवधि 15 वर्ष हुआ करती थी।

15 वर्ष तक सेना को अपनी सेवाएँ देने के बाद रामजी राव सेवा-मुक्त हो गए। अब उनके सामने रोजगार की समस्या थी। वे महू से सतारा आ गए। सतारा में वे बहुत दिनों तक रोजगार की तलाश में भटकते रहे, अंततः 50 रुपए मासिक पर उन्हें एक कंपनी में चौकीदारी का काम मिल गया।

पारिवारिक पृष्ठभूमि

बड़े जाँबाज और प्रतापी नवरत्न प्रदान कर जिस जिले ने अपना नाम सार्थक किया, उस रत्नागिरि जिले में मंडनगढ़ से 5 मील की दूरी पर आंबबड़े नाम का एक देहात स्थित है। यही है आंबेडकर घराने का मूल गाँव। इस घराने का कुलनाम है सकपाल, कुलदेवी है भवानी। कुलदेवी की पालकी रखने का सम्मान इसी महार घराने का था। गाँव के वार्षिक उत्सव के अवसर पर उन्हें अपने गाँववालों में विशेष सम्मान प्राप्त होता था और उस परिवार को वह दिन बड़ा महत्त्वपूर्ण लगता था।

भारत के सारे अस्पृश्य समाज में महार जाति मजबूत कद-काठीवाली, जोरदार आवाजवाली, चमकीली चमड़ीवाली, गुण से बुद्धिमान और वृत्ति से लड़ाकू दिखाई देती है। समय का ध्यान रखते हुए सावधानी से बरताव करनेवाली अगर कोई जाति है तो वह महारों की ही।

कुछ विद्वानों का कहना है कि महाराष्ट्र के मूल निवासी महार ही थे और उनके ही नाम पर इस राज्य को 'महाराष्ट्र' नाम प्राप्त हुआ। कुछ लोग 'महार' शब्द की व्युत्पत्ति इस तरह करते हैं—महा + अरि = बड़ा दुश्मन। दक्षिण भारत में राजनीतिज्ञों की सत्ता के लिए षड्यंत्र शुरू होने पर उन्होंने बड़ी होशियारी से उस उपेक्षित, लेकिन जुझारू महार जाति के साथ प्रथमतः संबंध जोड़ लिये। ईस्ट इंडिया कंपनी की बंबई

सेना के लिए कंपनी सरकार ने महारों के दल बनाए थे। उस समय बिहार के दस्यु और मद्रास राज्य की पेरिया दलित जातियों को लेकर उसी तरह सैनिक टुकड़ियों का गठन कंपनी सरकार ने किया था।

आंबेडकर के पितामह का नाम मालोजी सकपाल था। वे एक सेवानिवृत्त फौजी थे। उनकी संतानों में थे मीराबाई नामक बेटी और रामजी नामक बेटा। दोनों की ही कुछ जानकारी प्राप्त होती है। मालोजी के तीन और बेटे थे; परंतु प्रतीत होता है कि सेना की नौकरी में बार-बार तबादला होने के कारण परिवार के साथ मालोजी का विशेष संबंध नहीं रहा होगा।

रामजी सकपाल भी सेना में भरती हो गए। पहले जवानों को फौजी छावनी में स्कूल द्वारा अनिवार्य शिक्षा देने का प्रबंध ईस्ट इंडिया कंपनी सरकार ने किया था। इसके अतिरिक्त फौजी सिपाहियों की संतानों के लिए दिन के स्कूल थे। उनके प्रौढ़ रिश्तेदारों के लिए भी रात के स्कूल थे। ऐसी ही शिक्षा रामजी सकपाल को मिली थी।

रामजी सकपाल जिस सैनिक टुकड़ी में शामिल थे, उसी में इत्तफाक से सूबेदार मेजर धर्मा मुरबाड़कर का तबादला हुआ। मुरबाड़कर भी महार परिवार के थे। वे ठाणे जिले के मुरबाड़ के निवासी थे। मुरबाड़कर का कुलनाम पंडित था। उस परिवार के सात भाई भी सेना में मेजर थे। स्वाभाविक ही उनका रहन-सहन अमीरी ठाट-बाट का था। वे नाथ संप्रदाय के अनुयायी थे। उनके परिवार का वातावरण धार्मिक था। महार बस्ती और गाँव में उनका घराना बड़ा इज्जतदार माना जाता था।

आगे चलकर यह परिवार पनवेल के नजदीक उलवा बंदरगाह पर निवास करने चला गया। फिर कुछ दिनों बाद वह परिवार पनवेल जाकर बस गया।

धर्माजी की बेटी भीमाबाई और रामजी सकपाल का विवाह वर्ष 1865 के दौरान संपन्न हुआ। विवाह के समय रामजी की उम्र लगभग 22 वर्ष थी और भीमाबाई की उम्र लगभग 13 वर्ष। भीमाबाई बड़ी बातूनी, स्वाभिमानी, हठीली और धर्म-परायण महिला थीं। उनका रंग गोरा था, आँखें बड़ी-बड़ी और आकर्षक थीं। बाल घुँघराले थे। ललाट चौड़ा था और नाक तनिक चपटी थी। विवाह के बाद भीमा की माँ ही उसका हाल-चाल पूछती रहती थीं। परिवार के अन्य सदस्य रूठे ही रहते थे। इस बात से आहत होकर भीमाबाई ने दृढ़तापूर्वक कहा, ''अब मैं तभी मायके जाऊँगी जब जेवरों से भरी-पूरी अमीरी प्राप्त कर लूँगी।''

रामजी सकपाल उद्यमी, महत्त्वाकांक्षी और स्वतंत्र वृत्ति के आदमी थे। वे क्रिकेट, फुटबॉल आदि खेलों में प्रवीण थे। उत्साह के साथ निरंतर उद्यम करनेवाले पुरुष को

सम्मान और संपत्ति प्राप्त होती है। रामजी की उद्यमिता, प्रवीणता और कर्तव्यपरायणता को देखकर सेना का एक अधिकारी उनसे काफी प्रभावित हुआ। उसने रामजी की सिफारिश कर उन्हें पुणे के पंतोजी स्कूल में दाखिल करवा दिया। वहाँ अध्यापन के पेशे की शिक्षा लेते समय उन्हें जो वेतन मिलता था, उसमें से कुछ भाग वे पत्नी के लिए बंबई भेज देते थे। उन दिनों उनकी सैन्य टुकड़ी बंबई के निकट सांताक्रूज में थी। भीमा जीविका चलाने के लिए कंकड़-पत्थर पसारने का कार्य किया करती थीं। अपनी व्यावहारिक बुद्धि के कारण वे रूखी-सूखी में ही प्रसन्न रहती थीं।

जिस समय रामजी की शिक्षा-दीक्षा संभव हुई थी उस समय दलित शिक्षा की कल्पना भी नहीं की जा सकती थी। रास्तों की सफाई, सिर पर मैला ढोना, मृत पशुओं का चमड़ा निकालना—इसी प्रकार के कार्य करके उन्हें अपने अस्तित्व को बनाए रखना पड़ता था। ऐसी दशा में स्कूल और शिक्षा की बात सोची भी नहीं जा सकती थी।

रामजी का सौभाग्य था कि उन्हें शिक्षा की सुविधाएँ मिली थीं। उनके पिता जब सेना में थे तब कंपनी सरकार ने सेना की छावनियों में स्कूल शुरू किए थे। इन स्कूलों में सैनिकों के बच्चों को अनिवार्य रूप से शिक्षा दी जाती थी। इसके अलावा, सैनिकों को भी शिक्षा दी जाती थी। इसी तरह के एक स्कूल में पढ़कर वे प्रवीण हुए थे।

पुणे के पंतोजी स्कूल की परीक्षा में उत्तीर्ण होकर रामजी अध्यापक बने और धीरे-धीरे फौजी छावनी के स्कूल में प्रधानाध्यापक बन गए। उन्होंने प्रधानाध्यापक का कार्य लगभग 14 वर्षों तक किया। आखिर उनकी इस सेवा की कद्र कर उन्हें सूबेदार मेजर के ओहदे तक की पदोन्नति प्राप्त हुई। उनकी अध्यापन कला सराहनीय थी। शिक्षा के प्रति अपनी कर्तव्यनिष्ठा, प्रेम और लगाव से उन्होंने यह सिद्ध किया कि दलित भी अच्छा अध्यापक बन सकता है। साथ ही भीमाबाई का प्रण, जो उन्होंने अपनी संपन्नता हेतु किया था, वह भी पूरा हो गया। उनके परिवार को अब काफी प्रतिष्ठा प्राप्त हो चुकी थी।

सन् 1890 तक सूबेदार रामजी और भीमाबाई की तेरह संतानें पैदा हुईं। उनमें से बालाराम, गंगा, रमाबाई, आनंदराव, मंजुला, तुलसी ही जीवित रहे। शेष संतानें बचपन में ही मर गईं। इतने बच्चों की पैदाइश, आतिथ्य-सत्कार, कई व्रत-उपवास, पूजा-अर्चना और चिकित्सा की व्यवस्था न होने से भीमाबाई की सेहत बिगड़ती गई। कई छावनियों में तबादला होने के बाद रामजी का स्थान मध्य प्रदेश के महू

जिले की लश्करी छावनी में स्थिर हो गया।

महू में एक असाधारण घटना घटी। रामजी के एक चाचा बैरागी हो गए थे, जिनका काफी समय तक पता ही नहीं था। इत्तफाक से एक बार वे साधु जत्थे के साथ महू आ गए और उन्होंने रामजी को आशीर्वाद दिया—'अपने कर्तव्य से इतिहास पर चिरंतन छाप डालकर अपने वंश का नाम अमर करनेवाला एक पुत्र तुम्हारे घर में पैदा होगा।'

यह आशीर्वाद सुनकर सूबेदार दंपती को काफी प्रसन्नता हुई। दोनों धर्मपरायण और व्रतधारी थे। अपने घर कुलदीपक जैसा पुत्र हो, इसलिए दान, व्रत, उपासना, आराधना और धार्मिक कृत्यों की ओर उनका ध्यान बड़ी उत्कटता से बढ़ गया। साधना के अंत में संत का वरदान फलीभूत हुआ। चौदहवीं संतान के रूप में भीम का जन्म हुआ।

इसके साथ मत खेलना, यह अछूत है

सतारा में आंबेडकर परिवार ने किसी तरह जीवन-यापन का सहारा तो ढूँढ़ लिया, परंतु बच्चों के लिए समाज में घुलने-मिलने की समस्या खड़ी हो गई। भीमराव के भाई-बहन तो अब तक यह जान चुके थे कि वे अछूत हैं और उन्हें समाज के लोग इज्जत नहीं देते। लोग उनसे बचते हैं, उनकी छुई हुई चीज नहीं खा सकते। ब्राह्मण तो इतना परहेज करते हैं कि यदि उनका साया भी छू जाए तो 'शिव-शिव' करते हुए फिर से स्नान करने के लिए भाग खड़े होते हैं।

भीमराव अभी छोटे थे। जब उन्होंने यह सब देखा तो उन्हें बड़ी हैरानी हुई। बचपन से ही भीमराव में चीजों को गहराई से देखने और जानने की इच्छा तीव्र थी। वे जब घर से बाहर निकलते तो सब ओर देखते और दौड़कर सब बच्चों के साथ घुल-मिल जाना चाहते थे।

वे इस सच्चाई से अनजान थे कि जिस परिवार और जाति में उन्होंने जन्म लिया है, वह भारतीय समाज में अछूत है और उन्हें सब बच्चों के साथ हिल-मिलकर खेलने तथा उन्हें दोस्त बनाने का अधिकार नहीं है।

वे नहीं जानते थे कि ऊँची जाति और नीची जाति क्या होती है, निम्न वर्ण और उच्च वर्ण क्या होता है। जब वे बाहर होते और कोई बच्चा उन्हें दिखाई दे जाता तो उसकी ओर दौड़ पड़ते। वे उसके साथ खेलना चाहते थे; परंतु जैसे ही वे बच्चे के पास पहुँचते और बच्चा हँसकर उनकी ओर आने को होता, तभी दो हाथ पीछे से उस बच्चे को उठा लेते थे और भीमराव के कान में आवाज पड़ती, 'चल-चल···यह

अछूत है, इसके साथ मत खेलना।'

भीमराव ठगे से देखते रह जाते। उनको लगता, जैसे किसी ने मुँह में जाते-जाते उनके हाथ से निवाला छीन लिया है।

कहीं बच्चों का झुंड खेल रहा होता और भीमराव दौड़कर खेल में शामिल होना चाहते तो कोई बड़ा बच्चा झट से आगे आता और कहता, 'नहीं, तू हमारे साथ नहीं खेल सकता, तू अछूत है। जा, यहाँ से, भाग…।'

भीमराव रोते हुए घर लौट आते। माँ पूछती, 'क्या हुआ भीम, तू रो क्यों रहा है?' तो वे माँ से सिसकते हुए कहते, 'माँ, मुझे कोई अपने साथ नहीं खिलाता। सब बच्चे मुझे भगा देते हैं, कहते हैं–तू अछूत है!'

'हाँ बेटा!' माँ भीमराव को सीने से लगाती हुई कहतीं, "वे ठीक कहते हैं। हम अछूत हैं। वे बड़ी जाति के बच्चे हैं। तू उनके साथ खेलने की जिद क्यों करता है? क्यों जाता है तू वहाँ, बेटा? तू अपनी बहनों और भाइयों के साथ खेल लिया कर।'

सिसकते-सिसकते भीमराव चुप हो जाते और प्राय: माँ की गोद में सो जाते।

स्कूल में दाखिला : जूतों की जगह बैठते थे

भीमराव जब छह वर्ष के हुए तो माँ भीमाबाई रामजी राव के पीछे पड़ गईं, "भीम अब बड़ा हो गया है। उसका दाखिला क्यों नहीं करवाते?"

रामजी कुछ-न-कुछ बहाना बनाकर टाल देते। कभी कहते, "बात की है," कभी कहते, "हो जाएगा, थोड़ा सब्र करो।"

उस दिन जब भीमाबाई ने उन्हें टोका तो वे बिफर गए, "कहाँ कराऊँ इसका दाखिला? कैसे कराऊँ? तुम्हें क्या पता, मैं कहाँ-कहाँ नहीं गया हूँ। पूरे आठ दिन की छुट्टी ले चुका हूँ अब तक इस लड़के को दाखिल कराने के चक्कर में, परंतु कोई राजी नहीं होता, अछूत जो है। हाँ, स्कूल में ऊँची जाति के बच्चे पढ़ते हैं, कहते हैं, हमारा बच्चा उनके साथ नहीं बैठ सकता। मैं पूछता हूँ, जब साथ ही नहीं बैठ सकता तो पढ़ेगा कैसे?"

भीमाबाई शांत रहीं। आगे बढ़कर उन्होंने पीछे से पति के कंधों पर अपने दोनों हाथ रख दिए और बोलीं, "आपका दु:ख मैं समझती हूँ। पता नहीं क्यों, इतना कह गई। सूबेदार की पत्नी होने का गुरूर आ जाता है तो भूल जाती हूँ कि हम अछूत हैं।"

अपने आँसू पीने के बाद रामजी राव ने पत्नी की ओर मुँह फेरा और उसके

आँसुओं को अपनी दोनों हथेलियों में समेटते हुए बोले, "बड़ी मुश्किल से एक हेडमास्टर ने 'हाँ' की है, पर उसकी एक शर्त है।"

"शर्त?" कहते हुए भीमाबाई ने पति की आँखों में देखना चाहा।

रामजी राव ने आँखें झुकाते हुए कहा, "कहता है, आपके लड़के को दाखिल तो कर लूँगा, परंतु उसे बड़ी जाति के बच्चों के साथ नहीं बैठा सकता। उसे अपनी टाट-पट्टी साथ लानी पड़ेगी और कक्षा की दहलीज पर जहाँ बच्चे अपने जूते उतारते हैं, वहाँ बैठकर पढ़ना पड़ेगा।"

"नहीं!" भीमाबाई ने अपना माथा पीटा, "भगवान् इतना अन्यायी नहीं हो सकता।"

उसी समय भीमराव बाहर निकल आए। वे माता-पिता के बीच हुई बातचीत सुन चुके थे। उनकी आँखों में दृढ़ निश्चय की चमक थी। वे बोले, "माँ, मेरा दाखिला करवा दो। मैं जूतों की जगह बैठकर पढ़ लूँगा।"

फूल से कोमल बच्चे के मुँह से जब ये शब्द निकले तो पिता रामजी राव उनमें छिपी विवशता पर रो पड़े और अपने आँसुओं को छिपाने के लिए वहाँ से हट गए।

माँ भीमाबाई ने बेटे की आँखों के भाव पढ़े और उसे सीने से लगाकर सामने दीवार की तरफ देखने लगीं, "हाँ बेटा, कल से तू स्कूल जाएगा। जूतों की जगह ही सही, तू पढ़ेगा।"

जब वे ये शब्द धीमी आवाज में कह रही थीं तो उनकी आँखों में आँसुओं की बरसात हो रही थी। उन्होंने दोनों हाथों से दबाकर भीमराव को चिपटा रखा था। वे नहीं चाहती थीं कि नन्हा भीम उनके आँसू देखे।

अगले दिन भीमराव का दाखिला हो गया। अपने घर से पुरानी टाट-पट्टी लेकर वे अपने पिता के साथ स्कूल गए और जूतों को एक ओर हटाकर टाट-पट्टी बिछाकर दहलीज पर बैठ गए।

भीमराव ने देखा कि जब वे जूतों को हटाकर अपने बैठने के लिए जगह बना रहे थे तो कुरसी पर बैठे आचार्य और सामने बेंचों पर बैठे छात्र उन्हें ऐसे देख रहे थे मानो सोच रहे हों, इसके छूने से कहीं हमारे जूते तो अपवित्र नहीं हो गए।

भीमराव की पढ़ाई शुरू हो गई। रोजाना अपनी टाट-पट्टी साथ लाना, उसे बिछाकर जूतों की जगह बैठना, आधी छुट्टी में सबसे अलग बैठकर दूसरी ओर मुँह करके अपना खाना खत्म करना और इस बात का पूरा ध्यान रखना कि स्कूल का कोई विद्यार्थी या किसी की कोई चीज भूल से भी छू न जाए।

माँ बीमार पड़ीं, इलाज करने कोई वैद्य नहीं आया

भीमराव पढ़ने में बहुत तेज थे। घर से स्कूल दूर था। वे सवेरे घर से निकलते और शाम को लौटते थे। खूब मन लगाकर पढ़ते थे।

पहली कक्षा से वे पाँचवीं में आ गए। पूरे स्कूल में उन्होंने नाम कमा लिया। पिता जब उनके स्कूल जाते तो हेडमास्टर भीमराव की बड़ी तारीफ करते। पिता को बहुत अच्छा लगता। घर लौटकर वे भीम की माँ को बताते तो भीमाबाई की आँखों में खुशी के आँसू आ जाते। ईश्वर को धन्यवाद देते हुए वे कहतीं, "किसी ने सच ही कहा है, दाता! तेरे घर देर तो है, पर अँधेर नहीं है।"

भीमराव के भविष्य के बारे में सुनहरे स्वप्न देखते-देखते भीमाबाई एक दिन बीमार पड़ गईं। जब वे बीमार हुईं तो रामजी राव चिंतित हो उठे। उनके बल पर तो पूरी गृहस्थी चल रही थी।

उन्होंने बड़ी देखभाल की, बहुत भाग-दौड़ की, परंतु सब बेकार गई। उन्हें ऐसे वैद्य की जरूरत थी, जो उनके रोग को समझ सके और उन्हें अच्छी दवा देकर रोग के चंगुल से निकाल सके; परंतु उनके घर आने के लिए कोई वैद्य तैयार नहीं हुआ। सभी अच्छे वैद्य ऊँची जाति के थे। एक महार के घर भला वे कैसे आ सकते थे। उनका तो धर्म भ्रष्ट हो जाता।

वैद्य अपना धर्म बचाते रहे और सही दवा के अभाव में भीमाबाई का जीवन चला गया। वे मृत्यु की गोद में हमेशा-हमेशा के लिए सो गईं।

माँ की मृत्यु पर भीमराव बहुत रोए। भीमराव भूख के बहुत कच्चे थे। माँ इस बात का बहुत खयाल रखती थीं कि उनका बेटा भूखा न रहे। वही माँ जब अपने बच्चे को रोता-बिलखता छोड़कर चली गई तो भीमराव ने कई दिन तक रोटी को हाथ नहीं लगाया।

पिता ने बहुत समझाया, "बेटा, तुम ऐसे ही रोते रहोगे तो तुम्हारी माँ के सपनों का क्या होगा? तुम नहीं जानते, तुम्हारी माँ की आँखों में तुम्हें बड़ा आदमी बनाने के कितने सपने थे। बेटा, उठो, खाना खाओ और पढ़ाई में लग जाओ। वे सपने तुम्हें पूरे करने हैं।"

भीमराव ने पिता के चेहरे की तरफ देखा और रो पड़े। पिता ने पुत्र को गले से लगा लिया। अगले दिन से भीमराव नियमित स्कूल जाने लगे। माँ की आँखों के सपने अब उनके दिल में थे।

भूखे रहकर पढ़ाई की

माँ के लाडले भीमराव अब बिना खाना साथ लिये स्कूल आने लगे। ग्यारह-बारह साल के बच्चे ने निश्चय किया कि खाना मिले या न मिले, पढ़ाई का नुकसान नहीं होने दिया जाएगा।

हेडमास्टर ब्राह्मण थे, अछूतों के प्रति सामाजिक व्यवहार के सामने विवश थे; परंतु भीम को भूखा देखा तो उन्हें दया आ गई। उसके लिए दोपहर की रोटी का इंतजाम उन्होंने स्वयं किया।

जब सब बच्चे दोपहर का भोजन करने बैठते तो वे एक कोने में सब्जी भीमराव के सामने रख देते और दूर से ही रोटियाँ उसके हाथों में फेंक देते थे। भीमराव अपनी भूख मिटा लेते थे।

दूसरे लड़के ज़ब हेडमास्टर को ऐसा करते देखते तो कानाफूसी करते। वे कहते, "लगता है, हेडमास्टर का दिमाग खराब हो गया है। ब्राह्मण होते हुए भी वे महार जाति के अछूत लड़के को रोटी बनाकर खिलाते हैं।"

भीमराव लड़कों के मुँह से जब ऐसी बातें सुनते, अपने प्रति उनके चेहरों पर और आँखों में घृणा देखते तो सहम जाते। कभी-कभी उन्हें बहुत गुस्सा आता, परंतु वे सब सह जाते।

एक बार उनकी माँ ने उनके कान में कहा था, 'एक बात याद रखना बेटा, हम लोग महार जाति के हैं, अछूत हैं। तुम्हें समाज में बहुत अपमान सहना पड़ेगा, बड़ी घृणा झेलनी पड़ेगी। इस सबकी परवाह न करना। किसी से उलझना नहीं। तुम्हें पढ़ना है। बिना पढ़े कोई बड़ा आदमी नहीं बन सकता। जब तुम बड़े आदमी बन जाओ तब समाज की इस गंदी रीति को बदल डालना। यही मेरी इच्छा है।'

माँ तो नहीं रहीं, पर उनके वे शब्द भीमराव के लिए प्रेरणा के स्रोत बन गए। जब-जब उन्हें कोई घृणा की नजर से देखता, जब-जब कोई उनका अपमान करता, माँ मानो उनके कानों में कह देतीं, 'धड़कना नहीं, सहते जाना है। रुकना नहीं, आगे बढ़ते जाना है।' भीमराव वैसा ही करते, अपमान और घृणा सहते-सहते उन्हें चुप रहने की आदत पड़ गई। अब कोई कुछ भी कहे, कुछ भी करे, वे प्रायः चुप रहते और अपने काम में लगे रहते।

परात गिरवी रखकर हाई स्कूल में दाखिला दिलाया

कक्षा पाँच पास करने के बाद भीमराव को दूसरे स्कूल में दाखिला लेना था। भीमराव के पिता पत्नी के देहांत के बाद बहुत परेशान रहने लगे थे। नौकरी करना और घर में बच्चों की देखभाल करना, इस तरह उन्हें दोहरी जिम्मेदारी निभानी पड़

रही थी। घर में अभाव की स्थिति पैदा हो गई थी।

भीमराव के दाखिले के लिए वे एलफिंस्टन हाई स्कूल में बात कर आए। भीमराव के अंक बहुत अच्छे थे। दाखिला मिल गया, परंतु पैसा! वह कहाँ से आए? और कोई चारा न देखकर रामजी राव ने अपनी पीतल की भारी परात निकाली और उसे लेकर सेठ सूरजमल के पास जा पहुँचे।

सूरजमल सूदखोर और लालची था। वह जेवर या कीमती चीजें गिरवी रखकर ब्याज पर पैसा देता था और उससे लिखवा लेता था कि समय पर नहीं छुड़ा पाया तो चीज उसकी हो जाएगी।

उसने रामजी राव की भारी परात रख ली और उनसे लिखवा लिया कि यदि निश्चित तिथि तक पैसा ब्याज सहित वापस नहीं किया तो परात उसकी हो जाएगी।

पुत्र को पढ़ाने की धुन में रामजी राव ने परात की परवाह नहीं की। उस सूदखोर का ब्याज इतना अधिक था कि भीमराव के पिता दी हुई तिथि तक पैसा न जुटा सके और उनकी गाढ़ी कमाई से खरीदी गई परात उस सूदखोर की हो गई।

इस देश में मजबूरी का फायदा उठानेवाले इनसानियत का लबादा पहनकर इनसानियत का खून करनेवालों की कमी नहीं रही है। ऐसे लोगों ने अपने छोटे से स्वार्थ के लिए दूसरों के बड़े-बड़े नुकसान किए हैं। दुर्भाग्य तो यह है कि कोई कानून इस क्रूर हृदय और कुटिल बुद्धिवाले हैवानों की गरदनें नहीं नाप सका।

कानून बनाकर अछूतों का उद्धार करूँगा

परात गिरवी रखने के बाद रामजी राव प्राइमरी स्कूल के हेडमास्टर के पास पहुँचे। भीमराव की मदद करने के लिए उन्होंने हेडमास्टर को बार-बार धन्यवाद दिया और प्रमाण-पत्र माँगा।

हेडमास्टर भीमराव से बहुत खुश थे। उसने प्रथम श्रेणी पाई थी और बहुत अच्छे अंक प्राप्त किए थे।

हेडमास्टर ने प्रमाण-पत्र तैयार कर दिया और उसमें उनका पूरा नाम लिखा—'भीमराव आंबेडकर।'

पिता रामजी राव ने इसे ब्राह्मण का आशीर्वाद मानकर खुशी जाहिर की। उस दिन उन्हें क्या पता था कि यह 'आंबेडकर' शब्द एक दिन भारतीय इतिहास की शानदार धरोहर बनेगा और पूरे देश की पिछड़ी जातियों के लोग इससे प्रेरणा, आश्रय और सामाजिक न्याय प्राप्त करते हुए उन अधिकारों को प्राप्त करने में सफल होंगे, जिनका संकल्प लेकर भीमराव ने अपमान पर अपमान सहते हुए अपनी पढ़ाई जारी

रखी थी और जिन अधिकारों के सपने मरते समय उनकी माँ की आँखों में थे।

प्रमाण-पत्र लेने के बाद जब भीमराव चलने लगे तो श्रद्धा और भावुकता से भरकर वे गुरुजी के चरण छूने के लिए झुके; परंतु तभी उन्हें याद आ गया कि चरण तो वे छू नहीं सकते। छूने से गुरुजी अपवित्र हो जाएँगे। अत: चरणों की ओर बढ़ते उनके हाथ ठिठक गए। भीमराव ने कहा, ''क्षमा करें गुरुजी, मैं आपके चरण नहीं छू सकता, अछूत जो हूँ।''

''मेरा आशीर्वाद सदा तुम्हारे साथ है।'' भावुक स्वर में गुरुजी बोले, ''लेकिन बेटा, बहुत दिनों से एक बात मेरे मन में घूम रही है। अब पूछ ही लेता हूँ। तुमने पाँच साल तक दूसरे छात्रों के जूतों की जगह बैठकर पढ़ाई की। भूखे रहकर, बार-बार अपमान सहकर भी जरा विचलित नहीं हुए। ऐसा क्या है इस पढ़ाई में, जो तुम इतना सबकुछ सह रहे हो? आगे पढ़ोगे तो और भी अपमान झेलना पड़ेगा। फिर तुम महार जाति के हो, तुम्हें ऊँची नौकरी भी नहीं मिल पाएगी। तुम यह सब क्यों कर रहे हो?''

भीमराव के कदम रुक गए। वे मुड़े और सम्मानपूर्वक ब्राह्मण के सामने खड़े होकर बोले, ''गुरुजी, मैं सबकुछ सहूँगा, मुझे पढ़-लिखकर एक दिन एक नया कानून बनाना है। एक ऐसा कानून, जो अछूतों को समाज में ऊँचा स्थान दिलाएगा। देखिएगा, सरकार इस कानून को मानेगी।''

सन् 1904 में एलफिंस्टन हाई स्कूल में भीमराव को दाखिला मिल गया। अब वे 'भीमराव आंबेडकर' कहलाने लगे थे। लड़के उन्हें केवल 'भीम' कहा करते थे।

एक दिन कक्षा में एक अजीब घटना घटी। अध्यापक भूगोल पढ़ा रहे थे। तभी उन्होंने भीमराव से कहा, ''भीम, तुम उठो और चॉक से ब्लैकबोर्ड पर भारत का मानचित्र बनाओ।''

भीमराव उठे। चॉक लेकर जैसे ही वे ब्लैकबोर्ड की ओर बढ़े, लड़के आपस में खुसुर-फुसुर करने लगे।

तभी एक लड़का खड़ा हुआ और बोला, ''ठहरो भीम!''

लड़के की आवाज सुनकर भीमराव ठिठक गए।

अध्यापक ने डाँटकर लड़के से कहा, ''क्या बात है?''

''सर, ब्लैडबोर्ड के पीछे हम लोग अपना खाना रखते हैं। भीम महार है। इसने हमारा खाना छू लिया तो वह अपवित्र हो जाएगा और हमें फेंकना पड़ेगा।''

''ओह!'' अध्यापक के मुँह से केवल एक शब्द निकला।

तभी सब लड़के उठे और उन्होंने अपने-अपने खाने के डिब्बे वहाँ से उठा लिये।

इसके बाद भीमराव आगे बढ़े और उन्होंने ब्लैकबोर्ड पर भारत का मानचित्र बनाया।

कक्षा समाप्त होने के बाद सब लोग बाहर निकल गए। यह खाना खाने का समय था। सब लड़के एक झुंड में बैठकर खाना खाने लगे।

भीमराव आंबेडकर अपना खाना लेकर एक ओर चले गए। खाना खाते समय वे सोच रहे थे, एक अपमान और सही, अब तो इस सबकी आदत डालनी होगी।

सन् 1906 में पिता ने भीखू वालंगकर की बेटी रमाबाई के साथ भीमराव की शादी कर दी।

सन् 1907 में उन्होंने 750 में से 382 अंक लेकर हाई स्कूल की परीक्षा पास की।

बड़ौदा नरेश की कृपा

हाई स्कूल पास करने के बाद भीमराव के सामने आगे पढ़ने की समस्या आ खड़ी हुई। पिता पहले से ही बहुत दुखी थे। वे बूढ़े हो चले थे। घर में गरीबी थी।

हाई स्कूल में भीमराव को दाखिला दिलाने के लिए उन्होंने जो परात गिरवी रखी थी, उसे भी नहीं छुड़ा पाए थे। उधार तक कोई नहीं देता था। अब पैसा कहाँ से आ सकता था।

हाई स्कूल में पढ़ते समय भीमराव की एक ईसाई लड़के से अच्छी दोस्ती हो गई थी। उसका नाम था—कैलुस्कर। भीमराव ने अपनी समस्या कैलुस्कर के सामने रख दी। कैलुस्कर ने उसे बड़ौदा नरेश से मिलने की सलाह दी।

अगले ही दिन भीमराव बड़ौदा नरेश की कोठी पर जा पहुँचे। उन्होंने अपनी पूरी कहानी बिना झिझक के बड़ौदा नरेश को सुना दी।

राजा साहब जानते थे कि एक महार जाति के लड़के के लिए उच्च शिक्षा लेना बहुत कठिन है। पैसे की कमी और अपमान दोनों ही बाधक बनते हैं। परंतु भीमराव का भोलापन, ईमानदारी और राजा पर विश्वास काम कर गया। उन्हें लगा कि यह जानते हुए भी कि 'मैं उच्च जाति का हूँ', यह अछूत लड़का मुझ पर विश्वास करके मेरे पास आया है, तो मुझे इसकी मदद अवश्य करनी चाहिए।

उन्होंने एलफिंस्टन में भीमराव का दाखिला करा दिया और खजाने से 25 रुपए महीना छात्रवृत्ति दिलवा दी।

भीमराव ने लगन से पढ़ना शुरू कर दिया। उन्हीं दिनों उनके पिता रामजी राव भी बंबई आ गए। उनकी सतारा की नौकरी छूट गई थी, अतः वे परिवार को लेकर बेटे के साथ ही रहने लगे।

नवंबर 1912 में भीमराव की पत्नी रमाबाई ने एक पुत्र को जन्म दिया। उसका नाम रखा गया 'यशवंत'। जनवरी 1913 में भीमराव ने बी.ए. पास कर लिया। जब उन्होंने देखा कि पिता गरीबी की मार झेल नहीं पा रहे हैं तो उन्होंने बड़ौदा के महाराजा से कहकर सेना में नौकरी के लिए आवेदन कर दिया।

वे बी.ए. पास कर चुके थे। महाराज की सिफारिश थी, अतः भीमराव को लेफ्टिनेंट के पद पर नियुक्ति मिल गई।

पिता से मिलने के लिए नौकरी छोड़ दी

रमाबाई अपने पुत्र यशवंत की देखभाल तथा घर-परिवार का दायित्व निभा रही थीं और भीमराव लेफ्टिनेंट के रूप में बड़ौदा की रियासत को अपनी सेवाएँ दे रहे थे। एक दिन उन्हें घर से तार मिला। लिखा था—'पिता बहुत बीमार हैं, छुट्टी लेकर आ जाओ।'

पिता की बीमारी से भीमराव विचलित हो गए।

वे बड़े अफसर के पास पहुँचे और तार दिखाकर छुट्टी के लिए प्रार्थना की, परंतु अफसर ने छुट्टी देने से इनकार कर दिया।

भीमराव अपनी माता को तो खो चुके थे, अब पिता ही उनके लिए सबकुछ थे। वे उन्हें खोना नहीं चाहते थे। उन्हें संदेह था कि यदि अच्छा इलाज नहीं मिला तो पिता की भी उसी तरह मृत्यु हो जाएगी, जिस तरह माँ की हो गई थी। अतः उन्होंने अफसर से बार-बार प्रार्थना की, परंतु अफसर अड़ गया और बोला, "छुट्टी किसी हालत में नहीं मिल सकती।"

"तो फिर मेरा इस्तीफा स्वीकार कीजिए।" भीमराव ने कहा।

अफसर बोला, "हाँ, यह हो सकता है। आप नौकरी छोड़ दीजिए और अपने घर चले जाइए।"

भीमराव ने एक क्षण नहीं गँवाया, तुरंत इस्तीफा दिया और पिता की सेवा के लिए उनके पास चले गए।

पिता को जब यह पता चला कि उनका बेटा नौकरी छोड़ आया है तो उन्हें बड़ा दुःख हुआ।

भीमराव ने उनका बहुत इलाज कराया, परंतु वे ठीक न हो सके। फरवरी 1913

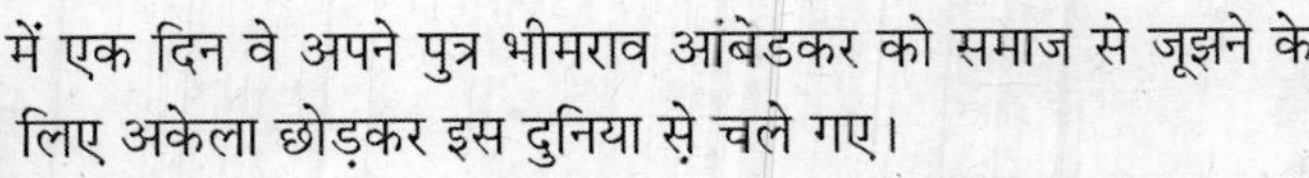

में एक दिन वे अपने पुत्र भीमराव आंबेडकर को समाज से जूझने के लिए अकेला छोड़कर इस दुनिया से चले गए।

उच्च शिक्षा के लिए अमेरिका रवानगी

अपने मित्र कैलुस्कर की सलाह पर एक बार फिर भीमराव बड़ौदा नरेश महाराज गायकवाड़ की कोठी पर जा पहुँचे। उस समय महाराज भारत के अनेक होनहार युवकों को उच्च शिक्षा के लिए अमेरिका भेजने की योजना में लगे थे।

भीमराव ने महाराज को पूरी कहानी सुनाई और अमेरिका जाकर पढ़ाई करने की इच्छा व्यक्त की। महाराज इस बात से अत्यंत प्रभावित हुए कि भीमराव ने पिता की सेवा करने के लिए लेफ्टिनेंट की नौकरी को लात मार दी थी। वे स्वाभिमानी युवकों को बहुत पसंद करते थे। उन्होंने भीमराव को अनुमति दे दी। उन्हें एक शपथ-पत्र पर हस्ताक्षर करने पड़े, जिसमें लिखा था कि शिक्षा पूरी करने के बाद वे 10 वर्ष तक रियासत की सेवा करेंगे।

4 जून, 1913 को भीमराव अपनी पत्नी रमाबाई और छह माह के बेटे यशवंत से विदा लेकर जहाज द्वारा बंबई से अमेरिका के लिए रवाना हुए।

जुलाई 1913 में उन्होंने कोलंबिया विश्वविद्यालय, न्यूयॉर्क में राजनीति संकाय में दाखिला लिया।

अमेरिका का वातावरण उन्हें बहुत अच्छा लगा। वहाँ जो खुला स्वागत उन्हें मिला, उससे वे हैरान रह गए। भारत की जमीन पर उन्हें कदम-कदम पर अपमान सहना पड़ता था; परंतु अमेरिका जातिवाद के विष से मुक्त था। वहाँ व्यक्ति को महत्त्व दिया जाता था।

दिन भर लड़के-लड़कियाँ उनसे खुलकर मिलते थे और खूब बातें करते थे। अमेरिका प्रवास के दौरान उनकी मुलाकात भारतीय छात्र नवल भटेना से हुई। उस समय भीमराव हॉस्टल में अलग कमरा लेकर रह रहे थे। नवल भटेना की भीमराव से दोस्ती हो गई। उन्होंने भीमराव को अपने ही कक्ष में बुला लिया। फिर वे दोनों एक ही कक्ष में साथ-साथ रहने लगे। साथ-साथ रहते, खाते-खेलते, बातें करते और जमकर पढ़ाई करते। भीमराव को इतने सुख का अनुभव होने लगा, जिसकी उन्होंने कल्पना भी नहीं की थी। हॉस्टल और पढ़ाई का पूरा खर्चा बड़ौदा नरेश दे रहे थे।

भीमराव सबके साथ घुल-मिलकर रहते हुए भी सबसे अलग थे। वे सादा जीवन पसंद करते थे, सादा खाना खाते थे और खूब जमकर पढ़ते थे। दूसरे लड़कों

की तरह फिल्म देखने अथवा गप्पें लड़ाने में उनकी तनिक भी रुचि नहीं थी।

अपनी माँ का स्वप्न, अपने पिता की प्रेरणा, अपनी जाति के उद्धार की चिंता, सदा उन्हें लक्ष्य से बाँधे रहती थी। भटकने और गुम हो जाने के लिए उनके जीवन में कोई गुंजाइश नहीं थी। कौम और देश के नाम समर्पित पूरा जीवन।

दो वर्ष की अवधि में उन्होंने इतनी पुस्तकें खरीद लीं कि उनका अपना छोटा पुस्तकालय ही बन गया। वे रात-दिन पढ़ते रहते थे और बहुत कुछ जानना चाहते थे, बहुत कुछ करना चाहते थे।

5 जून, 1915 को भीमराव आंबेडकर ने कोलंबिया विश्वविद्यालय से एम.ए. की परीक्षा पास की।

इसके बाद वे पी-एच. डी. के लिए अध्ययन और लेखन में जुट गए। उन्होंने 'द नेशनल डिवाइडेंड ऑफ इंडिया : ए हिस्टोरिकल एंड एनेलिटिकल स्टडी' नामक विषय पर थीसिस लिखी और उसे पी-एच. डी. के लिए कोलंबिया विश्वविद्यालय को दे दिया।

पी-एच. डी. का काम समाप्त होते ही जून 1916 में लंदन स्कूल ऑफ इकोनॉमिक्स एंड पॉलिटिकल साइंस में अध्ययन करने के लिए वे लंदन चले गए।

1917 में कोलंबिया विश्वविद्यालय ने उन्हें पी-एच.डी. की उपाधि प्रदान की।

जून 1917 में जब वे लंदन में अर्थशास्त्र में एम.एस-सी. की डिग्री के लिए अध्ययन कर रहे थे तब उन्हें सूचना मिली कि महाराज गायकवाड़ ने उनकी छात्रवृत्ति बंद कर दी है और अब उन्हें लिखित शर्त के अनुसार भारत लौट आना चाहिए और 10 वर्ष तक रियासत की सेवा करनी चाहिए।

आंबेडकर के सामने भारत लौटने के सिवाय कोई चारा न था। उन्होंने लंदन में अपनी उच्च स्तरीय पढ़ाई बंद कर दी और भारत लौट आए।

ऊँचा पद तो मिला, परंतु अपमान ने पीछा नहीं छोड़ा

बड़े उदास मन से आंबेडकर भारत लौटे। उनके आगमन से महाराज गायकवाड़ को बड़ी प्रसन्नता थी। वे उन्हें कोई बड़ा पद देना चाहते थे। अपने ही संसार में मस्त रहनेवाले महाराज यह नहीं जान पाए कि आंबेडकर को अपनी पढ़ाई बीच में रोककर आना पड़ा है। महाराज के सलाहकारों ने उन्हें यही सूचना दी थी कि आंबेडकर की पढ़ाई पूरी हो चुकी है। अब उनकी छात्रवृत्ति बंद कर दी जाए और रियासत की सेवा

के लिए वापस बुला लिया जाए।

महाराज ने अपने अधिकारियों को आदेश दिया था कि आंबेडकर को बड़े सम्मानपूर्वक एक जुलूस के साथ रियासत में लाया जाए; परंतु अधिकारियों को यह बात अच्छी नहीं लगी। ऊँची जाति के गर्व में चूर वे एक महार जाति के व्यक्ति का फूल-मालाओं से स्वागत भला कैसे कर सकते थे। उन्होंने मिलकर यह फैसला लिया कि महाराज की इस आज्ञा को न माना जाए।

भीमराव आंबेडकर दूसरी मिट्टी के बने थे। उन्हें अपने कर्तव्य-पालन और जातीय अधिकारों की लड़ाई लड़ने के सिवाय कुछ और सूझता ही नहीं था। मान-अपमान की दुनिया से तो वे बहुत ऊपर उठ चुके थे।

बड़ौदा रेलवे स्टेशन पर उतरने के बाद उन्होंने अपना सामान उठाया और पैदल ही सीधे राजभवन जा पहुँचे।

महाराज गायकवाड़ के प्रति उनके मन में बहुत श्रद्धा थी। उनकी पढ़ाई बीच में छूट गई थी, इस बात का दुःख तो जरूर था, परंतु महाराज ने उच्च जाति के अधिकारियों के विरोध के बावजूद उनकी इतनी सहायता की थी, इसके लिए उनका रोम-रोम महाराज का आभारी था।

महाराज के सामने पहुँचते ही उन्होंने झुककर उन्हें प्रणाम किया। महाराज ने उन्हें आशीर्वाद दिया और अमेरिका तथा ब्रिटेन के बारे में बहुत सी बातें कीं। भीमराव ने उन्हें बताया कि अमेरिका के लोग भारतीयों का आदर करते हैं, परंतु ब्रिटेन के लोग बराबरी का दर्जा नहीं देते। वे भारतीयों को अपना गुलाम मानते हैं।

महाराज को आंबेडकर से पश्चिमी दुनिया के बारे में बहुत कुछ जानने को मिला। आंबेडकर को महाराज का व्यवहार बहुत अच्छा लगा। महाराज ने सैन्य सचिव के पद पर भीमराव की नियुक्ति का आदेश जारी कर दिया। दो हजार रुपए महीने वेतन और सभी सुविधाएँ।

परंतु उनकी रियासत में ऊँच-नीच और जातिवाद का जो जहर पहले से ही मौजूद था, उसके आगे वे भी विवश थे। बड़ी जाति के अधिकारियों व कर्मचारियों को भीमराव के अंदर छिपा महान् इनसान उतना नजर नहीं आता था जितना यह सत्य नजर आता था कि वे महार जाति के हैं और उनके छूने से वे अपवित्र हो जाएँगे।

सैन्य सचिव के कार्यालय का चपरासी तक उन्हें छोटी जाति के कारण अपने से नीचा समझता था। वे सैन्य सचिव की कुरसी पर बैठकर जब कोई फाइल मँगवाते तो चपरासी दूर से ही उनकी मेज पर फेंककर चला जाता था। जब वह फाइल रखने

के लिए वापस लेने आता तब भीमराव से उसी तरह दूर से ही फाइल लेता था।

चपरासी के जाने के बाद वे अपना सिर दोनों हाथों से थाम लेते और दूर अमेरिका में उनका मन जा पहुँचता। वे सोचते, अमेरिका ने कितनी तरक्की की है। वहाँ के लोग जाति-पाँति के भेदभाव से बहुत ऊपर केवल काम करते हैं और खूब तरक्की करते हैं। जाति के कारण आदमी के हाथों आदमी का अपमान कितनी बड़ी बुराई है। इतना महान् अतीत, इतनी सुंदर संस्कृति और यह जातीय घृणा! हमारे देश के लोगों को आखिर हो क्या गया है? एक ओर तो हम मानवता की बातें करते हैं, मानव मूल्यों की बातें करते हैं, सारी दुनिया को एक महान् जीवन-शैली देने की बातें करते हैं, दूसरी ओर अपने ही भाइयों के साथ इतना अपमानजनक व्यवहार करते हैं कि उनसे छू जाएँ तो अपने आपको अपवित्र मानने लगते हैं। कितना बड़ा ढोंग, कितना बड़ा छलावा!

अब समझ में आता है, हमने तरक्की क्यों नहीं की। अब समझ में आता है, हम गुलाम क्यों हैं। यह जातीय घृणा लोगों के दिलों से नहीं निकाली गई तो यह देश नष्ट हो जाएगा। जो कौमें अपने ही एक भाग को पराया समझने लगती हैं, वे बच नहीं पातीं।

एक दिन उन्होंने चपरासी से पानी लाने के लिए कहा; परंतु चपरासी ने पानी देने से साफ मना कर दिया। बड़ी जाति का चपरासी छोटी जाति के सैन्य सचिव जैसे ऊँचे अधिकारी को भला पानी कैसे पिला सकता था। उसका सारा जातीय दर्प टुकड़े-टुकड़े हो जाता। वह धर्म-भ्रष्ट हो जाता।

भीमराव जानते थे कि चपरासी ऊँची जाति में जनमा है, इसलिए वह उच्च शिक्षा प्राप्त और उच्च पद प्राप्त आंबेडकर की तुलना में बहुत बड़ा है। उसे किसी भी तर्क से झुकाया नहीं जा सकता।

उन्होंने अधिक तर्क नहीं किया। कार्यालय में पानी के घड़े थे; परंतु क्योंकि वे अछूत थे, अत: घड़े से स्वयं पानी लेने का उन्हें अधिकार नहीं था। अत: कोई चारा न देखकर वे कुरसी से उठे और बाहर चले गए।

उनके बाहर जाते ही कार्यालय के सब कर्मचारी इकट्ठे हो गए। वे चपरासी को शाबाशी देने लगे। इतने बड़े अधिकारी का आदेश न मानकर उसने अपना जातीय स्वाभिमान बचा लिया था। सभी बड़ी जातिवालों के लिए यह गौरव की बात थी।

फिर सब राजा साहब की आलोचना करने लगे—विद्वान् है तो क्या हुआ, है तो महार जाति का। उसका छुआ कोई खा नहीं सकता। उसकी सेवा करे तो धर्म

जाता है। यह तो सरासर गलत है। नीची जाति के आदमी को ऊँचे पद पर नहीं होना चाहिए।

आंबेडकर कहीं से पानी पीकर अपने कार्यालय में लौटे तो कर्मचारी भागे; परंतु उनके चेहरों पर लिखे संदेश को आंबेडकर ने पढ़ लिया और फैसला कर लिया कि इस प्रकार रियासत की सेवा न हो सकेगी।

अपमान के इस घूँट को वे पी गए। फिर भी एक और समस्या उनके सामने आ खड़ी हुई। वे अपने परिवार को बड़ौदा लाना चाहते थे, परंतु अछूत होने के कारण कोई उन्हें मकान किराए पर देने के लिए तैयार नहीं था। जब उन्हें कहीं जगह नहीं मिली तो उन्होंने एक पारसी सराय में कमरा किराए पर ले लिया।

अभी कुछ ही दिन हुए थे इस कमरे में आए कि लोगों को उनकी जाति का पता लग गया। बस, सारे मुहल्ले के लोग इकट्ठे हो गए और उन्हें पीटने की तैयारी करने लगे। उनकी दृष्टि में यह सबसे बड़ा अपराध था कि एक अछूत उनकी सराय में उनके बीच आकर रहने लगे और उनका धर्म भ्रष्ट कर दे।

बहुत कहा-सुनी हुई। भीमराव ने हाथ जोड़कर सब लोगों से माफी माँगी और वचन दिया कि वे तुरंत कमरा खाली कर देंगे और इस सराय में कभी दिखाई नहीं देंगे।

जो अपमान हुआ था, उससे वे अंदर तक दहल गए। उन्होंने राजा साहब को एक पत्र लिखा। राजा साहब ने उनकी व्यथा को समझा और दीवान से कहा कि आंबेडकर के रहने का इंतजाम करें।

दीवान ने अकेले में आंबेडकर को समझाया, "बाबू, तुम अमेरिका और लंदन में पढ़े हो, यह तो ठीक है; पर इस समाज का मैं क्या करूँ? तुमसे राजा साहब को बहुत हमदर्दी है, मुझे भी कम नहीं है; पर क्या करूँ, ये लोग सुधरनेवाले नहीं। बाबू, यह मामूली बात नहीं है, छुआछूत का मामला है। मैंने तुम्हारे रहने का इंतजाम किया तो लोग मुझे जीने नहीं देंगे। हम हर चीज से निबट सकते हैं, पर छुआछूत से नहीं।"

दीवान की बात सुनकर आंबेडकर सन्न रह गए। फिर बोले, "तो फिर मैं क्या करूँ? मैं तो प्रतिज्ञा-पत्र से बँधा हूँ, नहीं तो इस्तीफा देकर चला जाता।"

दीवान धीरे से बोला, "प्रतिज्ञा-पत्र की बात भूल जाओ बाबू, मैं महाराज को सब समझा दूँगा।"

भीमराव समझ गए, दीवान को इस्तीफे से ऐतराज नहीं है, बल्कि वह तो यही चाहता है कि किसी तरह पीछा छूटे।

उन्होंने तुरंत त्याग-पत्र लिखा और दीवान को पकड़ा दिया।

महाराज से मिले बिना ही वे सीधे रेलवे स्टेशन चले गए और वहाँ से रेलगाड़ी पकड़कर बंबई जा पहुँचे।

प्रोफेसर आंबेडकर : घड़ा छूने पर हंगामा

बंबई (अब मुंबई) आते ही वे अपने ईसाई मित्र कैलुस्कर से मिले। उसने सीडेहॉम कॉलेज ऑफ कॉमर्स एंड इकोनॉमिक्स, बंबई में अर्जी देने के लिए कहा। कॉलेज में राजनीतिक अर्थशास्त्र के प्रोफेसर की जगह खाली थी। आंबेडकर को कॉलेज में नौकरी मिल गई। नवंबर 1918 में उन्होंने उस कॉलेज में राजनीति और अर्थशास्त्र पढ़ाना शुरू कर दिया।

उनके पढ़ाने का ढंग इतना अच्छा था कि देखते-ही-देखते वे छात्रों के बीच लोकप्रिय हो गए। ज्यों ही उनके साथियों को पता लगा कि वे लंदन व अमेरिका से पढ़कर आए हैं तो उनका आदर और बढ़ गया। छात्रों के बीच उनकी बड़ी तारीफें होने लगीं; परंतु लोकप्रियता के साथ-साथ अछूत जाति में पैदा होने का अभिशाप भी बराबर उनका पीछा कर रहा था।

कॉलेज में पानी पीने के लिए मिट्टी के बड़े-बड़े घड़े रखे गए थे। सब लोग घड़ों से गिलास में पानी लेकर पिया करते थे। एक दिन आंबेडकर जब कक्षा से पढ़ाकर निकले तो प्यास से उनका गला सूख रहा था। वे एक क्षण के लिए यह भूल गए कि घड़े से पानी लेने का हक दूसरों की तरह उन्हें प्राप्त नहीं है। वे झट घड़े के पास पहुँचे और एक गिलास में पानी डालकर पीने लगे। उसी समय वहाँ से एक गुजराती प्रोफेसर गुजरे। उनकी निगाह आंबेडकर पर पड़ गई। उन्हें ध्यान आया कि आंबेडकर तो अछूत हैं। उन्होंने घड़े से पानी ले लिया। उन्होंने घड़ा छू लिया। यह तो अनर्थ हो गया। अछूत चाहे कितना ही विद्वान् क्यों न हो जाए, पर घड़ा तो नहीं छू सकता।

वे डॉ. आंबेडकर की ओर झपटे और गुस्से से बोले, "यह आपने क्या किया? क्या आप भूल गए कि आप अछूत हैं? आपने घड़ा अपवित्र कर दिया। अब इसका पानी कोई नहीं पी सकता।"

डॉ. आंबेडकर के हाथ से गिलास गिर गया। वे थोड़ा सा पानी पी पाए थे, बाकी नीचे बिखर गया। हैरान थे—इतने बड़े कॉलेज के इतने विद्वान् प्रोफेसर और इतने छोटे खयालात।

तभी शोर सुनकर अनेक अध्यापक उस ओर दौड़ पड़े। प्रिंसिपल महोदय भी आ गए। अनेक छात्र भी पीछे आ खड़े हुए। मजमा लग गया।

डॉ. आंबेडकर ठगे से उन तमाम चेहरों को देख रहे थे, जो उन्हें इस महापाप के लिए घूर रहे थे।

अचानक आंबेडकर बोले, "प्रिंसिपल महोदय, आप ही इन लोगों को समझाइए। मेरा खयाल है, ये सब बहुत पढ़े-लिखे और विद्वान् प्रोफेसर हैं। मेरे छूने से घड़ा अपवित्र नहीं हुआ है। यह बिलकुल वैसा ही स्वच्छ और पवित्र है जैसा मेरे छूने से पहले था। विश्वास रखिए, मेरे हाथ बिलकुल साफ हैं।"

"इतने भोले मत बनो प्रोफेसर आंबेडकर!" प्रिंसिपल ने उन्हें झिड़का, "माना आप विलायत में पढ़े हैं, परंतु आपके बाप-दादा तो इस देश की अछूत जाति के ही थे। क्या आप नहीं जानते?"

डॉ. आंबेडकर की ऊपर की साँस ऊपर और नीचे की साँस नीचे रह गई। उन्होंने कुछ कहने के लिए ज्यों ही मुँह खोलना चाहा, प्रिंसिपल ने चपरासी को आदेश दिया, "यह घड़ा उठाओ और इसे फोड़ दो।"

चपरासी ने तुरंत आज्ञा का पालन किया।

अगले ही क्षण वह घड़ा उठाकर जमीन पर पटक दिया गया और उसके पानी के छींटे आंबेडकर के कपड़ों पर आ गए।

आंबेडकर बहुत देर तक उस टूटे हुए घड़े को खड़े-खड़े देखते रहे। दुःख और उदासी की अनेक रेखाएँ उनके चेहरे पर खिंचीं और विलीन हो गईं।

फिर वे मुसकराए, मानो कह रहे हों, 'यह सब कोई नई बात नहीं है आंबेडकर। यह तो वह अपमान-गाथा है, जो हर अछूत के जीवन का अभिन्न अंग है। पढ़-लिख जाने मात्र से समाज तुम्हें तुम्हारा अधिकार देनेवाला नहीं। इसके लिए लंबी लड़ाई लड़नी होगी, लंबी और लगातार लड़ाई बिना धैर्य खोए, बिना क्रोध किए इस त्रासदी का मुकाबला करो और चाहे कुछ भी करना पड़े, अछूतों के जीवन से इस महाकलंक को धो डालो। मत भूलो कि तुम्हारी माँ को छोटी सी बीमारी के कारण इसलिए मरना पड़ा कि कोई अच्छा वैद्य अछूत के घर में आने के लिए तैयार नहीं हुआ। पिता जब बीमार हुए तो ऊँची जाति के अफसर ने इसलिए छुट्टी देने से इनकार कर दिया कि तुम अछूत हो, और नौकरी से इस्तीफा देकर ही तुम पिता का इलाज करा सकते हो। पग-पग पर संघर्ष करना है। बहुत कुछ खोना पड़ेगा। अपने आपको तिल-तिल कर मिटाना पड़ेगा, तब कहीं जाकर मंजिल हाथ आएगी।'

उनका सीधा हाथ पेन पर पहुँच गया था। अंदर के स्वाभिमान ने कहा था, इस्तीफा दो और चलो यहाँ से। परंतु डॉ. आंबेडकर ने पेन पर से अपना हाथ हटा लिया। संकल्प के साथ हुंकार भरी, 'नहीं, मैं इस्तीफा नहीं दूँगा। मैं मुकाबला करूँगा। देखूँगा, यह समाज मुझे कितना सताता है।'

इस घटना के बाद कॉलेज में एक ऐसा सन्नाटा छा गया जैसे कोई बहुत बड़ी दुर्घटना हो गई हो। डॉ. आंबेडकर के साथ घुल-मिलकर बातें करनेवाले छात्र अपने आप में सिमटे हुए से कटे-कटे रहने लगे। अध्यापक भी डॉ. भीमराव आंबेडकर से बात नहीं करते थे। वे सब खाली समय में अलग समूह में बैठकर ठहाके लगाते और आंबेडकर अकेले बैठकर कुछ पढ़ते रहते, कुछ सोचते रहते।

कुछ दिन बाद थोड़ा बदलाव आया। आपसी बातचीत का सिलसिला बढ़ा। छात्र भी करीब आने लगे। आंबेडकर ने फैसला कर लिया था कि किसी चीज को हाथ नहीं लगाएँगे।

इस प्रकार उन्होंने डेढ़ वर्ष इस कॉलेज में गुजारे। इस बीच 31 जनवरी, 1920 को उन्होंने 'मूकनायक' नामक एक मराठी अखबार का प्रकाशन शुरू किया। इसका उद्‌देश्य पिछड़ों की लड़ाई लड़ना था। इसमें आंबेडकर ऐसे लेख लिखा करते थे, जिनमें जातिवाद और छुआछूत की कठोर आलोचना की जाती थी।

मार्च 1920 में उन्होंने कॉलेज की नौकरी से इस्तीफा दे दिया और नवल भटेना व महाराज गायकवाड़ के सहयोग से अपने शोध पर अध्ययन करने के लिए लंदन चले गए। सितंबर 1920 में उन्होंने लंदन स्कूल ऑफ इकोनॉमिक्स में दाखिला ले लिया और अध्ययन कार्य में जुट गए। वे 18 घंटे अध्ययन और लेखन के काम में लगे रहते थे।

डॉ. आंबेडकर ने लंदन विश्वविद्यालय से एम.एस-सी. (अर्थशास्त्र) की डिग्री प्राप्त की और रुपए की समस्या पर थीसिस लिखी। साथ ही उन्होंने कानून की पढ़ाई पूरी की।

वे तीन वर्ष तक लंदन में रहे। इस पूरी अवधि में महाराज गायकवाड़ ने उनकी आर्थिक मदद की। उन्हें अपनी पत्नी और बेटे यशवंत की बहुत याद आती थी। वे उन सबको पत्र लिखकर अपनी सफलता के समाचार देते रहते थे।

अप्रैल 1923 में महाराज गायकवाड़ और अपने मित्र नवल भटेना को उन्होंने लिखा कि उनका अध्ययन पूरा हो चुका है, अब वे भारत वापस आना चाहते हैं। भारत में रहकर वे वकालत करेंगे और अछूतों तथा दलितों के हितों के लिए संघर्ष करेंगे।

उनके पत्र से उनके मित्रों तथा परिजन में खुशी की लहर दौड़ गई।

अप्रैल के महीने में जब आंबेडकर जलयान से भारत पहुँचे तो बंबई बंदरगाह पर पहुँचकर उनके मित्रों ने उनका भव्य स्वागत किया।

वकालत शुरू : पहले मुकदमे में ही अछूतों को न्याय दिलाया

भारत पहुँचने के बाद आंबेडकर ने एक वर्ष तक सारी स्थिति को ठीक से समझा, वकालत के लिए अनुमति प्राप्त की तथा दलित उद्धार के लिए रचनात्मक आंदोलन की रूपरेखा तैयार की।

जून 1924 में उन्होंने वकालत का काम शुरू कर दिया।

अब तक भीमराव आंबेडकर अछूतों के हितों के लिए संघर्ष करनेवाले नेता के रूप में थोड़े-थोड़े चर्चित होने लगे थे।

उनके पास पहला मुकदमा तीन अछूत लेकर आए। उन्होंने एक ऐसी पुस्तक लिखी थी, जिसमें अछूत प्रथा के लिए उच्चवर्ग को दोषी ठहराया गया था। पुस्तक प्रकाशित होते ही उच्चवर्ग के लोग भड़क उठे। उन्होंने उन तीनों पर मानहानि का दावा ठोंक दिया।

सेशन जज की अदालत में जब इस मुकदमे पर आंबेडकर ने बचाव पक्ष के वकील के रूप में बहस की तो उसे सुनने के लिए अदालत में लोगों की भीड़ उमड़ पड़ी।

सामाजिक न्याय दिलाने के लिए समर्पित डॉ. आंबेडकर रात-दिन कानून की किताबें पढ़ते और जब बोलते तो विपक्षी वकीलों की बोलती बंद हो जाती। जज उनके तर्कों से हैरान रह जाता।

उन्होंने साबित कर दिया कि किताब में दलितों के साथ उच्च वर्ग की ज्यादती की जो दर्द भरी कहानी लिखी गई है, वह समाज का जीता-जागता सच है। इसे पुस्तक के रूप में देश व समाज के सामने लाकर तीनों अछूत लेखकों ने कोई अपराध नहीं किया है। अपराधी तो ये लोग हैं, जिन्होंने छुआछूत और जातिवाद का विष फैलाकर समाज को गुमराह किया है।

तीनों अछूत मुकदमा जीत गए। इस मुकदमे ने आंबेडकर को दलितों के मसीहा के रूप में स्थापित कर दिया। दलित समाज ने उन्हें हाथोहाथ उठा लिया।

विशाल महार सम्मेलन

अब डॉ. आंबेडकर ने फैसला कर लिया कि वे सामाजिक न्याय के लिए खुला संघर्ष करेंगे। वैचारिक आंदोलन की भूमिका वे बहुत पहले बना चुके थे। मुकदमा जीतने के बाद पिछड़ों की भीड़ अपने चारों ओर उमड़ते देखकर डॉ. आंबेडकर समझ गए कि जिस मौके का उन्हें इंतजार था, वह यही है।

19-20 मार्च, 1926 को उन्होंने बंबई के निकट जिला कुलाबा में एक विशाल

महार सम्मेलन का आयोजन किया।

उनके इशारे पर महार तथा अन्य पिछड़ी जातियों के सैकड़ों युवक व्यवस्था में जुट गए। पूरे महाराष्ट्र व गुजरात क्षेत्र में पोस्टर लगाए गए, प्रचार किया गया।

सम्मेलन में एक लाख से अधिक लोग इकट्ठे हुए। 'डॉ. आंबेडकर जिंदाबाद' के नारों से विशाल पंडाल गूँज रहा था। देश के इतिहास में पहली बार इतना बड़ा सम्मेलन केवल दलितों के चंदे से हुआ था।

आंबेडकर जब मंच पर आए तो तालियों की गड़गड़ाहट और नारों से पंडाल गूँज उठा। आंबेडकर ने कहा, "अब समय आ गया है कि जिन्हें समाज अछूत कहता है, जिनका बात-बात पर अपमान करता है, वे सब एक हो जाएँ। तुम किसी से डरते क्यों हो, क्यों दबते हो? तुम तो देश और समाज के आधार हो। तुम्हारी मेहनत से उगाए गए अन्न से सबका पेट भरता है।

"मैं कहता हूँ, दबना छोड़ो। अपने स्वाभिमान को इकट्ठा करो और उठकर खड़े हो जाओ, एकजुट हो जाओ। तुम सबसे बड़ी ताकत हो। परंतु हाँ, एक बात याद रखो—कोई भी गलत काम न करो। मेहनत की खाओ। मुर्दा पशुओं का मांस खाना बंद कर दो। काम कोई बुरा नहीं होता, परंतु बुरी आदतों को अपने अंदर नहीं आने देना चाहिए। दृढ़ इरादा कर लो, हमें अपने आपको बदलना है और अपने प्रति समाज का रवैया बदलना है। जो बात मुँह से निकालो वह ठोस होनी चाहिए, उसमें वजन होना चाहिए। जब अपनी बात कहो तो आवाज में दम होना चाहिए। दबी जबान में बोलना छोड़ दो। तुम्हारे साथ अगर गलत होता है, अन्याय होता है तो ऊँची आवाज में उसका विरोध करो।"

डॉ. आंबेडकर के विचार युवाओं में गहराई तक उतर गए। बूढ़े और कमजोर लोग तो थोड़ा डरे। उन्हें लगा कि यह आंबेडकर हमें मरवाएगा। बड़ी जातिवालों से टकराएँगे तो वे हमें और हमारे बच्चों को चैन से जीने नहीं देंगे। परंतु युवा वर्ग में उत्साह की लहर दौड़ गई। युवाओं ने घोषणा की, "आज से आंबेडकर का वाक्य हमारे लिए ब्रह्म वाक्य है। हम अपना सबकुछ न्योछावर कर देंगे। हम आंबेडकर के पीछे एकजुट होकर खड़े होंगे और सामाजिक न्याय लेकर रहेंगे।"

इसी सम्मेलन में डॉ. आंबेडकर ने सरकार से माँग की कि वह अछूतों और पिछड़ों को सेना में भरती करे। उन्होंने ब्राह्मणों से निवेदन किया कि वे अछूतों को इनसान समझें और छुआछूत बंद करें।

अगले दिन जन-समूह के साथ डॉ. आंबेडकर चौदार तालाब पहुँचे। उस तालाब के पानी को छूने का अछूतों को अधिकार नहीं था। आंबेडकर ने नारा लगाया, "तालाब

पर हमारा भी उतना ही हक है जितना दूसरों का।'' और उन्होंने दोनों हाथों की अंजलि बनाकर तालाब से पानी लेकर पी लिया। उन्होंने कहा, ''अधिकार भीख में नहीं मिल सकते, इन्हें आगे बढ़कर हासिल किया जाता है।''

महारों का मनोबल बढ़ गया। तालाब से पानी पीने के बाद जुलूस की शक्ल में आंबेडकर आगे बढ़े और उन्होंने वीरेश्वर मंदिर में प्रवेश किया। उनके साथ सैकड़ों दलितों ने मंदिर में प्रवेश किया और पूजा की।

इस सबसे ऊँची जाति के लोग चिढ़ गए। उन्हें लगा कि अछूत सीमा पार कर रहे हैं। इन्हें सबक सिखाया जाना चाहिए।

जैसे ही जुलूस पंडाल में पहुँचा, भालों और लाठियों से उन्होंने दलितों पर हमला कर दिया। उच्च वर्ग के इशारे पर किए गए हमले में बहुत लोग घायल हुए। वे निहत्थे थे। कुछ लोग बुरी तरह सहम गए।

परंतु आंबेडकर और उनके उत्साही साथी अड़ गए। आंबेडकर ने घोषणा की, ''कल हम चौदार तालाब पर कब्जा करेंगे। देखें, पानी पीने के अधिकार से हमें कौन रोकता है!''

अगले दिन ऊँची जाति के लोगों ने पूरी ताक़त लगा दी। पुलिस का भी सहारा लिया। परंतु दलितों में इतना जोश था कि किसी के रोके न रुके। आगे बढ़ते ही गए और उन्होंने तालाब पर कब्जा कर लिया।

जॉन साइमन से मुलाकात

उन्हीं दिनों साइमन कमीशन भारत आया हुआ था। आंबेडकर ने एक स्मरण-पत्र तैयार किया और अपने साथ दलितों का एक जत्था लेकर कमीशन के नेता जॉन साइमन से मिले। उन्होंने दलितों की दर्द भरी कहानी जॉन साइमन को सुनाई तो वे हैरान रह गए।

जॉन साइमन ने आंबेडकर से वादा किया कि वे दलितों को पूरा न्याय दिलाएँगे। छुआछूत खत्म की जाएगी, अछूतों व दलितों के प्रतिनिधि को मंत्रिमंडल में शामिल किया जाएगा।

गोलमेज सभा, लंदन में जॉर्ज पंचम से भेंट

आंबेडकर ने अपने निजी प्रयासों से दोनों गोलमेज सम्मेलनों में भाग लिया। पहले सम्मेलन में उन्होंने दलितों की ओर ब्रिटेन के सम्राट् का ध्यान खींचा और दूसरे सम्मेलन में उन्होंने दलितों की ओर से माँग-पत्र पेश किया।

माँग-पत्र में उन्होंने दलितों के लिए अलग निर्वाचन क्षेत्र तथा जनसंख्या के अनुसार प्रांतों की विधानसभाओं तथा केंद्र सरकार में प्रतिनिधित्व की माँग की।

गांधीजी से समझौता

इन माँगों से गांधीजी घबरा गए। इन माँगों का अर्थ था दलितों के लिए देश का बँटवारा। महात्मा गांधी इसके विरोध में आमरण अनशन पर बैठ गए। पूरे देश में तहलका मच गया। कांग्रेसी नेता जवाहरलाल नेहरू, मदन मोहन मालवीय आदि ने आंबेडकर को मनाने की बहुत कोशिश की। वे चाहते थे कि आंबेडकर गांधीजी से मिलें और उनका अनशन तुड़वाएँ, गांधीजी के साथ समझौता करें।

आंबेडकर इसके लिए तैयार नहीं थे। वे जानते थे कि गांधीजी की बात उन्हें माननी पड़ जाएगी और फिर दलितों को उनके अधिकार अलग से नहीं मिल पाएँगे।

जब गांधीजी की हालत बहुत खराब हो गई तो उनकी पत्नी कस्तूरबा गांधी आंबेडकर के पास गईं। उन्होंने आंबेडकर को समझाया कि वे गांधीजी से मिलकर बात करें और उनकी जान बचाएँ।

अब आंबेडकर समझ गए कि गांधीजी से मिलना ही पड़ेगा। उनके मन में गांधीजी के प्रति बहुत प्यार था; परंतु दलितों को न्याय दिलाने के लिए वे कृत संकल्प थे।

उन्होंने 10 सूत्री माँग-पत्र तैयार किया और गांधीजी के सामने रख दिया। इस माँग-पत्र में वे सभी शर्तें शामिल की गई थीं, जिन्हें मान लेने से दलित समाज देश की मुख्यधारा में शामिल हो सकता था और उसकी अगली पीढ़ियाँ सिर उठाकर चल सकती थीं।

गांधीजी ने वे माँगें मान लीं और दोनों नेताओं के बीच ऐतिहासिक समझौता हो गया। आंबेडकर दलितों के उद्धार में जुटे थे और देश में स्वाधीनता का आंदोलन जोरों पर था। आंबेडकर को बहुत काम करना पड़ता था। वे घर-परिवार के लिए तनिक भी समय नहीं निकाल पाते थे।

मई 1935 में उनकी पत्नी रमाबाई की मृत्यु हो गई। वे बीमार थीं। समाज के काम में जुटे होने के कारण वे पत्नी के इलाज के लिए अधिक भाग-दौड़ नहीं कर सके।

पत्नी की मृत्यु ने उन्हें झकझोरकर रख दिया। वे अपने आपको उनकी मौत के लिए जिम्मेदार मानकर बहुत दुखी रहने लगे।

भारत की आजादी और संविधान

पत्नी की मृत्यु के बाद बहुत दिनों तक डॉ. आंबेडकर ने कुछ नहीं किया। पूरा दलित समाज उनके पीछे था और स्वाधीनता का आंदोलन जोरों पर था। उनके कंधों पर दो-दो जिम्मेदारियाँ थीं—आजादी की लड़ाई में महात्मा गांधी जैसे महान् नेता का साथ देना तथा दलितों व पिछड़ों को उनके अभिशाप से मुक्त कराना, उन्हें सामाजिक न्याय दिलाना।

1935 से 1947 तक का 12 वर्ष का समय आंबेडकर के लिए विविध संघर्षों और चुनौतियों से भरा था।

इस बीच कभी वे दलितों के संगठनों का नेतृत्व करते, उन्हें राष्ट्र की मुख्यधारा में लाने के लिए भाषण देते, लेख लिखते, किताबें लिखते; कभी देश-विदेश में आयोजित सभाओं में अपने विचार रखते। गांधीजी के साथ मिलकर आजादी की योजना पर काम करते और उन्हें बार-बार याद दिलाते कि दलितों के साथ न्याय अवश्य होना चाहिए।

इधर उच्च समाज की चोटें उन्हें लगातार सहन करनी पड़ रही थीं। रात-दिन व्यस्त रहने के कारण स्वास्थ्य पर भी विपरीत प्रभाव पड़ने लगा था।

फिर भी वे थके नहीं। उन्होंने दलितों को संगठित करने के लिए बड़े काम किए। उनके पीछे लाखों शक्तिशाली लोगों की टोली बन गई। एक के बाद एक काम होते गए। समस्या आती गई, समाधान होते रहे।

15 अगस्त, 1947 को देश आजाद हो गया। इसके साथ ही देश दो भागों में बँट गया—भारत और पाकिस्तान।

देश में सांप्रदायिक तनाव पैदा हो गया। लाखों की संख्या में हिंदुओं और मुसलमानों की जानें गईं। करोड़ों बेघर-बार हुए।

भारत में संविधान की रचना के लिए सभा का गठन हुअ। डॉ. भीमराव आंबेडकर को संविधान-रचना का कार्य सौंपा गया।

बहुत मेहनत से रात-दिन लगकर उन्होंने संविधान समिति के सहयोग से भारत के लिए श्रेष्ठ संविधान का निर्माण कर लिया।

26 जनवरी, 1950 को उन्होंने यह संविधान भारत सरकार को सौंप दिया।

शारदा कबीर से मुलाकात : दूसरा विवाह

दलितों को न्याय दिलाने के लिए संघर्ष करते-करते तथा भारत के संविधान की रचना में रात-दिन जुटे रहने के कारण आंबेडकर थक गए और बुरी तरह बीमार पड़ गए। उन्हें मधुमेह के रोग ने जकड़ लिया।

पुत्र यशवंत राव ने उन्हें इलाज के लिए बंबई के मापलकर अस्पताल में दाखिल करा दिया।

वहाँ उनकी मुलाकात लेडी डॉक्टर शारदा कबीर से हुई। शारदाजी आंबेडकर के व्यक्तित्व से इतनी प्रभावित हुईं कि उन्होंने उनसे शादी का प्रस्ताव रख दिया। डॉ. शारदा जाति से ब्राह्मण थीं। आंबेडकर शारदाजी की सेवा से बहुत प्रभावित हुए थे। उन्होंने 'हाँ' कर दी और दोनों ने विवाह कर लिया। विवाह के बाद शारदाजी सविता आंबेडकर बन गईं।

हिंदू कोड बिल : मंत्रिमंडल से इस्तीफा

सविताजी की सेवा और स्नेह-वर्षा से डॉ. आंबेडकर शीघ्र ही ठीक हो गए। वे उस समय नेहरूजी की सरकार में कानून मंत्री थे।

स्वस्थ होते ही वे 'हिंदू कोड बिल' नामक प्रस्ताव पर काम करने में जुट गए। प्रस्ताव तैयार होते ही उन्होंने उसे संसद् में पेश कर दिया।

उस बिल की बड़ी आलोचना हुई। राष्ट्रपति डॉ. राजेंद्र प्रसाद, श्यामा प्रसाद मुखर्जी, सरदार पटेल आदि सभी वरिष्ठ नेता उस बिल का विरोध करने लगे। नेहरूजी इसके समर्थन में थे, परंतु अकेले वे कुछ न कर सके।

डॉ. राजेंद्र प्रसाद ने यहाँ तक धमकी दे दी कि "यदि नेहरू सरकार ने यह बिल पास किया तो मैं अपने पद से इस्तीफा दे दूँगा।"

डॉ. आंबेडकर ने इसे अपना सरासर अपमान माना। काफी सोच-समझकर एक दिन वे उठे और उन्होंने प्रधानमंत्री को अपना इस्तीफा सौंप दिया।

बौद्ध धर्म की दीक्षा ली

अब आंबेडकर का मन राजनीति से भर गया था। वे इतने उखड़ गए कि सविताजी का स्नेह भी उन्हें न बाँध सका।

संविधान के अनुसार दलित व पिछड़े वर्गों को आरक्षण मिल गया था। संसद् तथा विधानसभाओं में उन्हें प्रतिनिधित्व मिल गया था। निःशुल्क शिक्षा और नियुक्तियों में आरक्षण की बात मान ली गई थी। फिर भी, व्यवहार में पिछड़ों का अपमान करने की परंपरा जारी थी।

आंबेडकर ने एक दिन एक विशाल बौद्ध सम्मेलन बुलाया और उसमें उन्होंने हिंदू धर्म छोड़कर बौद्ध धर्म की विधिवत् दीक्षा ली। उनके साथ हजारों दलितों ने बौद्ध धर्म की दीक्षा ली। पूरे देश में दलितों पर उसका भारी प्रभाव पड़ा। बड़ी संख्या में युवकों ने हिंदू धर्म छोड़ दिया और वे बौद्ध बन गए।

दीक्षा के समय उन्होंने ब्रह्मा, विष्णु और महेश को ईश्वर न मानने की घोषणा की। राम, कृष्ण आदि अवतारों को ईश्वर न मानने का संकल्प लिया। ब्राह्मणों से कोई रस्म न करवाने और देवी-देवताओं को न मानने की प्रतिज्ञा की।

संसार से वैराग्य : महानिर्वाण की कामना

बौद्ध धर्म स्वीकार करने के बाद वे बहुत गंभीर रहने लगे। हिंदू धर्म की विसंगतियों और आडंबरों से उनका जी ऊब गया था। उन्होंने प्रतिज्ञा की कि वे हिंदू धर्म में पैदा हुए, यह उनके हाथ में नहीं था; परंतु अब वे हिंदू धर्म में मरेंगे नहीं। यह उनके हाथ में है और वे ऐसा करके दिखाएँगे। अपनी धारणा के आधार पर उन्होंने हिंदू धर्म का त्याग कर दिया और बौद्ध बन गए।

अब वे बंबई स्थित अपने घर में ही रहते थे। उनकी पत्नी सविता आंबेडकर तथा पुत्र यशवंत उनका बहुत खयाल रखते थे।

धीरे-धीरे उनका रोग बढ़ता ही गया। अस्पताल जाने से उन्होंने इनकार कर दिया। वे कहा करते थे, ''जब एक बहुत अच्छी डॉक्टर मेरी जीवन साथी बनकर मेरी सेवा कर रही है तो मैं अस्पताल में पड़कर क्या करूँगा!''

अब वे बिस्तर के बंदी हो गए। शरीर कमजोर होता गया और स्वभाव में चिड़चिड़ाहट का भाव बढ़ता गया।

एक दिन उनका मन बंबई से उचट गया। पत्नी सविता आंबेडकर तथा पुत्र यशवंत राव के साथ वे अपने दिल्ली स्थित निवास 26, अलीपुर रोड आ गए।

सविताजी रात-दिन उनकी सेवा में लगी रहती थीं। बड़े-बड़े नेता उनके स्वास्थ्य के बारे में जानकारी लेने के लिए आते-जाते रहते थे।

6 दिसंबर, 1956 को प्रातःकाल महामानव आंबेडकर ने अचानक अपना शरीर त्याग दिया। दलितों को सामाजिक न्याय दिलवाने वाले, भारतीय समाज से अछूत प्रथा को समाप्त करानेवाले महाविद्वान् आंबेडकर की शव-यात्रा में जन-समूह उमड़ पड़ा था।

प्रस्तुत पुस्तक में महामानव आंबेडकर के प्रेरक विचारों को संकलित किया गया है।

अनुक्रमणिका

	अपनी बात	*5*
1.	सामाजिक चिंतन	35
2.	धार्मिक चिंतन	45
3.	आर्थिक चिंतन	52
4.	राजनीतिक चिंतन	56
5.	भारत में जाति-प्रथा	66
6.	विविध चिंतन	82
7.	आर्य	97
8.	हिंदू धर्म की मान्यताएँ	99
9.	कांग्रेस और गांधी ने अस्पृश्यों के लिए क्या किया?	113
10.	धर्मांतरण क्यों?	125
11.	भारत-विभाजन	133
12.	पत्रों में व्यक्त विचार	138
13.	संविधान विषयक विचार	141
14.	भारत में शूद्र	155
	संदर्भ ग्रंथ	*160*

सामाजिक चिंतन

व्यक्ति समूह समाज से बनता है—यह एक बहुत ही मजबूत व स्थूल तथा सामान्य वक्तव्य है। समाज कभी भी व्यक्ति से नहीं बनता है, वह वर्ग से बनता है। प्रत्येक समाज के वर्ग अस्तित्व में होते हैं। वर्ग-रचना के मूल में निहित कारण अलग-अलग होते हैं। कभी आर्थिक, कभी बौद्धिक, कभी राजनीतिक कारणों से वर्ग बनते हैं।

~***~

व्यक्ति किसी-न-किसी वर्ग की इकाई बनकर ही जीना चाहता है। यह एक वैश्विक सत्य है। हिंदू समाज भी इसके लिए अपवाद नहीं है। इस नियमानुसार वर्ग जाति में रूपांतरित होते गए। वास्तव में जाति व वर्ग आमने-सामने रहनेवाले पड़ोसी की तरह होते हैं। बहुत मामूली भेदों के कारण उनका अस्तित्व भिन्न-भिन्न लगता है। वस्तुतः जाति स्वयं एक मर्यादित वर्ग है।

~***~

यह संभव है कि मैं गलती पर होऊँ, परंतु मैंने यही हमेशा मुनासिब समझा है कि दूसरों के पथ-निर्देशन और आदेशों को मानने तथा मौन बैठे रहने और स्थितियों को बिगाड़ देने की अपेक्षा त्रुटियाँ करना कहीं श्रेयस्कर है।

~***~

मुझसे ज्यादा योग्य विद्वानों ने जाति के रहस्यों को खोलने का प्रयास किया है। किंतु यह दुःख की बात है कि यह अभी तक व्याख्यायित नहीं हुआ है और हम लोगों को इसके बारे में अल्प जानकारी है। मैं जाति जैसी संस्थाओं की जटिलता के प्रति सजग हूँ और इतना निराशावादी नहीं हूँ कि यह पहेली अगम, अज्ञेय है; क्योंकि मेरा विश्वास है कि इसे जाना जा सकता है। जाति की समस्या में

असैद्धांतिक और व्यावहारिक रूप में एक विकराल समस्या है।

***

भारतीय समाज रक्त की शुद्धता की बात चाहे जितनी ही क्यों न करे, पर यह शुद्ध रक्तवाला समाज नहीं है। इस भारतीय समाज में आर्यों, द्रविड़ों, मंगोलों, शकों, हूणों, आभीरों, नागों, यक्षों आदि न जाने कितनों का सम्मिश्रण है। अनेक जातियाँ देश-विदेश से घूम-घूमकर यहाँ पहुँचीं, बसीं और इस देश की सांस्कृतिक धारा में मिलकर एक हो गईं। ये जातियाँ अपने पूर्ववर्तियों को धकेलकर इस देश की मंत्र बन गईं। इनके परस्पर सतत संपर्कों और संबंधों के कारण एक समन्वित संस्कृति का सूत्रपात हुआ।

***

भारतीय समाज के विषय में यह बात कहना असंगत है कि वह विभिन्न जातियों का संकलन है। संकलन से सजातीयता उत्पन्न नहीं होती। यदि रक्तभेद की दृष्टि से देखा जाए तो भारतीय समाज विजातीय है। हाँ, यह संकलन सांस्कृतिक रूप से अत्यंत सघन है। इसी आधार पर मेरा कहना है कि इस प्रायद्वीप को छोड़कर संसार का कोई देश ऐसा नहीं है, जिसमें इतनी सांस्कृतिक समरसता हो। हम केवल भौतिक दृष्टि से ही संगठित नहीं हैं, बल्कि हमारी सांस्कृतिक एकता भी अविच्छिन्न एवं अटूट है, जो पूरे देश में चारों दिशाओं में व्याप्त है। इसी सांस्कृतिक एकरूपता के कारण जाति प्रथा इतनी विकराल बन गई है कि उसकी व्याख्या करना कठिन कार्य है। कठिन कार्य इसलिए है कि यहाँ सजातीय समाज में भी जाति प्रथा की घुसपैठ है।

***

भारतीय जाति समस्या को यूरोपीय नृविज्ञान और समाजशास्त्र के सिद्धांतों से हम कभी नहीं समझ सकते हैं। भारत में जाति प्रथा का अर्थ है समाज को कृत्रिम हिस्सों में विभाजित करना, जो रीति-रिवाज और शादी-विवाह की भिन्नताओं से बँधे हों। परिणाम स्पष्ट है कि सजातीय विवाह एकमात्र लक्ष्य है, जो जाति प्रथा की विशेषता है, और यदि हम यह जानने में सफल हो जाएँ कि सजातीय विवाह ही क्यों होते हैं, तो हम व्यावहारिक रूप से साबित कर सकते हैं कि जातियों की उत्पत्ति कैसे हुई और उनका ताना-बाना क्या है।

***

वह कौन सा पहला वर्ग था, जो जाति में रूपांतरित हुआ? दूसरे शब्दों में, जाति-व्यवस्था का जनकत्व किसे दिया जाए? आरंभ में ब्राह्मण वर्ग जाति में

बदल गया। इसी ब्राह्मण वर्ग ने इस देश में जाति प्रथा को बढ़ाया।

***

मनु जाति व्यवस्था के जन्मदाता नहीं हैं। मनु के सैकड़ों वर्ष पूर्व जाति प्रथा अस्तित्व में थी। मनु इस व्यवस्था को एक दार्शनिक धरातल पर ले जाकर शास्त्रीय रूप देते हैं। जाति से संबंधित नियमों का मनु ने केवल संकलन किया है।

***

विषम समाज-रचना भावुकता अथवा आदर्शों से समाप्त होनेवाली नहीं है। इसके लिए परिवर्तित मानसिकता और ठोस कार्यक्रमों की आवश्यकता है।

***

आज तक (1935 तक) के आंदोलन केवल पारिवारिक प्रश्नों को सुलझाने के लिए ही किए गए। संपूर्ण समाज की पुनर्रचना की बात कहीं की ही नहीं गई। विधवा विवाह, वृद्ध विवाह, बाल विवाह आदि सुधारों का संबंध व्यक्ति और परिवार के साथ है, समाज-रचना के साथ नहीं है। उच्च वर्णियों की रुचि पारिवारिक सुधारों में अधिक रही। इसी में उनके हित सुरक्षित रहे। जाति या वर्गभेद से टकराने की आवश्यकता उन्हें महसूस नहीं हुई। जाति और वर्ग के कारण उपेक्षित समाज की यातना को वे समझ ही नहीं पाए। विधवा विवाह एवं वृद्ध विवाह आदि स्त्री विषयक समस्याएँ उनके अपने वर्ग में ही अधिक कठोर थीं, इस कारण उनके सामाजिक सुधार उनके वर्ग तक ही सीमित रहे। समाज-रचना में परिवर्तन हो, इस प्रकार की दृष्टि अंग्रेजों की सत्ता यहाँ आने के बाद ही पनपी।

***

जाति संस्था केवल श्रम-विभाजन नहीं है, वह श्रमिकों का भी विभाजन है। केवल श्रमिकों का विभाजन करके ही वह रुकती नहीं, अपितु श्रमिकों के एक-एक समूह को वह नीचे से ऊपर की ओर एक-दूसरे पर रखती जाती है—और सबसे भयंकर बात यह है कि वह क्रमबद्धता जन्म के आधार पर हमेशा के लिए चिपका दी जाती है। विश्व के किसी भी देश में श्रम-विभाजन को श्रमिकों के सामाजिक स्तर के साथ चिपकाया नहीं गया है।

***

भारतीय समाज सीढ़ी-दर-सीढ़ी का ऐसा सिलसिला है, जिसमें हर ऊपर की सीढ़ी का आदमी हर नीचे की सीढ़ी के आदमी को घृणा की दृष्टि से देखता है और नीचे की सीढ़ी का आदमी अपने ऊपर की सीढ़ी के आदमी को आदर की

दृष्टि से। ऐसे समाज में समता, बंधुत्व और प्रजातंत्र को पनपने की रत्ती भर भी गुंजाइश नहीं है। प्रजातंत्र में असमानता कैसी, भेदभाव कैसा, राग-द्वेष कैसा?

***

जाति-व्यवस्था से जो दोष हिंदू समाज में आए, क्या इन दोषों से हम मुक्त हो सकते हैं? जाति-व्यवस्था नष्ट कैसे की जा सकती है? जाति व्यवस्था नष्ट करने के लिए सर्वप्रथम उपजातियाँ नष्ट करनी चाहिए, ऐसी एक धारणा है, जो गलत है। इस प्रयत्न में जातियाँ अधिक मजबूत और शक्तिशाली बनेंगी।

***

जाति-व्यवस्था तोड़ने के लिए एकमात्र उपाय है अंतर्जातीय विवाह। रक्त संबंधों के कारण ही आत्मीयता की भावना तैयार होती है। इसी कारण अंतर्जातीय विवाह अधिकाधिक होते जाएँ तो जाति के बंधन कमजोर हो जाएँगे। एक-दूसरे के प्रति अधिक आत्मीयता तैयार होती जाएगी।

***

जाति व्यवस्था मनुष्य निर्मित है। वह नष्ट को जा सकती है। जाति एक काल्पनिक व्यवस्था है, एक मानसिक वृत्ति है। किसी की मानसिकता को बदलना आसान काम नहीं है।

***

हिंदू जाति का पालन करते हैं। इसका एकमात्र कारण कि हिंदू मूलत: धार्मिक हैं। जाति को धार्मिकता से जोड़ा गया है। मूल धर्म ने ही एक गलत अमानवीय व्यवस्था का समर्थन किया है। शास्त्रों की हँसी उड़ाकर यह व्यवस्था खत्म होनेवाली नहीं है। इसके विरोध में प्रखर बुद्धिवाद और समतावादी समाज-रचना के प्रति प्रतिबद्धता की नितांत आवश्यकता है।

***

जाति और वर्ण अधिष्ठित लोगों के आचरण के मूल में हजारों वर्षों से चली आई श्रद्धाएँ हैं।

***

अंतर्जातीय विवाह का नारा लगाकर काम नहीं होगा। केवल नई व्यवस्था से बात नहीं बननेवाली है।

***

जब-जब आदर्श और काल्पनिक वर्ण-व्यवस्था का प्रचार किया जाता है

तब-तब जातियों से चिपककर जीने का एक बहुत बड़ा आधार जनसामान्य को मिल जाता है।

****

वर्ण-व्यवस्था की जड़ों पर ही आघात करने की आवश्यकता है। जब तक जड़ें तोड़ी नहीं जाएँगी तब तक अस्पृश्यता की मानसिकता समाप्त नहीं हो सकती। अस्पृश्यता और जातिभेद नष्ट करने के लिए शास्त्रों से आधार ढूँढ़ने का प्रयत्न वास्तव में कीचड़ से कीचड़ साफ करने का हास्यास्पद प्रयत्न है।

****

मेरे लिए यह चातुर्वर्ण्य, जिसमें पुराने नाम जारी रखे गए हैं, घिनौनी वस्तु है, जिससे मेरा व्यक्तित्व विद्रोह करता है। लेकिन मैं यह नहीं चाहता कि मैं केवल भावनाओं के आधार पर चातुर्वर्ण्य के प्रति आपत्ति करूँ। इसका विरोध करने के लिए मेरे पास ठोस कारण है। इस आदर्श की अच्छी तरह जाँच-परख के बाद मुझे पूरी तरह विश्वास हो गया है कि चातुर्वर्ण्य सामाजिक संगठन प्रणाली के रूप में अव्यावहारिक, घातक और अत्यंत असफल रहा है। व्यावहारिक दृष्टि से भी चातुर्वर्ण्य से ऐसी अनेक कठिनाइयाँ उत्पन्न होती हैं, जिन पर इसके समर्थकों ने ध्यान नहीं दिया।

****

जाति का आधारभूत सिद्धांत वर्ण के आधारभूत सिद्धांत से भिन्न है; न केवल मूल रूप से भिन्न है, बल्कि मूल रूप से परस्पर विरोधी भी है।

****

यदि कोई व्यक्ति गुण के आधार पर नहीं, जन्म के आधार पर ऊँची हैसियत प्राप्त कर गया हो तो आप क्या करेंगे? उसे कैसे झेलेंगे? इसका उत्तर यही है कि आपको जाति प्रथा का उन्मूलन करना पड़ेगा।

****

प्लेटो ने भी लोगों को प्रकृति के आधार पर तीन भागों में बाँटा था, पर हुआ क्या? वे असफल रहे। चातुर्वर्ण्य सिद्धांत मानव प्रकृति के प्रतिकूल है। वह चल नहीं सकता। चाहे उसे पालन कराने के लिए दंड-विधान की नीति को ही क्यों न अपनाया जाए।

****

आदर्श मानव समाज वह है जो समता, स्वतंत्रता और भ्रातृत्व की भावना पर खड़ा हो और जिसकी नींव त्याग, बलिदान, समर्पण, सत्य, अहिंसा और प्यार पर

रखी गई हो। यदि किसी समाज में ऐसा नहीं है तो वह समाज कैसे हो सकता है? समाज तो मानवकृत है और उसने उसे सबकी सुविधा और विकास के लिए बनाया है।

***

तुम्हारे मुखमंडल की दयनीय दशा देखकर और तुम्हारी निराशामयी आवाज सुनकर मेरा हृदय विदीर्ण हो उठा है। कितने समय से तुम अत्याचारों की चक्की में पिसते आ रहे हो और फिर भी तुममें साहसहीनता व अंधविश्वास त्यागने का विचार पैदा नहीं होता है। तुम लोग जन्म लेते ही क्यों नहीं मर जाते? तुम अपने दयनीय, घृणित और उपेक्षित जीवन से धरती का बोझ क्यों बढ़ाते हो? यदि तुम नवजीवन नहीं अपना सकते और अपनी स्थिति नहीं बदल सकते तो इस जीने से मरना कहीं बेहतर है। यकीनन भोजन, वस्त्र और आवास प्राप्त करना तुम्हारा जन्मसिद्ध अधिकार है। यदि तुम ससम्मान जीना चाहते हो तो तुम्हें स्व-सहायता पर विश्वास करना चाहिए, क्योंकि वही सर्वोपरि सहायता है।

***

हिंदू समाज एकवर्णी समाज बनेगा तो ही उसमें अपनी रक्षा करने की शक्ति का निर्माण होगा। इस आंतरिक शक्ति के अभाव में स्वराज्य स्वतंत्रता की सीढ़ी न बनकर गुलामी की एक सीढ़ी सिद्ध होगा।

***

यह मान भी लिया जाए कि चातुर्वर्ण्य व्यावहारिक है, फिर भी मैं यह निश्चयपूर्वक कहूँगा कि यह एक बहुत ही दोषपूर्ण व्यवस्था है कि ब्राह्मण द्वारा विद्या का संवर्धन किया जाना चाहिए, वैश्य को व्यापार करना चाहिए और शूद्र को सेवा करनी चाहिए। यद्यपि यह एक श्रम-विभाजन ही था, लेकिन क्या इस सिद्धांत का उद्देश्य यह था कि शूद्र को धन-संपत्ति आदि अर्जित ही नहीं करनी चाहिए? यह बहुत ही दिलचस्प प्रश्न है। चातुर्वर्ण्य के समर्थक इसका पहला अर्थ यह निकालते हैं कि शूद्र को धन-संपत्ति आदि अर्जित करने का कष्ट क्यों उठाना चाहिए, जबकि तीनों वर्ण उसकी सहायता के लिए मौजूद हैं। यदि शूद्र को लिखने-पढ़ने की जरूरत होगी तो वह ब्राह्मण के पास जा सकता है। शूद्र को शस्त्र धारण करने की चिंता क्यों करनी चाहिए, जबकि क्षत्रिय मौजूद हैं। इस अर्थ में समझे गए चातुर्वर्ण्य सिद्धांत के अनुसार कहा जा सकता है कि शूद्र एक आश्रित व्यक्ति है और तीनों वर्ण उसके संरक्षक।

***

शिक्षा प्रत्येक व्यक्ति को मिलनी चाहिए, रक्षा के साधन सभी लोगों के पास होने चाहिए। प्रत्येक व्यक्ति के आत्म-परीक्षण के लिए ये परम आवश्यकताएँ हैं। किसी अशिक्षित और निरस्त्र व्यक्ति को उस व्यक्ति से क्या सहायता मिल सकती है कि उसका पड़ोसी शिक्षित और सशस्त्र है?

~***~

इसमें कोई संदेह नहीं है कि व्यवहार में चातुर्वर्ण्य का संबंध वस्तुत: मालिक और नौकर का था। ब्राह्मण, क्षत्रिय और वैश्यों के आपसी संबंध भी संतोषजनक नहीं थे; फिर भी वे मिलकर कार्य करने में सफल हुए। ब्राह्मण ने क्षत्रियों की खुशामद की और दोनों ने वैश्य को जीवित रहने दिया, ताकि वे उसके सहारे जीवित रह सकें। लेकिन तीनों शूद्र को पददलित करने के लिए सहमत हो गए। उसे धन-संपत्ति अर्जित करने की अनुमति नहीं दी गई, क्योंकि कहीं ऐसा न हो कि वह तीनों वर्गों पर निर्भर ही न रहे। उसे विद्या प्राप्त करने से रोका गया कि कहीं ऐसा न हो कि वह अपने हितों के प्रति सजग हो जाए। उसके लिए शस्त्र धारण करना निषिद्ध था कि कहीं ऐसा न हो कि वह उनकी सत्ता के विरुद्ध विद्रोह करने का साधन प्राप्त कर ले।

~***~

हमारे चिंतन का मूलाधार 'मनुष्य' होना चाहिए। मनुष्य को उसके मानवीय अधिकार उपलब्ध करा देना हमारी नैतिक जिम्मेदारी है। शास्त्रों ने मानवीय अधिकार ही छीन लिये, इसलिए उनको नकारना जरूरी है।

~***~

धर्म मनुष्य के लिए है, मनुष्य धर्म के लिए नहीं।

~***~

जब तक अछूत हिंदू समाज में बना रहेगा तब तक उसका जीवन-स्तर नहीं सुधरेगा और वह निरंतर शोषित, अपमानित तथा अकिंचन बना रहेगा। उसका भाग्य कभी नहीं बदलेगा। कोई भी उसे आर्थिक प्रलोभन देकर अपने धर्म में परिवर्तित कर लेगा। इसका परिणाम यह होगा कि हिंदू समाज में कभी भी क्रांति नहीं हो सकेगी। निरंतर रूढ़ियों के अँधेरे में पड़ा-पड़ा सड़ता जाएगा। हिंदू समाज में रहना है तो उसमें क्रांतिकारी परिवर्तन किए बिना जीवन-स्तर नहीं सुधारा जा सकेगा।

~***~

अछूतों की आर्थिक स्थिति दिन-प्रतिदिन गिरती जा रही है। उसके पास सामाजिक सौहार्द, सहजीवन की पकड़ और प्रतिष्ठा कुछ भी नहीं है। फिर भी वह हिंदू समाज की इकाई कैसे बना हुआ है? सामाजिक सम्मान प्राप्त किए बिना कोई वर्ग किसी समाज में सम्मान से नहीं रह सकता। यह सब सामाजिक क्रांति लाए बिना संभव नहीं है।

***

गांधीजी के 'हरिजन' पारिभाषिक शब्द पर मुझे आपत्ति है, जिसके मुख्य बिंदु इस प्रकार हैं—

(1) क्या अछूत हरिजन हो जाने से अछूत नहीं रहेगा?

(2) क्या वह मैला नहीं उठाएगा या झाड़ू नहीं लगाएगा?

(3) क्या इससे अन्य वर्णवाले उसे गले से लगा सकेंगे?

(4) क्या उसे हिंदू समाज अछूत न समझकर सवर्ण जैसा मान लेगा?

(5) क्या इससे उसे सामाजिक समता का अधिकार मिल सकेगा?

***

यदि नहीं, तो फिर हरिजन की चक्रव्यूह रचना क्यों? जिस मार्ग से अछूतों का आत्मसम्मान नहीं लौट सके, उस मार्ग से क्या लाभ? हरिजन तो हर एक है।

***

कोई मनुष्य अपमानित होकर धन्यवाद नहीं दे सकता, कोई नारी अपना शील भंग हो जाने पर धन्यवाद नहीं दे सकती और कोई राष्ट्र अपनी स्वाधीनता खोकर धन्यवाद नहीं कर सकता। फिर अछूत 'हरिजन' के वज्र से अपमानित होकर कैसे गांधीजी का धन्यवाद अदा कर सकता है।

***

जब तक अस्पृश्य वर्ग हिंदू समाज में समाविष्ट है, तब तक उसकी उन्नति नहीं होगी। सच्ची स्वतंत्रता प्राप्त करने के लिए उनको हिंदू समाज त्याग देना चाहिए। इससे उन्हें कपड़े बनाने, खाने-पीने, नौकरी करने, शिक्षा प्राप्त करने, सुसंस्कृत समाज में रहने आदि की स्वतंत्रता मिल जाएगी।

***

तुम्हारे पास गँवाने के लिए कुछ भी नहीं है। तुम सिर्फ अपनी बेड़ियों को ही गँवानेवाले हो। धर्मांतरण से तुम्हें अनेक लाभ होनेवाले हैं। सामाजिक दृष्टि से धर्मांतरण की समस्या का ओर देखा जाए तो यह संघर्ष ऊपरी तौर से भले ही सामाजिक प्रतिष्ठा प्राप्त करने की दिखाई पड़े, तो भी वह प्राय: वर्गीय संघर्ष ही

है। हकदार और बेरोजगार में जो निरंतर संघर्ष चल रहा है, वरिष्ठों का जुल्म उसका एक हिस्सा ही है। संघर्ष फलदायक करने के लिए अस्पृश्यों में तीन आवश्यक शक्तियों की कमी है। वे शक्तियाँ हैं—मनुष्य शक्ति, द्रव्य शक्ति और बौद्धिक शक्ति। ये तीनों शक्तियाँ जब तक तुम हिंदू समाज में रहोगे, तब तक तुम प्राप्त नहीं कर सकोगे।

***

कुछ लोग कहते हैं, धर्मांतरण करने से तुम्हें क्या लाभ होनेवाला है। उसका उत्तर है कि स्वराज से भारत को क्या लाभ होगा? हिंदुस्तान को जितनी स्वराज की आवश्यकता है उतनी ही अस्पृश्यों को धर्मांतरण की। धर्मांतरण और स्वराज का अंतिम प्रयोजन है स्वतंत्रता-प्राप्ति।

***

समान नागरिक कानून होना चाहिए। हिंदू, मुसलमान और ईसाई धर्म पर आधारित कानूनों में जहाँ मतभेद के मुद्दे हैं, उन्हें विचारपूर्वक हल करना चाहिए। सब में कुछ समानता के मुद्दे भी हैं। केवल मतभेदों के मुद्दों से बात नहीं बन सकती। समान नागरिक कानून हमारा अंतिम उद्देश्य है।

***

सामाजिक समता की लड़ाई में दलितों को ही महत्त्वपूर्ण भूमिका निभानी है। कमजोर वर्ग को संगठित होकर अपने अधिकार खुद ही प्राप्त करने चाहिए। मैं अस्पृश्य समाज को राजनीतिक अधिकार दिलाने के लिए प्रयत्नशील हूँ। देश की स्वतंत्रता के लिए काम करनेवाले बहुत से युवक हैं। दलित युवकों का उद्धार उन्हें स्वयं करना है। दलितों के उद्धार से ही सचमुच का स्वराज्य संभव है।

***

समाज के सबसे शोषित और उपेक्षित व्यक्ति तक शिक्षा की सुविधाएँ पहुँचनी चाहिए। परंपरा से नकारे गए वर्ग को उच्च शिक्षा हेतु अधिक खर्च न करना पड़े, ऐसी व्यवस्था होनी चाहिए। स्नातक और छात्रों के पाठ्यक्रम द्वारा सांस्कृतिक प्रगति करने के लिए महाविद्यालयों और विश्वविद्यालयों को एकत्र होना जरूरी है।

***

शिक्षा दोधारी शस्त्र के समान है। चरित्रहीन और विनयहीन शिक्षित व्यक्ति पशु से भी अधिक भयंकर होता है। सुशिक्षित मनुष्य का ज्ञान और उसकी शिक्षा जनहित के विरोध में जा रही हो, तो ऐसा व्यक्ति समाज के लिए अभिशाप साबित होता है।

***

अछूत और सवर्ण में वंशभेद नहीं है। पूर्व में यह अंतर विजित और विजेता का था। वंशभेद की भाँति व्यवसाय-भेद भी उसका आधार नहीं है।

***

चांडाल और अस्पृश्य पर्यायवाची शब्द नहीं हैं। ये भिन्न हैं। चांडालों का वर्ग धर्म सूत्र के कालखंड में हुआ, परंतु इसके कई वर्षों बाद करीब 400 ई.पू. में अस्पृश्यों की निर्मिति हुई।

***

अस्पृश्य सारे देश में गाँव से बाहर क्यों रहते हैं? आदिम काल में टोलियों में सतत लड़ाइयाँ हुआ करती थीं। वे टोलियाँ जब स्थायी समाज में परिवर्तित हो गईं तब अन्य स्थानों से विस्थापित टोलियों के लोग उनके यहाँ आए। तब उन्हें गाँव के भीतर प्रवेश नहीं दिया गया। उन्हें गाँव के बाहर ही रहना पड़ा। न केवल भारत में अपितु दुनिया के कई देशों में विस्थापितों को गाँव के बाहर ही रुकना पड़ता था।

***

आर्य-द्रविड़ों के संघर्ष में विजयी आर्यों ने द्रविड़ों को जीत लिया और उन्हें शूद्र बनाया। इसके पूर्व द्रविड़ों ने आदिवासियों को जीता था और उन्हें अस्पृश्य बनाया था। द्रविड़ और नाग इन दोनों को एक ही वंश का सिद्ध करते हैं। बाद में आर्यों ने नाग लोगों को विस्थापित कर दिया और उन्हें अस्पृश्य घोषित किया।

***

विभिन्न वंशों संबंधी निर्णय लेने का आधार अगर वांशिक शरीर रचना शास्त्र है तो उसके आधार पर अगर हम हिंदू समाज की विभिन्न जातियों का मूल्यांकन करें (सिर, आँख, नासिका, ठुड्डी, रंग, कद आदि) तो यह सिद्ध हो जाएगा कि अस्पृश्य, आर्य और द्रविड़ वंश एक-दूसरे से कहीं पर भी भिन्न नहीं हैं। शरीर के विभिन्न अंगों के मापन से यह सिद्ध होगा कि ब्राह्मण और अस्पृश्य एक ही वंश के लोग हैं। अगर ब्राह्मण आर्य हैं तो अस्पृश्य भी आर्य हैं।

***

अछूत बौद्ध थे। शंकराचार्य ने दार्शनिक स्तर पर जाकर जब हिंदू धर्म की पुनर्स्थापना की तब समाज का बहुत बड़ा तबका सत्ता, शक्ति और आतंक के कारण हिंदू धर्म में लौटा; परंतु एक तबका ऐसा था, जो बौद्ध धर्म छोड़ने के लिए तैयार नहीं था। इन्हें ही तिरस्कृत कर अस्पृश्य बना दिया गया। यह षड्यंत्र ढाई हजार वर्ष पूर्व से आरंभ हुआ था।

□

धार्मिक चिंतन

धर्म ही जीवन है और जीवन ही धर्म है। जैसे जीवन क्या है, इस पर मानव का चिंतन निरंतर चलता आया है और भविष्य में भी अहर्निश चलता रहेगा। ठीक वैसे ही धर्म के संबंध में सोचना कभी बंद नहीं होगा। जहाँ आज धर्म है वहाँ धर्म के झगड़े हैं। ये झगड़े आज से नहीं, सदियों से हैं।

~***~

इस धरती पर कोई भी धर्म ऐसा नहीं, जो उत्थान-पतन के दौर से न गुजरा हो। किसी भी धर्म में मानवतावादी दृष्टिकोण का रत्ती भर भी विरोध नहीं है। फिर भी बराबर धर्मयुद्ध होते रहे हैं और धर्म ने अपने सहज स्वभाव को छोड़कर अधर्म का रूप अपना लिया है। धर्म कब अधर्म बनकर भी धर्म बना रहता है, इसके किस्से धर्म के इतिहास में हजारों मिल जाएँगे। कारण भी मालूम पड़ जाएँगे। फिर भी धर्म अपने इस इतिहास को दोहराने से पीछे नहीं हटेगा।

~***~

धर्म मानव इतिहास की सबसे शक्तिशाली चालक शक्तियों में से एक है और इसकी गुणवत्ता केवल उसके द्वारा प्रस्तुत सामाजिक आदर्श की रोशनी में ही समझी जा सकती है प्रत्येक धर्म को न्याय और उपयोगिता द्वारा ही परखा जाना चाहिए। न्याय, मुक्ति, समानता और भाईचारे का ही दूसरा नाम धर्म है।

~***~

धर्म और समाज की विषमता की जड़ चातुर्वर्ण्य व्यवस्था है। चातुर्वर्ण्य ही अस्पृश्यता की जननी है। जाति-भेद और अस्पृश्यता विषमता के अन्य रूप हैं। अगर इस जड़ को नष्ट नहीं किया गया तो अस्पृश्य वर्ग इस धर्म को निश्चित ही त्याग देगा।

~***~

ऐसे लोगों की कमी नहीं है, जो अस्पृश्यों की दयनीय स्थिति से दुखी होकर, चिल्लाकर अपना जी हलका करते फिरते हैं कि हमें अस्पृश्यों के लिए कुछ करना चाहिए। लेकिन इस समस्या का जो हल करना चाहते हैं, उनमें से शायद ही कोई ऐसा व्यक्ति हो जो यह कहता है कि हमें स्पृश्य हिंदुओं को बदलने के लिए भी कुछ करना चाहिए। यह धारणा बनी हुई है कि अगर किसी का सुधार करना है तो वह अस्पृश्यों का ही करना है। अगर कुछ किया जाना है तो वह अस्पृश्यों के प्रति किया जाना है और अगर अस्पृश्यों को सुधार दिया जाए, तब अस्पृश्यता की भावना मिट जाएगी।

***

जाति-व्यवस्था मनुष्य को संवेदनहीन बनाती है। यह बाँझपन की एक प्रक्रिया है। शिक्षा, संपत्ति एवं श्रम प्रत्येक व्यक्ति के लिए, जो स्वतंत्र एवं पूर्ण मनुष्यत्व प्राप्त करना चाहता है, आवश्यक है। समाज के पिछड़ेपन और अकर्मण्यता का दोष मुख्यतः अप्राकृतिक एवं अवैज्ञानिक सामाजिक व्यवस्था को दिया जा सकता है।

***

क्या कोई यह विश्वास कर सकता है कि पुरुष नाम का कोई ऐसा पशु भी है, जिसके स्पर्श मात्र से पानी गंदा हो जाता है, मनुष्य भ्रष्ट हो जाता है और ईश्वर पूजनीय नहीं रहता। यदि ऐसा पुरुष है तो वह अछूत है, जिसे हिंदू धर्म और हिंदू समाज में पशु से भी बुरा समझा जाता है।

***

एक धर्म में सबको समानता का अधिकार नहीं हो तो वह धर्म कहाँ रह जाता है।

***

अगर हिंदू समाज एकवर्णी समाज बनेगा तो ही उसमें अपनी रक्षा करने की शक्ति निर्मित होगी। इस आंतरिक शक्ति के अभाव में स्वराज्य हिंदुओं की स्वतंत्रता की सीढ़ी न बनकर गुलामी की एक सीढ़ी सिद्ध होगी।

***

धर्म-दर्शन को मैं प्राकृतिक धर्म-दर्शन के रूप में देखता हूँ, जिसमें देव और ईश्वर का अस्तित्व है, जो प्रकृति से अभिन्न रूप से आबद्ध है। धर्म-दर्शन के तीन आधार हैं—

(अ) ईश्वर की सत्ता है और वही प्रकृति का निर्माता है, जिसे हम जगत् कहते हैं।

(आ) ईश्वर जगत् का नियामक अर्थात् नियंत्रक है तथा

(इ) ईश्वर अपनी सत्ता मानवता पर नैतिकता के माध्यम से लागू करता है।

***

धर्म कोई विवरणात्मक या वैज्ञानिक विश्लेषण की विषय-वस्तु नहीं है। धर्म-दर्शन तब तक विश्लेषणात्मक दर्शन होता है जब तक कि वह अपने में तर्क को समाहित नहीं करता है। परंतु तर्क को समाहित करने के बाद वह मानक वैज्ञानिक विषय-वस्तु हो जाता है।

***

मैं नहीं समझता कि सभी धर्म-दर्शनों का एक ही दर्शन होगा, बल्कि सभी धर्मों में कुछ-न-कुछ अलगाव होता है। किसी भी आंदोलन या संस्था का एक क्रांतिकारी दर्शन होता है, जिसके आधार पर वह संस्था संचालित होती है।

***

प्राचीन समाज में मानव अपने राजनीतिक ईश्वर को सबकुछ के रूप में मान्यता देता था। आधुनिक समाज मानवीय पहलुओं को मान्यता देता है, जो ईश्वर को समाज के बाहर रख देता है, तथापि वह ईश्वर को स्वीकार करता है।

***

प्राचीन समाज में प्रत्येक समुदाय के लिए एक ईश्वर था। आधुनिक समाज की भाँति एक संयुक्त विश्वव्यापी ईश्वर नहीं था। वह मानवता को सामूहिकता में रखकर सोचता था। वह मात्र अपने समुदाय के विषय में सोचता था, जबकि वर्तमान में ऐसा नहीं है।

***

एक क्रांति का उद्‌भव जीवन में निरपेक्षता की दशा लाता है और जीवन को उस दिशा में ले जाता है, जहाँ धर्म की सत्ता सफल होती है। दूसरे में धार्मिक विचारों की 'वास्तविक सत्ता' की स्थापना होती है, जिसमें दैवी शक्ति द्वारा अच्छी मानवता के सृजन की अवधारणा होती है। यह प्रणाली बदलाव की है।

***

हालाँकि समूचे विश्व का सारा-का-सारा समाज पूर्णतया किसी एक ही वर्ग का नहीं बना है और प्रत्येक समाज में वर्ग होते हैं, फिर भी अन्य समाज के वर्ग मूलतः भिन्न होते हैं।

***

व्यक्ति के जीवन में धर्म की आवश्यकता के चार कारण हैं—

(1) समाज की स्थिरता और नियंत्रण के लिए नीति की आवश्यकता होती है।

इनमें से किसी एक के अभाव में समाज रसातल को जा सकता है। इसलिए किसी भी समाज को धर्म की आवश्यकता पड़ती है।

(2) धर्म को बनाए रखने के लिए उसे बुद्धि-प्रमाण्यवादी होना चाहिए। विज्ञान बुद्धि-प्रमाण्यवादी है।

(3) केवल नीति की संहिता का अर्थ धर्म नहीं है। धर्म की नीति संहिता में स्वतंत्रता, समता और बंधुत्व इन मूलभूत तत्त्वों को मान्यता मिलनी चाहिए।

(4) दरिद्रता को पवित्र मानने का आग्रह किसी भी धर्म को नहीं करना चाहिए अथवा दरिद्रता का उदारीकरण भी नहीं होना चाहिए।

***

प्रारंभ में चार वर्ण थे। वर्ण का आधार धर्म था और बाद में ये चार वर्ण जातियाँ बन गईं। चार वर्ण एक ही अंश के विभाजन थे। यह कार्य नहीं अपितु श्रम का विभाजन था। इस धर्म द्वारा जातिवाद का उद्भव हुआ है। इसने एक व्यक्ति को उसकी जाति तक सीमित कर दिया है। इससे मानव की विशेषताएँ गौण हो गई हैं। जहाँ जाति है वहाँ कोई पवित्रता नहीं है।

***

धर्म मानव की अंतरात्मा की मुक्ति का व्यक्तिगत प्रकरण है।

***

धर्म नीतियों के आधार पर मानव के पारस्परिक सौहार्द और सदाचार को कायम रखता है।

***

धर्म मन, आत्मा और शुद्धि की पहचान का मार्ग है।

***

धर्म की नींव सदाचार और शुद्ध आचरण पर रखी जानी चाहिए।

***

धर्म ग्रंथों से नहीं, व्यवहार से मुखरित हो।

***

धर्म भेदभाव को नहीं मानता है।

***

धर्म करुणा, प्रेम और त्याग का दूसरा नाम है। धर्म घृणा को प्रेम से जीतता है और वह बिना शासक हुए लोगों के दिलों पर राज करता है।

***

धर्म विचलित मन को शांति प्रदान करने का अमृत व्यवहार है। जो धर्म में है,

वह प्रेम में है और सबके साथ समान है।

****

धर्म असहाय, गूँगों और अपाहिजों का सहायक है। वह उनकी अमिट शक्ति है।

****

ये धर्म अपरिवर्तनशील होने का दावा करते हैं। प्रचलित सभी धर्मों की नींव श्रद्धा है। ईश्वरीय कृपा प्राप्त करना और उसकी अनुकंपा से बचना, यह धर्म मार्ग है। इसके सिवा बाकी सभी अधर्म और निषिद्ध माना गया है। परमेश्वर और श्रद्धा के सिवाय अन्य तथ्य का धर्म में कोई स्थान नहीं है। मनुष्य की बुद्धि, उसके विचार आदि को किसी भी धर्म ने मान्यता नहीं दी है। इस कारण ईश्वर के अस्तित्व के संबंध में चर्चा ही नहीं की जा सकती है।

****

सिद्धार्थ नामक व्यक्ति ने जिस धर्म की बात की है, वह प्रचलित कमजोरियों से परे है। एक व्यक्ति द्वारा अनेक व्यक्तियों से कहा गया, यह धर्म है। इस धर्म में गूढ़ता का कोई स्थान नहीं हैं। ईश्वरीय कृपा, वरदान या शाप, स्वर्ग या नरक—इसकी कोई गुंजाइश इस धर्म में नहीं है। सिद्धार्थ का यह धर्म हाड़-मांस के व्यक्ति, उसके ही जैसे हाड़-मांस के लोगों के लिए, उनकी ही भाषा में उनके ही शब्दों में कहा गया है। इस धर्म में श्रद्धा के लिए कोई स्थान नहीं है। उलटे किसी व्यक्ति या वस्तु पर अंधश्रद्धा न रखने का इसमें आग्रह किया गया है। प्रश्न करो, शंकाएँ उपस्थित करो, विचार करो और अगर विवेक को मान्य हो तो ही स्वीकार करो—ऐसा इस धर्म का आग्रह है। सिद्धार्थ कहीं भी सर्वज्ञाता होने का दावा नहीं करते। वे किसी को संदेश देने की बात नहीं करते। वास्तव में, वे किसी प्रकार का आदेश नहीं देते, कोई निर्णायक बात नहीं करते। यही इस धर्म की शक्ति है।

****

एक सच्चा धर्म समान अवसरों की घोषणा व प्रतिष्ठापना करता है, बाकी सभी धर्म झूठे व बकवास हैं।

****

मेरे बाद मेरे अनुयायियो, बौद्ध धर्म तुम्हारे द्वारा संचालित होगा। यदि तुम इसको आत्मसात् नहीं कर सकोगे तो इसका पालन कैसे करोगे? धर्म मानसिक शुद्धता को तुम्हारे आचरण में स्वीकार करता है।

****

धर्म नैतिक चरित्र का निर्माण स्वतंत्रता, समानता, बंधुत्व के आधार पर करना

चाहता है और मात्र बुद्धवाद ही मानवता के समान अवसरों की बात करता है।

~***~

आध्यात्मिक दृष्टि से धर्मांतरण की समस्या की ओर देखा जाए तो कहा जा सकता है कि वैयक्तिक विकास सच्चे धर्म का उद्देश्य है। मुझे धर्म की वह परिभाषा स्वीकार है, जिससे सारी प्रजा की धारणा होती है। उसके लिए धर्म को बंधुता, समता और स्वतंत्रता—इन सद्‌गुणों की सीख देनी चाहिए।

~***~

अस्पृश्यता के कारण तुम्हारे गुण व्यर्थ साबित हुए। अस्पृश्यता की वजह से तुम्हें सेना में, पुलिस विभाग में या नाविक दल में प्रवेश नहीं मिलता। अस्पृश्यता तुम्हारे लिए अभिशाप है। अस्पृश्यता ने तुम्हें जगत् के सही जीवन से, सम्मान से और प्रतिष्ठा से वंचित किया है। अस्पृश्यों के लिए कानूनी स्वतंत्रता की अपेक्षा सामाजिक स्वतंत्रता की आवश्यकता है।

~***~

धर्म-परिवर्तन की बात करना बच्चों का खेल नहीं है। जिस प्रकार कोई नौका खेनेवाला यात्रियों की संख्या का अनुमान लगाकर नौका खेता है और तदुपरांत अपना शेष साजो-सामान सँभालता है, उसी प्रकार हमें भी करना होगा।

~***~

जब 'दीर्घनिकाय' आदि पालि ग्रंथों को आधार बनाकर भगवान् बुद्ध की जीवनी लिखने की चेष्टा करते हैं, तब हमें कठिनाई है और उनमें उपदेशों की सुसंगत व वैज्ञानिक अभिव्यक्ति बहुत कठिन लगने लगती है। वास्तविकता तो यह है कि संसार में जितने भी धर्म-संस्थापक हुए हैं, उनमें भगवान् बुद्ध की चर्चा या विवरण हमारे सामने अनेक ऐसी समस्याएँ लाता है, जिनका समाधान असंभव नहीं तो दूभर अवश्य है।

~***~

बुद्ध का धम्म परलोक के संबंध में नहीं है। इसका संबंध मात्र इहलोक से है। यह न स्वर्गवादी है, न नरकवादी है; यह पृथ्वीवादी है। यह धम्म व्यक्तिगत मोक्ष की बात नहीं करता है, सामाजिक मुक्ति का संदेश देता है। यह धम्म निरीश्वरवादी, अनात्मवादी, भौतिकवादी और बुद्धिवादी है। यह धम्म अपरिवर्तनीय नहीं है, परिवर्तनवाद इसका अंग है। यह धर्म विचारशील है।

~***~

अन्य धर्मों की तुलना में बुद्ध धम्म आधुनिक, वैज्ञानिक निष्कर्षों पर, कसौटियों

पर श्रेष्ठ सिद्ध होता है। आधुनिक युग की चुनौतियों को यह स्वीकार करता है। बुद्धिवाद, सामाजिक सत्ता, बंधुता, स्वतंत्रता, समाजवाद, जनतंत्र आदि आधुनिक मूल्यों के आलोक में इस धम्म को सिद्ध किया जा सकता है।

***

बुद्ध की विचार-पद्धति और तर्क मीमांसा में शब्द प्रमाण्य, ग्रंथ प्रमाण्य का कहीं भी स्थान नहीं है। प्रत्यक्ष प्रमाण और अनुमान पर ही बुद्ध का तर्कशास्त्र खड़ा है। बुद्ध ने एक स्थान पर कहा कि—'मेरे शब्दों को प्रमाण न मानो। तुम्हारी बुद्धि अथवा अनुभव से जो बात जँचती हो, उसे ही सत्य मानो। इस विश्व में अंतिम और अपरिवर्तनीय कुछ भी नहीं है। परिवर्तन और सतत परिवर्तन, यही सत्य है।'

***

बुद्ध का धम्म व्यक्तिनिष्ठ न होकर सामाजिक है। बौद्ध दर्शन धर्म अथवा 'रिलीजन' शब्द के अंतर्गत नहीं आता। वह धम्म अथवा फिलॉसफी ऑफ रिलीजन है। बौद्ध धम्म ने नैतिकता को सर्वाधिक महत्त्व दिया। यह नैतिकता ही धम्म है। उसके अभाव में समाज-धारणा संभव ही नहीं है। यह नैतिकता, स्वतंत्रता, समता और बंधुता के संवर्धन के लिए ही होनी चाहिए।

***

जो धर्म हमारी चिंता करता हो, हमें अवसर प्रदान करना हो, उसके लिए हम प्राण तक देने को तैयार हैं। जो धर्म हमारी परवाह ही नहीं करता, उसकी परवाह हम क्यों करें?

***

समाज धर्म और मजिस्ट्रेट दोनों का चुनाव कर सकता है—जितने अंश में समाज धर्म का पालन करे उतने अंश में धर्म और जहाँ धर्म का पालन न करे वहाँ मजिस्ट्रेट।

***

धर्म तो मनुष्य की अपनी मनुष्यता के लिए है। धर्म के विकास के लिए उसे बलि का बकरा नहीं बनाना चाहिए।

***

सिद्धांत के प्रतिकूल कोई समझौता नहीं किया जा सकता। ऐसा किया भी नहीं जाना चाहिए।

□

आर्थिक चिंतन

माता-पिता शिशु को केवल जन्म देते हैं, भविष्य नहीं। भविष्य तो उनके बनाने से बनेगा।

~***~

भारत में समस्त दरिद्र और दलित लोग गौरव को ईश्वर का दंड मानकर चुपचाप भोग रहे हैं। वे उसे समाज और अर्थव्यवस्था को रुग्णता नहीं मान रहे हैं। वे शोषण, अन्याय, अत्याचार, बलात्कार सहन कर लेंगे; परंतु न्यायालय नहीं जाएँगे।

~***~

रुपए व पाउंड के संबंधों का लाभ उठाकर अंग्रेज भारतीयों का शोषण कर रहे हैं। मुद्रा विनिमय से भारत लाभान्वित नहीं हो सकता है। इसे स्वर्ण स्तर की देशीय स्थिरता प्राप्त न होने के कारण लागू नहीं करना चाहिए।

~***~

मेरा जन्म सर्वसाधारण जनता की जिम्मेदारी लेने के लिए हुआ है। मैं एक गरीब परिवार में पैदा हुआ और बंबई की इंप्रूवमेंट ट्रस्ट चॉल में गरीब लोगों की तरह ही बड़ा हुआ। मुझे आपकी शिकायतें मालूम हैं। खोली प्रणाली आपके खून का शोषण कर रही है। खोली की पद्धति खत्म होनी चाहिए। ऐसा होने पर आपको शांति एवं समृद्धि प्राप्त होगी। अपना ध्येय प्राप्त करने तक यह आंदोलन आपको इसी तरह चालू रखना चाहिए।

~***~

महार महाराष्ट्र की सबसे प्रमुख अछूत जाति है। निम्न तथ्य महार व स्पृश्य हिंदुओं के अधिकारों को परिभाषित करते हैं—

(1) महार प्रत्येक गाँव में पाए जाते हैं।

(2) महाराष्ट्र के सभी गाँव दीवारों से घिरे हैं। बारी-बारी से उनकी पहरेदारी महारों को करनी पड़ती है।

(3) महारों के कुछ अधिकार भी थे, जिनमें वतन भूमि का अधिकार भी था।

महारों के पास वतन भूमि का होना एक पुराना रिवाज था, जिसकी पुष्टि हिंदू-मुसलिम अभिलेखों द्वारा की गई थी; परंतु अंग्रेज सरकार ने इस भूमि पर बहुत अधिक कर-भार लाद दिया था, जिससे महारों पर मुसीबतें काफी ज्यादा बढ़ गईं।

***

क्रूर वतन व्यवस्था के कारण ही महार जाति गुलामी की ओर अग्रसर हुई।

***

वतन प्रणाली का उन्मूलन किया जाए, अर्थात् प्रत्येक गाँव में हिंदुओं की गुलामी कर रहे महारों को अन्य नागरिकों की भाँति आजाद किया जाए। महारों द्वारा की जानेवाली पुश्तैनी सेवा को सशर्त किया जाए। भूमि को स्थायी भूमि मानकर महारों को मालिकाना हक दिया जाए।

***

यद्यपि वतनदारों का कार्य, सेवा, समय और मेहनताना अनिश्चित था, तथापि वतनदारों को सरकारी नौकर माना जाए। वतनदारों के पूरे परिवार न केवल सरकार की अपितु ग्रामीण अधिकारियों तथा पाटील और कुलकर्णी की भी, किसी भी समय सेवा के लिए मजबूर थे। चूँकि वतनदारों को अपने से ऊपर की श्रेणी को संतुष्ट करना होता है, जो भूमि हथियाने के लिए आजाद थे, अतः उनके वतन की कोई सुरक्षा नहीं थी।

***

इतिहास हमेशा एक उद्धरण नहीं होता है, बल्कि वह प्रायः चेतावनी भी होता है।

***

अव्वल तो श्रमिक संहिता भार नहीं है, दूसरे इनके होने से मालिक एवं श्रमिक दोनों के न्यायोचित संबंधों का विकास संभव है। यही तो वास्तविक अवस्था है, जब श्रम कानून बनकर लागू हो जाना चाहिए।

***

सबको अवसर मिले, यह तो कदापि संभव नहीं है कि सब समान होने का एक सार्वजनिक फॉर्मूला तैयार कर लें। आवश्यकताओं तथा शक्ति के अनुसार सदैव

परिवर्तन आते रहेंगे।

***

लोकतांत्रिक समाज यंत्रीकरण के प्रतिकूल नहीं हो सकता। आधुनिक सभ्यता का नारा यंत्र है। यंत्रों के कारण बुराइयाँ नहीं पैदा हुई हैं और न आधुनिक सभ्यता की वजह से। वे गलत सामाजिक संरचना का प्रतिफल हैं।

***

हमारे दल का यह ठोस मत है कि खेती की भूमि के टुकड़े करना और उस पर बढ़ती हुई जनसंख्या का बोझ डालना—ये सही अर्थों में किसानों की दरिद्रता के कारण हैं। पुराने धंधे फिर से चलाना और नए शुरू करना, यही सुधारात्मक उपाय है। जनता की उत्पादन शक्ति और उसकी कार्य-क्षमता में वृद्धि हो, इसलिए तकनीकी शिक्षा विषयक धंधे और जहाँ जरूरत हो वहाँ शासन के स्वामित्व व नियंत्रण में चलनेवाले धंधे शुरू करने का बड़ा विस्तृत कार्यक्रम हम हाथ में लेने वाले हैं।

***

मालिकों द्वारा चलाए गए शोषण से काश्तकारों का बचाव हो। मालिक उन्हें भगा न दें, इसलिए काश्तकारों को सुरक्षा का आश्वासन दिया गया था। श्रमिकों के हित में कानून बनाए जाएँगे। नौकरियाँ देने, पदच्युत करने, कारखानों में बोनस देने, काम के घंटे तय करने, छुट्टी देने, आवास व्यवस्था करने आदि हेतु कानून बनाए जाएँ। बेकारी से मुक्ति के लिए भूमिहीनों को भूमि दी जाएगी।

***

शक्ति द्वारा औद्योगिक शांति संभव नहीं है। कानून के माध्यम से औद्योगिक शांति का प्रयास किया जा सकता है, परंतु गारंटी नहीं दी जा सकती है। यदि सामाजिक न्याय को औद्योगिक शांति का आधार बनाया जाता है तो उसके परिणाम सर्वथा अनुकूल हो सकते हैं, क्योंकि सामाजिक न्याय स्वामी और श्रमिक दोनों का हित-चिंतन करता है। यह लड़ाई तो इतनी सी है कि स्वामी चाहता है कि वह श्रमिक को कम-से-कम देकर अधिक-से-अधिक लाभ कमाए और श्रमिक की मंशा है कि उसे जीवनयापन के लिए यथोचित धन मिल जाए। मालिक का दृष्टिकोण पूँजीवादी होता है और श्रमिक का सदा समाजवादी। जब पूँजीवाद श्रमिकवाद या समाजवाद को निगलने का प्रयत्न करता है तो स्वामी और श्रमिक आमने-सामने आ जाते हैं। श्रम के शोषण के परिणामस्वरूप श्रमिक को आर्थिक विषमताओं से ही नहीं, मानसिक परेशानियों से भी जूझना पड़ता है, जिससे अशांति पैदा होती है, जो हड़तालों के रूप में सामने आती है। आखिर पूँजीपति के सोचने का आधार यह क्यों नहीं बनता है कि वह और श्रमिक

दोनों सामाजिक प्राणी हैं और उत्पादन दोनों के संयुक्त प्रयत्नों का परिणाम है। श्रम भी तो पूँजी है। वह श्रम पूँजी को नकारना क्यों चाहता है?

***

किशोरावस्था में मैंने अपने रिश्तेदारों को खाने के डिब्बे पहुँचाने का काम किया था, अत: मजदूर समस्याओं की कुछ जानकारी मुझे है। अब हमने आर्थिक कठिनाइयों को दूर करने की शुरुआत की है। आज तक हम दलित अस्पृश्य के रूप में इकट्ठा होते थे, आज हम मजदूर के रूप में एकत्रित हैं।

***

हमारे दलित वर्ग को जो अवसर मिलता है वह काफी सीमित है। ऐसे अनेक अवसर होते हैं, जहाँ अस्पृश्यता के कारण अवसर नहीं मिलता है। कपड़ा मिलों के कुछ विभागों में अस्पृश्य नौकर नहीं लिये जाते। रेलवे में वे गैंगमैन के रूप में सड़ते हैं। कुली भी नहीं बन पाते हैं। जब तक तात्त्विक दृष्टि से खुला अन्याय और पक्षपात है तो इनके पराभव के बिना मजदूर एकता कैसे संभव है?

***

हड़ताल करना एक दीवानी अपराध है, फौजदारी गुनाह नहीं। मनुष्य की इच्छा के विरुद्ध उससे काम कराना गुलामी है।

***

कांग्रेस ने ब्रिटिश साम्राज्यवाद के लिए पुकारा होता तो मैं उस दल से सहयोग अवश्य करता, लेकिन यथार्थ ऐसा नहीं है। कांग्रेस दल अपने हाथ में आए संविधानात्मक शासन-तंत्र को पूँजीवादी और निरंतर स्वार्थ में लिप्त लोगों के कल्याण के लिए ही कार्यान्वित कर रहा है। उसने किसान और मजदूर के कल्याण की बलि दे दी है।

***

यदि मुद्रास्फीति तथा अंदरूनी मूल्य-असंतुलन दूर करना है तो टकसाल बंद कर दी जाए। मुद्रा में लचीलेपन के लिए सोने के मूल्य का मानदंड बनाना चाहिए।

***

भूमि ही आज देश की आय का साधन है तो जो परजीवी की भाँति बहुत सारी भूमि अपने पास रखे हैं और उससे ढेर सारी आय प्राप्त कर रहे हैं, उन्हें अवश्य आय कर देना चाहिए।

□

राजनीतिक चिंतन

किसी भी जाति या समुदाय के अभिजन अपनी जाति एवं समुदाय की उन्नति में महत्त्वपूर्ण भूमिका अदा करते हैं; क्योंकि उनकी शक्तियाँ इतनी व्यापक होती हैं कि वे अपने वर्ग के लोगों की सहायता कर सकते हैं। उनसे यह भी अपेक्षा की जाती है कि वे अपने समुदाय की प्रगति में सहायक बनें। दलितों के अभिजन भी इसके अपवाद नहीं हो सकते हैं। चूँकि अभिजन अपने-अपने क्षेत्रों में मान्यता प्राप्त एवं प्रतिष्ठित सदस्य होते हैं, अतः दलित उनसे यह अपेक्षा करते हैं कि अभिजन उनके विकास में सहायक सिद्ध होंगे।

~***~

स्व-सरकार पर्याप्त नहीं है, बल्कि एक अच्छी सरकार चाहिए, जो समाज के सभी वर्गों के विकास के लिए प्रतिबद्ध हो।

~***~

मात्र स्व-सरकार पूर्ण नहीं मानी जा सकती, जबकि कांग्रेस की माँग स्व-सरकार की है। स्वतंत्रता तब पूर्ण मानी जाएगी जब समाज का प्रत्येक नागरिक इस स्वतंत्रता का अनुभव करे और सरकारी तथा गैर-सरकारी प्रतिष्ठानों में बराबर भागीदारी निभाए।

~***~

स्वराज्य जैसा ब्राह्मणों का जन्मसिद्ध अधिकार है वैसा ही महारों का भी है, यह बात कोई भी स्वीकार करेगा। इसलिए उच्च वर्ग के लोगों का यह प्रथम कर्तव्य है कि वे दलितों को शिक्षा देकर उनका मनोबल और सामाजिक स्तर ऊँचा करने की कोशिश करें। जब तक यह नहीं होगा तब तक भारत की स्वतंत्रता का दिन बहुत दूर रहेगा, इसमें संदेह नहीं।

~***~

भारत के स्वतंत्र होने से ही सबकुछ साध्य होगा, ऐसी बात नहीं। भारत एक ऐसा राष्ट्र बनना चाहिए, जिसमें प्रत्येक नागरिक के धार्मिक, सामाजिक, आर्थिक व राजनीतिक अधिकार समान हों और हर एक को व्यक्तित्व के विकास के लिए उचित अवसर प्राप्त हो।

****

बहुमत का राजनीतिक आदर्शवाद समाज का आदर्शवाद हो जाएगा।

****

जनतंत्र सार्वजनिक जीवन जीने की पद्धति है। जनतंत्र एक ऐसी प्रणाली है, जिसके द्वारा आर्थिक और सामाजिक क्षेत्र में बिना रक्त की एक बूँद बहाए क्रांतिकारी परिवर्तन लाए जा सकते हैं।

****

एक ही व्यक्ति अथवा एक ही राजनीतिक दल के हाथों यदि सत्ता सुरक्षित रह जाए तो ऐसी स्थिति में उस राष्ट्र में संसदीय प्रणाली और जनतंत्र की हत्या हो जाती है, अराजकता का राज्य पैदा हो जाता है। मजबूत विरोधी दल सत्ताधारियों पर नियंत्रण रखेगा तथा उसे व्यापक लोक-हित की ओर ले जाने का प्रयास करेगा।

****

भारतीय राजनीति में अभिजन की महत्त्वपूर्ण पकड़ है और यह लोक-कल्याणकारी नहीं रह गया है। इससे संपूर्ण समाज, किसान, मजदूर का भला नहीं होगा।

****

दुनिया में दो ही वर्ग हैं—गरीब और अमीर, शोषित और शोषक। एक और मध्य वर्ग है, वह वास्तव में बहुत छोटा है। किसान और मजदूर शोषित हैं और इसी कारण उनका संगठित होना आवश्यक है।

****

सभी भारतीयों को समता का व्यवहार मिलना चाहिए।

****

जीवन के सभी क्षेत्रों में ऐसी व्यवस्था हो, जिससे प्रत्येक व्यक्ति अपनी बुद्धि और प्रतिभा के बल पर विकास कर सके।

****

प्रत्येक भारतीय को आर्थिक, धार्मिक और राजनीतिक स्वतंत्रता मिले।

****

प्रत्येक भारतीय को समान अवसर मिलें।

~***~

प्रत्येक भारतीय अपनी जरूरतों से और भय से मुक्त हो जाए। उसे इन दोनों से मुक्त रखने की जिम्मेदारी राज्य की है।

~***~

एक व्यक्ति दूसरे व्यक्ति का, एक वर्ण दूसरे वर्ण का, एक राष्ट्र दूसरे राष्ट्र का शोषण न करे।

~***~

मुझे अच्छा नहीं लगता, जब कुछ लोग कहते हैं कि हम पहले भारतीय हैं और बाद में हिंदू और मुसलमान। मुझे यह स्वीकार्य नहीं है। धर्म, संस्कृति, भाषा आदि की प्रतिस्पर्धी निष्ठा के रहते हुए भारतीयता के प्रति निष्ठा नहीं पनप सकती। मैं चाहता हूँ कि लोग पहले ही भारतीय हैं और अंत तक भारतीय रहें, भारतीय के अलावा कुछ नहीं।

~***~

और मैं इस सदन में पूरे जोर से कहता हूँ कि जब कभी देश के हित और अस्पृश्यों के हित के बीच टकराव होगा तो मैं अस्पृश्यों के हित को तरजीह दूँगा। यदि कोई आततायी बहुमत देश के नाम पर बोलता है तो मैं उसका समर्थन नहीं करूँगा। मैं किसी पार्टी का समर्थन सिर्फ इसलिए नहीं करूँगा कि वह पार्टी देश के नाम पर बोल रही है। जो यहाँ हैं और जो यहाँ नहीं हैं, सब मेरी इस भूमिका को समझ लें। मेरे अपने हित और देश के हित के साथ टकराव होगा तो मैं देश को तरजीह दूँगा; लेकिन अगर देश के हित और दलित वर्ग के हित में टकराव होगा तो मैं दलित वर्ग के हितों को प्राथमिकता दूँगा।

~***~

दलितों की स्थिति में परिवर्तन समतामूलक समाज की स्थापना के लिए जरूरी है। जाति प्रथा न केवल मानव-गरिमा के विरुद्ध है, बल्कि समतामूलक समाज के स्वप्न के भी विरुद्ध है।

~***~

जाति-व्यवस्था मनु से पहले ही अस्तित्व में थी। इसलिए यह कहना भूल है कि जाति श्रेणियों का निर्माण शास्त्रों और स्मृतियों ने किया।

~***~

सांस्कृतिक एकता में कोई भी देश भारत का मुकाबला नहीं कर सकता।

इसमें न केवल भौगोलिक एकता है, अपितु उससे ज्यादा गहरी और मूलभूत एकता है सांस्कृतिक एकता, जो एक छोर से दूसरे छोर तक सारे देश में व्याप्त है। लेकिन इस संस्कृतिजन्य एकरूपता के कारण ही जाति की गुत्थी को सुलझाना अत्यंत कठिन हो जाता है।

***

जाति-व्यवस्था को तोड़ने के लिए सामाजिक क्रांति की आवश्यकता है। सामाजिक सुधार पर्याप्त नहीं हैं।

***

यूरोपीय समाज की वर्तमान व्यवस्था में संपत्ति ही शक्ति का प्रमुख स्रोत है। परंतु यह विचार धोखा है। भारत के संदर्भ में धार्मिक-दार्शनिक विचारधाराओं की बड़ी भूमिका रही है और यह भूमिका भारतीय मार्क्सवादी समझने में असमर्थ रहे हैं।

***

यह गौरतलब है कि मनु का धर्म अब किस स्वरूप में शेष रह गया है। यहाँ स्वीकार करना होगा कि कानून के रूप में, उन नियमों के रूप में, जो किसी न्यायालय के लिए विवादों के निर्णय के लिए बाध्यकारी होते हैं, मनु के धर्म की कोई प्रवर्तन-शक्ति नहीं रही है। इसमें अपवाद विवाह, उत्तराधिकार आदि जैसे मामले हैं, जिनका प्रभाव केवल व्यक्ति पर पड़ता है। सामाजिक आचरण तथा नागरिक अधिकारों को नियमित करनेवाले कानून के रूप में उसे लागू नहीं किया जा सकता है। लेकिन भले ही कानून के रूप में उसकी मान्यता नहीं रही है, प्रथा के रूप में वह अब भी विद्यमान है। कानून के मुकाबले प्रथा भी कोई नगण्य चीज नहीं है। यह सच है कि कानून को राज्य अपने पुलिस बल के माध्यम से लागू करता है और प्रथा यदि वैध नहीं है तो उसे राज्य लागू नहीं कर सकता है। लेकिन व्यवहार में इस अंतर का कोई महत्त्व नहीं है। राज्य जितनी शक्ति से कानून लागू करता है, उतनी ही शक्ति से प्रथा को जन-समूह लागू करता है। इसका कारण है कि संगठित जन-समूह की बाध्यकारी शक्ति राज्य की बाध्यकारी शक्ति से कहीं अधिक प्रबल होती है। मनु का धर्म तकनीकी दृष्टि से कानून नहीं रहा है, फिर भी इस कारण उसकी परिवर्तनीयता पर कोई आँच नहीं आई है।

***

भ्रष्टाचार संसदीय लोकतांत्रिक व्यवस्था के लिए हानिकारक है। भ्रष्टाचार बढ़ेगा तो लोकतंत्र नष्ट हो जाएगा।

मेरा राजनीतिक लक्ष्य है—

(1) पद-दलितों को राजनीतिक अधिकार दिला पाना।

(2) राजनीतिक, सामाजिक तथा आर्थिक संतुलन बनाते हुए इन क्षेत्रों में लोकानुभूति को लोकतंत्र का आधार देना।

(3) जाति व धर्म के भेद के बिना मजदूर मोर्चे का संगठन करना।

(4) राजनीतिक शक्ति से जन-शक्ति का आभास देना।

(5) शिक्षित, संगठित और संषर्घशीलता का लक्ष्य प्राप्त करना।

जनतांत्रिक व्यवस्था एक परिवर्तनशील व्यवस्था है।

किसी भी देश की समाज-रचना में निम्नलिखित बातें हों, तभी जनतंत्र की सफलता संभव होती है—

(1) उस समाज-व्यवस्था में विषमता अधिक मात्रा में न हो।

(2) वहाँ विरोधी दल का अस्तित्व हो।

(3) कानून और प्रशासन में समानता हो। वे एक-दूसरे के लिए पूरक हों, विरोधी नहीं।

(4) संवैधानिक नैतिकता के पालन की वृत्ति जनता तथा शासन में हो। असंवैधानिक आचरण न हो।

(5) बहुसंख्यकों की दादागीरी न चलती हो।

(6) उस समाज में नैतिक मूल्यों को सर्वोपरि महत्त्व दिया जाता हो। नैतिक मूल्य आचरण से निर्धारित हुए हों।

(7) जाग्रत् जनमत हो, देश के प्रशासन के प्रति संवेदनशील हो। देश में घटी घटनाओं, पारित कानूनों के प्रति उनमें अत्यधिक जागरूकता हो।

***

वर्गीय समाज-रचना जनतंत्र के लिए हमेशा खतरनाक साबित हुई है; क्योंकि वर्गीय समाज में एक ओर जुल्म, झूठ, अहंकार, लोभ और स्वार्थ फैला हुआ होता है तो दूसरी ओर स्वतंत्रता, अस्मिता, स्वाभिमान और प्रतिष्ठा का पूर्ण अभाव होता है। असुरक्षा, दरिद्रता तथा भय भी ऐसे समाज में होता है।

***

कानून व्यक्ति द्वारा व्यक्ति के लिए बनाए जाते हैं। कानून में सतत संशोधन की आवश्यकता होती है, परंतु यह संशोधन भी सबकी सहमति से होना चाहिए। सभी के लिए सभी स्थानों पर एक कानून हो। सभी में सभी का हित सुरक्षित होता है। कानून सामाजिक हो, मानवीय हो अर्थात् उसका परिणाम सार्वत्रिक हो। शिक्षा

द्वारा कानून का महत्त्व सिद्ध हो, उसका प्रचार-प्रसार हो, परंतु कानून की ओर जनता भय की दृष्टि से न देखे। कानून पाँच मानवीय मूल्यों पर आधारित होना चाहिए—

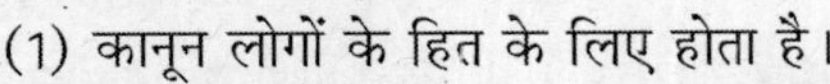

(1) कानून लोगों के हित के लिए होता है।

(2) कानून स्वच्छंद अथवा भय के स्रोत से मुक्त होता है।

(3) कानून धर्मनिरपेक्ष होता है।

(4) वह ईश्वर-प्रेरणा के बजाय मानव-प्रेरणा से बनता है।

(5) आवश्यकतानुसार वह परिवर्तनीय है।

राजनीति में समता और आर्थिक क्षेत्र में विषमता—इस दृश्य को बदलना जरूरी है। दोनों के मध्य की खाई को जल्दी-से-जल्दी समाप्त करना होगा। निकट भविष्य में यदि ऐसा नहीं हो सकता तो विषमता में जो छटपटा रहे हैं, ऐसे वर्ग जनतंत्र के मुखौटे का पर्दाफाश कर देंगे।

****

यद्यपि सरकार ने यह महसूस किया कि जमींदार लोग असहाय, गरीब एवं दलितों का खून चूस रहे हैं, फिर भी सरकार ने उन बुराइयों का अंत नहीं किया, जिनसे दलितों का जीवन सदियों से मुरझाया पड़ा है। सरकार के पास इन बुराइयों को समाप्त करने की कानूनी शक्ति है, पर उसने सामाजिक जीवन की वर्तमान संहिता को नहीं बदला।

****

ब्रिटिश सरकार से पूर्व हम अस्पृश्य थे। क्या ब्रिटिश सरकार ने उसकी समाप्ति के लिए कुछ किया है? ब्रिटिश सरकार से पूर्व हम कुओं से पानी नहीं भर सकते थे। क्या ब्रिटिश सरकार ने हमें यह अधिकार दिलाया है? ब्रिटिश सरकार से पूर्व हमें मंदिर, पुलिस एवं सेना में प्रवेश का अधिकार नहीं था। क्या इस सरकार ने हमें वह अधिकार दिलाया है? भले ही ब्रिटिश शासन के भारत में 150 वर्ष बीत चुके हैं, परंतु अस्पृश्यों की दशा ज्यों-की-त्यों है।

****

भारत में बौद्ध इतिहास का अध्ययन करनेवाले जानते हैं कि जिन लोगों ने बौद्ध धर्म का प्रचार आरंभ किया, वे नाग लोग थे। नाग लोग अनार्य थे। आर्यों और नागों के मध्य घोर शत्रुता थी। आर्य-अनार्यों के मध्य अनेक युद्ध लड़े गए। आर्य लोग अनार्यों को समूल नष्ट कर देना चाहते थे। इससे संबंधित पुराणों में अनेक कथाएँ मिलेंगी। आर्यों ने नागों को भस्म कर दिया। अगस्त्य मुनि ने एक नाग को

बचा लिया था और हम उसी नाग के वंशज माने जाते हैं।

****

हमारा लक्ष्य मात्र जल की प्राप्ति और मंदिर में प्रवेश नहीं है, बल्कि इससे आगे समतावाद तक है। हमारा लक्ष्य वर्ण-व्यवस्था को ध्वस्त करना है, जिसके कारण समाज में असमतावाद का जन्म हुआ है।

****

हिंदू धर्म का आधार वर्ण-व्यवस्था है, जो हमें अछूत बनाती है। मैं हिंदू व्यवस्था की वर्ण प्रथा और अस्पृश्यता को स्वीकार नहीं करता हूँ। मैं 'भगवद्‌गीता' के अलावा किसी अन्य सत्ता को स्वीकार नहीं करता। यद्यपि मैं वेदों की सत्ता को स्वीकार नहीं करता, फिर भी मैं एक सनातन हिंदू हूँ।

****

किसी भी कार्य की सफलता या विफलता जितनी साधन पर निर्भर करती है उतनी ही उसके नैतिक स्वरूप पर। यदि कार्य के मूल में सत्य निहित है तो परिणाम की चिंता की कोई आवश्यकता नहीं, क्योंकि सत्याग्रहियों में आत्मबल की आवश्यकता होती है। जिस कार्य में लोक-संग्रह होता है, वही सत्कार्य है। यही हमारी विचारधारा है। इसे हमने 'गीता' से लिया है। सत्याग्रह ही 'गीता' का मुख्य प्रतिपाद्य है। 'गीता' स्पृश्य-अस्पृश्य दोनों को स्वीकार्य है। हमारे आंदोलन का आधार लोक-संग्रह है।

****

किसी एक व्यक्ति का सत्याग्रह है या असत्याग्रह, यह उस आग्रह की सफलता के लिए उपायोजित साधनों पर निर्भर नहीं करता, बल्कि पूर्णतया उसके नैतिक स्वरूप पर निर्भर करता है। हिंसा-अहिंसा केवल उस आग्रह की सफलता के साधन हैं। यद्यपि कर्म या कर्ता के अनुरोध पर क्रिया का रूप परिवर्तित होता है, परंतु कुछ साधनों के अनुरोध से आग्रह का नैतिक स्वरूप परिवर्तित नहीं होता; क्योंकि किसी एक दुराग्रही ने अपना आग्रह सिद्ध करने के लिए अहिंसा का मार्ग स्वीकार किया तो उसके दुराग्रह को सत्याग्रह नहीं कहा जा सकता या किसी एक सत्याग्रही ने सत्याग्रह की सिद्धि के लिए हिंसा की तो उसके सत्याग्रह को दुराग्रह नहीं कहा जा सकता। यदि ऐसा कहा जाता तो 'गीता' में श्रीकृष्ण ने अर्जुन को सत्याग्रह सिद्धि के लिए जो हिंसा का मार्ग स्वीकारने के लिए विवश किया, उसे क्या कहा जाए।

****

सच कहा जाए तो अस्पृश्यता इतनी भयंकर बात है कि उसके निवारण के लिए कुछ लोगों के प्राण न्योछावर हों तो भी कोई हर्ज नहीं। जीना ही जगत् का पुरुषार्थ नहीं। जीने के अनेक तरीके हैं। गंदगी खाकर कौए भी बहुत साल जीते हैं, लेकिन यह कोई भी नहीं कहेगा कि उनके जीवन में पुरुषार्थ है। मृत्यु के बारे में रोना या घबराना किस लिए? नश्वर देह न्योछावर कर इससे भी शाश्वत कोई चीज प्राप्त करने के लिए, उदाहरणार्थ देश के लिए, सत्य के लिए, ध्येय के लिए, यश के लिए—अनेक महापुरुषों ने अनेक घटनाओं में कर्तव्य रूपी अग्नि में अपने प्राणों की आहुति दी है। महाभारत में वीर माता विदुला ने अपने बेटे को यह उपदेश दिया था कि सड़ते रहने या 100 व्यर्थ की जिंदगियाँ जीने की अपेक्षा कुछ क्षण वीरता की ज्योति दिखाकर बुझ जाएगा तो भी अच्छा है। इस तरह का उपदेश हर माता अपने पुत्र को दे, इस तरह का समय नजदीक आ गया है।

***

मैं यह मानता हूँ कि प्रत्येक माता को प्रसूति पूर्व और प्रसूति पश्चात् की अवधि में विशिष्ट समय तक विश्राम मिलना राष्ट्र-हित की दृष्टि से लाभदायक है।

***

भाषावार प्रांत रचना का तत्त्व इतना विशाल है कि उसे प्रत्यक्ष कार्यान्वित करना मुश्किल है। इस तत्त्व का छोर तक युक्तिवाद स्वीकारने से इतने नए प्रांत बनाने होंगे कि उनकी संख्या ही उस बात की अव्यावहारिकता सिद्ध करेगी।

***

एक भाषा, एक जीवन-पद्धति और एक धर्म राष्ट्रीयता के आधार हैं।

***

भावनात्मक सुसंवाद और राष्ट्रीय एकता हिंदी के कारण अधिक तेजी से हो सकती है।

***

भाषिक प्रांत रचना करनी है तो एक भाषा और अनेक राज्य—इस सूत्र को स्वीकार किया जाए, एक राज्य एक भाषा का नहीं। एक राज्य एक भाषा के कारण उत्तर भारत के एकत्रीकरण की और दक्षिण भारत के विभक्तीकरण की प्रक्रिया आरंभ हो जाएगी।

***

मुझे हिंदी के प्रति अधिक प्रेम है, परंतु हिंदी भाषी लोग ही हिंदी के सबसे बड़े शत्रु हैं और यही मेरे लिए चिंता की बात है।

***

ग्रीस, यूगोस्लाविया, रूमानिया, बुल्गारिया आदि में भी मुसलमान अल्पसंख्यक हैं; परंतु वहाँ तो पृथक् निर्वाचन क्षेत्र नहीं हैं। यूरोप में विभिन्न धर्म व पंथ के लोग संयुक्त मतदाता संघ का विरोध न करते हुए एक ही शासन के नीचे पड़ोसी के रूप में सुख-शांति से रहते हैं।

***

कुछ वर्गों को अलग प्रतिनिधित्व प्राप्त हो, इस मत का समर्थक होने पर भी मेरा मानना है कि इस प्रतिनिधित्व के लिए स्वतंत्र निर्वाचक मंडल हो, इस मत के मैं पूरी तरह खिलाफ हूँ। प्रादेशिक मतदाता संघ और अलग मतदाता संघ दो विपरीत ध्रुव हैं। जनतंत्र के अभाववाले अपने इस देश में जनतंत्र प्रधान राज्य-पद्धति का बीज बोने के लिए जिन मतदान योजनाओं का अवलंबन किया जाएगा, इन सारी योजनाओं से इन दो बातों को पूरी तरह टालना चाहिए। इन दोनों का समाधान है—आरक्षित सीटोंवाली संयुक्त निर्वाचक मंडल पद्धति।

***

यूरोप के राष्ट्रों में अलग वंशों के, अलग भाषाओं के, अलग धर्म-पंथों कें लोग भी स्वतंत्र राष्ट्र के रूप में अपना अस्तित्व रख सकते हैं। हिंदुस्तान के स्व-शासित राष्ट्र के रूप में अस्तित्व में रहने में कोई हर्ज नहीं। हिंदुस्तान की परिस्थिति की विभिन्नता को परिलक्षित कर संविधान बनाया जाना चाहिए। जिस तरह किसी भी राष्ट्र का दूसरे राष्ट्र पर शासन करने का अधिकार नहीं है, उसी तरह किसी विशिष्ट वर्ग को अन्य वर्ग पर अधिसत्ता जताने का अधिकार भी नहीं है। प्रत्येक व्यक्ति जीवन एक बार ही व्यतीत करता है। इसलिए इसी जीवन में अधिक-से-अधिक उन्नति का अवसर देने का मानवीय मूल्य स्वीकार करना जरूरी है। इस तरह की श्रद्धा रखना ही आधुनिक मानवीय जनतंत्र राज्य-पद्धति का मूलभूत सिद्धांत है।

***

स्वराज्य क्यों चाहिए, इसका अन्य कोई भी कारण आपको स्वीकार नहीं है तो भी दरिद्रता का कारण आपको स्वीकार करना चाहिए। हिंदुस्तान से तुलना करने लायक दरिद्रता विश्व के किसी भी कोने में आपको नहीं मिलेगी। ब्रिटिश शासनकाल के दौरान भारत में पिछली शताब्दी में कुल 31 अकाल पड़े, जिसमें लगभग ढाई से

तीन करोड़ लोग भूख से मारे गए। यह सच है कि अंग्रेजों ने हिंदुस्तान को परिष्कृत न्याय-पद्धति और सुव्यवस्था दी है, तथापि मनुष्य जाति केवल न्याय व सुव्यवस्था पर जीवित नहीं रहती। वह अन्न पर जिंदा रहती है। लोगों का कल्याण राजनीतिक अधिकार पाने के बाद ही होगा।

***

महात्मा कोई अमर व्यक्ति नहीं है, न ही कांग्रेस हमेशा रहने वाली है। भारत में ऐसे अनेक महात्मा हुए हैं, जिनका लक्ष्य मात्र छुआछूत मिटाना था, उनका सुधार करना था और उन्हें समाज में मिला देना था। लेकिन वे सभी अपने लक्ष्य में सफल नहीं हुए। अनेक महात्मा आए और चले गए, लेकिन अछूत जहाँ थे वहीं रह गए।

***

ग्राम पंचायतों के कारण ही जाति समूहीकरण बढ़ा है। एक आरक्षित देश में विकेंद्रित लोकतंत्र का अर्थ अधिनायकवाद है।

□

भारत में जाति-प्रथा

'जाति' जैसी प्राचीन सामाजिक व्यवस्था की पेचीदी उलझनों का पर्दाफाश करने का सामर्थ्य मुझमें है। मैं निराशावादी नहीं हूँ, जो 'जानने योग्य नहीं है' कहकर विषय की उपेक्षा कर बैठूँ। यह सब जाना जा सकता है। 'जाति की समस्या' सैद्धांतिक और व्यावहारिक रूप में बहुत बड़ी है। यह स्थानीय समस्या होने पर भी शरारतपूर्ण व्यवहार में अति समर्थ समस्या है।

~***~

जब तक भारत में जातिवाद है, हिंदू बाहरी लोगों से अंतर्जातीय विवाह और अन्य सामाजिक संबंध स्वीकार नहीं करेंगे। हिंदुओं के देशांतर-गमन से भारतीय जातिवाद विश्व-समस्या बन जाएगा। सैद्धांतिक रूप में तो इस व्यवस्था ने जातिवाद के मूल स्रोत को जानने में संलग्न दृढ़ निश्चयी विद्वानों का निरादर ही किया है।

~***~

प्रख्यात नृवंश-शास्त्रियों के अनुसार भारत की जनसंख्या आर्य, द्रविड़, मंगोल और स्काइथियन अंशों का मिश्रण है। सदियों पहले अपने विकास की आरंभिक दशा में ये जातियाँ भिन्न-भिन्न संस्कृतियाँ लेकर भिन्न-भिन्न दिशाओं से भारत में आईं। इन्होंने अपने पूर्ववर्ती लोगों से युद्ध कर बलपूर्वक इस देश में प्रवेश किया। इनमें से कुछ शांतिप्रिय पड़ोसियों की भाँति यहाँ बस गए। निरंतर संपर्क और आपसी संबंधों से इन्होंने अपनी मूल संस्कृतियों के ऊपर एक सामान्य संस्कृति को जन्म दिया। फिर भी इन समुदायों का भारतीयों में समग्र एकीकरण नहीं हो पाया और इसका स्पष्ट साक्ष्य एक यात्री अपनी भारत-यात्रा में पूर्व-पश्चिम और उत्तर-दक्षिण के लोगों की शरीर रचना और त्वचा के रंग से जान-पहचान सकता है। कुछ विद्वानों की परिकल्पना के आधार पर मिश्रण को समान जातीयता की कसौटी

नहीं माना जा सकता; क्योंकि नृवंश-शास्त्र के अनुसार, सभी व्यक्ति भिन्न जाति के हैं। यह तो सांस्कृतिक एकता है कि विश्व का कोई भी देश भारत प्रायद्वीप की इस सांस्कृतिक एकता को चुनौती नहीं दे सकता। यह मात्र भौगोलिक एकता नहीं है, एक गहन मौलिक एकता है—असंदिग्ध सांस्कृतिक एकता है, जिसने देश को एक छोर से दूसरे छोर तक बाँध रखा है। इस एकरूपता—एकता के कारण ही जाति की समस्या इतनी जीवित बन गई है कि इसकी विवेचना दुष्कर कार्य है। हिंदू समाज को अलग-अलग इकाइयों का समूह मान लेने से मामला बहुत आसान हो जाता है। किंतु 'जाति' तो स्वयं मिश्रणों का पुलिंदा है। 'जाति' की उत्पत्ति की विवेचना इस 'पुलिंदे' की विवेचना की ही प्रक्रिया है।

***

गवेषणा के क्षेत्र में प्रवेश से पूर्व 'जाति' के गुण-धर्म को समझ लेना उचित रहेगा। इस संदर्भ में कतिपय विद्वानों द्वारा की गई 'जाति' की परिभाषाओं की ओर ध्यान आकर्षित करना मैं प्रासंगिक मानता हूँ—

(1) एक फ्रांसीसी पेत्ता सीनारे महाशय ने जाति की परिभाषा की है—जाति एक ऐसा संकुचित निगम है, जो पैतृक सिद्धांतों, निर्धारित परंपराओं से युक्त एक प्रमुख और परामर्शदात्री समिति सहित स्वतंत्र संगठन है और जो कम या अधिक अधिकार-संपन्न सभाओं में एकत्रित होता है, निश्चित उत्सव-अवसरों पर मिलता है, समान व्यवसायों से बँधा है, जिसका संबंध विशेषतः विवाह, खान-पान और उत्सवी शुद्धि के प्रश्न से है, जो अपने अधिकार-क्षेत्र (जो बदलता रहता है) से अपने सदस्यों को शासित करता है तथा निश्चित दंड-विधान और विशेषकर समुदाय से बहिष्कृत करने की स्वीकृति से जाति को अपनी अधिकार-सत्ता के महत्त्व को अनुभव कराने में समर्थ है।

(2) नेस्फील्ड महोदय ने जाति की परिभाषा इस प्रकार की है—एक ऐसा जाति वर्ग, जो अपने वर्ग के अलावा किसी अन्य वर्ग से न तो वैवाहिक संबंध रखता है और न खान-पान का।

(3) सर एच. रिसले के अनुसार, जाति ऐसे परिवारों या परिवार दलों का समूह है, जिसका व्यवसाय विशेष का सूचक एक सामान्य नाम हो, जो अपनी उत्पत्ति किसी एक पौराणिक मानवीय या दैवी पुरुष से मानता हुआ उसी व्यवसाय-वृत्ति का अनुगमन करे, जो जाति-संरचना में अपना निर्णय देने

में सक्षम व प्रवीण व्यक्तियों द्वारा मान्य हो।

(4) डॉ. केलकर ने जाति की परिभाषा इस प्रकार दी है—जाति एक ऐसा सामाजिक़ वर्ग है, जिसके दो विशिष्ट गुण हैं, (1) सदस्यता अपने सदस्यों तथा सदस्यों से उत्पन्न जनों तक ही सीमित रहती है तथा (2) सदस्यों के लिए अंतर्जातीय विवाह दृढ़ सामाजिक कानूनों द्वारा निषिद्ध होता है।

***

इन परिभाषाओं का पुनरीक्षण हमारे लिए महत्त्वपूर्ण है। प्रथम तीन विद्वानों की परिभाषाओं का अलग-अलग विवेचन करने से स्पष्ट है कि इनमें 'बहुत कम' या 'बहुत अधिक' का समावेश है। कोई भी परिभाषा अपने आप में पूर्ण या सही नहीं है। सभी में जाति-प्रणाली की यांत्रिक संरचना के मुख्य मुद्दे नजरअंदाज कर दिए गए हैं। उनकी यह भूल निश्चित संबंधों के साथ समग्र जाति प्रणाली की अपेक्षा 'जाति' को एक प्रथम इकाई के रूप में परिभाषित करने की वजह से है। सामूहिक रूप में सभी परिभाषाएँ एक-दूसरे की पूरक हैं और इस बात पर बल देती हैं कि दूसरी में क्या 'गूढ़' या 'अस्पष्ट' है।

***

'शौच' जाति की विलक्षणता किसी भी प्रकार नहीं है। यह सामान्यत: पौरोहिती शिष्टाचार है और 'शुचिता' में विश्वास का एक विशेष मामला है। फलत: इसकी जाति से संबद्धता जाति की कार्यशीलता को क्षति पहुँचाए बिना ही अस्वीकार की जा सकती है। 'शौच का विचार' जाति संस्था से इसलिए जोड़ा गया है कि 'पुरोहित वर्ग' जाति का उच्चतम वर्ग होता है तथा 'पौरोहित' और 'शौच' पुरातन संगी हैं। इस प्रकार इसका निष्कर्ष यह हुआ कि 'शौच' धार्मिक कृत्यों में ही जातीय विलक्षणता का प्रतीक है।

***

जाति स्वयं में एक इकाई होने के कारण सामाजिक संसर्ग, सहभोज आदि स्वाभाविक रूप से अपने सदस्यों तक ही सीमित कर देती है। अत: बाह्य लोगों के साथ सहभोज का अभाव परोक्ष अथवा सुनिश्चित वर्जना न होकर जाति के अलगाव की वजह से है।

***

जाति-व्यवस्था के नाना विशिष्ट गुणों के समालोचनात्मक मूल्यांकन के अनंतर 'जाति' को ठीक से समझने के लिए अंतर्जातीय विवाह की वर्जना को जातिवाद

का सार कहा जा सकता है। कुछ विद्वान् इस विचार को मानव-शरीर रचना-शास्त्र के आधार पर अमान्य कर सकते हैं और इसका कारण ऐसे समुदायों का अस्तित्व है, जो 'जाति' की समस्या को बढ़ाए बिना ही इतर लोगों से सामाजिक संबंध नहीं स्थापित करते। सांस्कृतिक भिन्नता लिये दूरस्थ या समीपस्थ प्रदेशों-क्षेत्रों में रहते हुए तथा बाह्य लोगों से सामाजिक संसर्ग न करनेवाले समुदायों का अस्तित्व एक सामान्य सच और वास्तविकता है। इस मत की पुष्टि में संयुक्त राज्य अमेरिका के अमेरिकन इंडियंस नाम से ख्यात नीग्रो, श्वेतांग तथा अन्य विभिन्न जातीय समुदायों का उदाहरण दिया जा सकता है। हमें भारत की भिन्न स्थिति को ध्यान में रखते हुए 'विचार' में गड़बड़ नहीं करनी चाहिए। भारत की भिन्न-भिन्न जातियाँ भिन्न-भिन्न क्षेत्रों में वास करती हुई आपस में कम या अधिक एकरूप होकर सांस्कृतिक एकता प्राप्त कर बैठी हैं और यही एक समान जनसंख्या का प्रमाण है। एकरूपता या समानता का यह स्वरूप ग्रहण कर 'जाति' एक सर्वथा नई समस्या बन जाती है, जो उस स्थिति से पूर्णत: भिन्न है, जिसका प्रादुर्भाव जातीय या सामाजिक समुदायों के अंतर्जातीय विवाह निषेध या सगोत्रता के आधार पर हुआ है। भारत में 'जाति' का अर्थ जनसंख्या का उन सुदृढ़ एवं निश्चित इकाइयों में कृत्रिम विभाजन है, जिन पर आपसी वैवाहिक रीति-रिवाजों के माध्यम से एक-दूसरे में एकरूप होने का प्रतिबंध है। इसका निष्कर्ष हुआ—सजातीय विवाह ही एकमात्र लक्षण है, जो 'जातिवाद' का विशिष्ट गुण है। अब, यदि हम यह स्पष्ट करने में सफलता प्राप्त कर लें कि सजातीय विवाह का पालन कैसे किया जाता है तो निश्चय ही हम जातिवाद की उत्पत्ति और उसके यंत्रीकरण को व्यावहारिक रूप में सिद्ध कर पाने में समर्थ होंगे।

***

यह कहना अप्रासंगिक न होगा कि आज के युग का कोई भी सभ्य समाज भारतीय समाज की भाँति अपने आदिम 'अति जीवन' को प्रस्तुत नहीं करता। इसका गुण-धर्म शुद्ध रूप में प्राचीन है। इसकी जातीय आचार-संहिता समय और सभ्यता के उत्थान के उपरांत भी अपने पूर्वकालीन रूप में कार्य करती है। इन अति जीवनों में से एक सजातीय विवाह-रीति की ओर आपका ध्यान आकर्षित करना चाहता हूँ। प्राचीन विवाह-पद्धति की सर्व विदितता की अधिक विवेचना आवश्यक नहीं। इतिहास के विकास के साथ-साथ सजातीय विवाह ने विशिष्ट गुण गँवा दिया और निकटतम रक्त संबंध को छोड़ कोई अन्य प्रतिबंध विवाह के क्षेत्र में अब नहीं रहा। परंतु भारतीयों में विजातीय विवाह-कानून आज भी निषिद्ध है।

भारतीय समाज 'वंश' शेष न रहने पर भी 'वंश प्रणाली' का आनंद लेता है। यह सजातीय विवाह के सिद्धांत पर आधारित विवाह-कानून में देखा-जाना जा सकता है। बात यह नहीं कि सपिंड विवाह हो नहीं सकते। वस्तुतः सपिंड विवाह रक्त की पवित्रता को दूषित करनेवाला माना जाता है।

***

भारत के विभिन्न गोत्र गोत्रांतर विवाह करते हैं और करते रहे हैं। ऐसी ही रीति संबंध-सूचक समुदायों में पाई जाती है। यह कहना अतिशयोक्ति नहीं कि गोत्रांतर विवाह भारतीयों में स्वीकृत सिद्धांत है, जिसकी अवहेलना करने का कोई साहस तक नहीं कर सकता। जातियों में सगोत्र विवाह संबंध होने पर भी गोत्रांतर विवाह के नियम का दृढ़ता से पालन किया जाता है। सगोत्र विवाह-पद्धति को भंग करने की अपेक्षा गोत्रांतर विवाह-नीति का उल्लंघन करने पर कठोर दंडों का प्रावधान है। इस प्रकार आप देख सकते हैं कि गोत्रांतर विवाह का अर्थ एकरूपता—एक का दूसरे में पूरी तरह मिल जाना है और इस नियम के आधार पर किसी 'जाति' का अस्तित्व नहीं उभरता। लेकिन हम (भारतीयों) में जातियाँ हैं। परिणामतः अंतिम वाक्य विश्लेषण में जातियों की उत्पत्ति—जहाँ तक भारत का संबंध है—का अर्थ है विजातीय (गोत्रांतर) विवाह पर सगोत्र (सजातीय) विवाह संबंध की सर्वोपरि स्थिति। चूँकि मूल गोत्रांतरिक जनाधार में सगोत्रीय विवाह-रीति, जो 'जाति' की उत्पत्ति के समरूप है, एक गंभीर समस्या है।

***

गोत्रांतर विवाह पर सगोत्र विवाह की सर्वोपरि स्थिति का अर्थ-निष्कर्ष है 'जाति' की उत्पत्ति। लेकिन यह इतना सरल नहीं। इसके लिए हम स्वयं को 'जाति' में परिणत होने के आकांक्षी एक समुदाय की परिकल्पना करें और उन साधनों का विश्लेषण-विवेचन करें, जिन्हें वह सगोत्र विवाह-पद्धति के लिए उपयोग में लाता है। यदि एक समुदाय स्वतः को सजातीय बनाना चाहे तो बाह्य समुदायों के साथ वैवाहिक संबंधों के विपरीत औपचारिक निषेधादेश का कोई महत्त्व नहीं, विशेषकर यदि सगोत्र विवाह-रीति के परिचय से पहले सभी वैवाहिक संबंधों में गौत्रिक विवाह प्रथा एक नियम बन जाती है। फिर निकट संपर्क में रहनेवाले समुदायों में एक-दूसरे में एकीभूत होने की प्रवृत्ति पाई जाती है और इस एकीकरण से वे 'समान समाज' बन जाते हैं। जाति-रचना के हित में इस प्रवृत्ति को बलपूर्वक रोकने के लिए 'समान समाज' को एक दायरे में परिसीमित करना परम आवश्यक

है, ताकि लोग बाहर विवाह संबंध न करें।

***

बाह्य विवाह संबंधों को रोकने के लिए किया गया समाज-परिसीमन ऐसी आंतरिक समस्याएँ पैदा करता है, जिनका समाधान आसान नहीं है। माना कि एक सामान्य समुदाय में स्त्री व पुरुष कमोबेश बराबर हैं और उनमें आयु की समानता है (वस्तुत: वास्तविक समुदायों में यह समानता नहीं होती है)। अत: 'जाति' बनने के इच्छुक समुदाय के लिए स्त्री-पुरुष की संख्या में 'समानता' कायम रखना अंतिम लक्ष्य है। अन्यथा 'सजातीयता' कभी नहीं रह सकती। दूसरे शब्दों में, सगोत्र विवाह प्रथा की सुरक्षा के लिए आंतरिक विवाह के अधिकार देने होंगे, अन्यथा समुदाय के सदस्य अपनी इच्छानुसार स्वयं को सँभालने के लिए समुदाय से बाहर चले जाएँगे। वैवाहिक रीति की व्यवस्था करने पर भी 'जाति बनने के इच्छुक समुदाय' को विवाह के योग्य स्त्री-पुरुष इकाइयों की संख्या में समानता बनाए रखना आवश्यक है। मात्र इस समानता के आधार पर ही सगोत्र विवाह-रीति को अक्षुण्ण रखा जा सकता है। बहुत अधिक असमानता निश्चित रूप से सगोत्र विवाह नीति को खंडित कर देगी।

***

तब 'जाति' की समस्या विवाह योग्य स्त्री-पुरुष इकाइयों के मध्य असमानता समाप्त करने की प्रक्रिया में स्थिर हो जाती है। प्रकृति पर छोड़ने पर दंपती के एक साथ मरने पर दोनों इकाइयों में वांछित संख्या-समानता प्राप्त की जा सकती है। परंतु, ऐसा होना विरल घटना होगी। पति की मृत्यु से एक स्त्री 'अतिरिक्त' हो जाती है। उसका प्रबंध होना चाहिए, अन्यथा वह अंतर्जातीय विवाह कर सगोत्र विचार-नीति का उल्लंघन कर बैठेगी। उसी प्रकार पत्नी की मृत्यु से 'अवशिष्ट' पति का उसकी वियोग पूर्ति का समुदाय द्वारा सहानुभूतिपूर्वक प्रबंध किया जाना चाहिए, अन्यथा वह जाति से बाहर विवाह कर सगोत्र विवाह-नीति भंग कर देगा। इस प्रकार 'अवशिष्ट' स्त्री और पुरुष दोनों ही जाति के लिए खतरा हैं। यदि निर्धारित दायरे में उनके लिए योग्य पति अथवा पत्नी की व्यवस्था न की गई तो बहुत संभव है कि वे मर्यादा भंग करें, बाहर जाकर विवाह करें और ऐसी संतति को जन्म दें, जो 'जाति' के लिए बाह्य हो।

***

आइए, देखते हैं कि हमारा दलित समुदाय 'अवशिष्ट' पुरुष और स्त्री का क्या करता है। हम पहले अवशिष्ट स्त्री को लेते हैं। सगोत्रता के आधार पर उसका

निबटारा दो भिन्न-भिन्न ढंगों से किया जा सकता है। प्रथम उसे मृत पति की चिता में डालकर छुटकारा पा लिया जाए। लिंग-भेद की समस्या के हल का यह अव्यावहारिक स्वरूप कुछ-कुछ मामलों में कारगर साबित हो सकता है, कुछ में नहीं। नतीजतन, हर एक अवशिष्ट स्त्री का निपटान इस तरह नहीं किया जा सकता; क्योंकि यह समाधान आसान होते हुए भी कठिन साधन है। इस प्रकार निबटारा न होने पर अवशिष्ट (विधवा) स्त्री समुदाय में बनी रहेगी। उसके अस्तित्व से समाज को दोहरा खतरा बना रहता है। वह जाति से बाहर विवाह कर सगोत्रीयता का उल्लंघन करे या जाति के भीतर विवाह कर प्रतियोगिता के आधार पर जाति के संभाव्य विवाह-योग्य बंधुओं के लिए सुरक्षित अवसरों का अतिक्रमण करे। वह हर हालत में खतरा है। इसलिए, यदि पति के साथ उसका दहन नहीं किया जाता तो उसके लिए 'कुछ' करना आवश्यक है। दूसरा उपाय उस पर शेष जीवन के लिए वैधव्य लाद देना है। जहाँ तक उद्देश्य—परिणाम का संबंध है, अनिवार्य वैधव्य की अपेक्षा जला देना उत्तम समाधान है। विधवा-दहन से अवशिष्ट स्त्री के अस्तित्व से संभावित तीनों बुराइयों से छुटकारा मिल जाता है। मरकर समाप्त हो जाने पर वह जाति के भीतर या बाहर पुनर्विवाह की समस्या उत्पन्न नहीं करती। फिर भी, व्यावहारिक होने के नाते दाह की तुलना में अनिवार्य वैधव्य श्रेयस्कर है। इसके अतिरिक्त तुलनात्मक दृष्टि से मानवीय होने से यह दाह की भाँति पुनर्विवाह के दोषों के विपरीत सुरक्षात्मक होने पर भी समुदाय की नैतिकता की रक्षा में विफल है। अनिवार्य वैधव्य की सीमा में स्त्री रहती है, यहाँ तक तो ठीक है; किंतु भविष्य में नियमानुसार पत्नी के बनने के सामान्य अधिकार से वंचित कर दिए जाने की वजह से उसमें अनैतिक आचरण के प्रति प्रलोभन बढ़ जाता है। यह अजेय विघ्न नहीं है। इसे एक ऐसी हीन स्थिति में रखा जा सकता है, जिसमें वह चित्त आकर्षित करने का साधन बिलकुल न बन सके।

***

जाति बनने के आकांक्षी समुदाय में अवशिष्ट स्त्री की तुलना में अवशिष्ट पुरुष-विधुर की समस्या अधिक महत्त्वपूर्ण और कठिन है। आदिकाल से ही स्त्री की अपेक्षा पुरुष का हाथ ऊपर रहा है। प्रत्येक समुदाय में स्त्री-पुरुष संबंधों में उसकी अधिक प्रतिष्ठा-प्रधानता रही है। स्त्री वर्ग पर परंपरागत श्रेष्ठता के कारण पुरुष की इच्छाएँ सदा ही विचारणीय रही हैं। दूसरी ओर, स्त्री अन्यायपूर्ण धार्मिक, सामाजिक तथा आर्थिक निषेधों का शिकार रही है। इन निषेधाज्ञाओं का जनक

पुरुष इन सबसे ऊपर रहा है। ऐसी स्थिति में एक अवशिष्ट पुरुष के साथ अवशिष्ट स्त्री सरीखा व्यवहार नहीं किया जा सकता।

****

पुरुष को मृत पत्नी के साथ जलाने की युक्ति दो प्रकार से जोखिम भरी है। प्रथम तो यह पुरुष होने के कारण जलाया नहीं जा सकता; द्वितीय, यदि दहन किया गया तो जाति का एक बलिष्ठ व्यक्ति नष्ट हो जाएगा। अत: उसके सुविधाजनक निबटारे के निमित्त दो उपचार शेष रह जाते हैं। पुरुष समुदाय की संपत्ति है, यह मानकर ही मैंने 'सुविधाजनक' शब्द का प्रयोग किया है।

****

समुदाय के लिए जितना महत्त्वपूर्ण पुरुष है उतनी ही महत्त्वपूर्ण सगोत्र विवाह-नीति है। समाधान द्वारा दोनों सीमाओं का संरक्षण आवश्यक है। ऐसी परिस्थिति में पुरुष को स्त्री की भाँति शेष जीवन विधुर के रूप में व्यतीत करने के लिए बाध्य—मैं कहूँगा कि प्रेरित—किया जाए। यह समाधान कठिन नहीं, क्योंकि कुछ पुरुष तो बिना किसी विवशता के स्वत: ही अविवाहित जीवन बिताते हैं और कुछ जन तो एक कदम आगे जाकर संसार के सुख-भोगों को अस्वीकार कर देते हैं। यह समाधान, जैसा कि मानव स्वभाव है, कठिनता से संपन्न होगा। दूसरी ओर, यदि अवशिष्ट पुरुष समुदाय के लिए क्रिया-कलापों में भागीदार बनकर समाज में रहता है—जैसी कि संभावना है—तो वह समुदाय की नैतिकता के लिए खतरा है। भिन्न दृष्टि से ब्रह्मचर्य—जीवन जहाँ सफल हो जाए, सरल है। किंतु यह 'जाति' की गवेषणा में लाभकारी नहीं है। यदि वह (विधुर) ब्रह्मचर्य का पालन करे और सांसारिक भोगों से विरत रहे तो निस्संदेह उससे सगोत्रीयता और नैतिकता को किसी भी प्रकार का भय नहीं; क्योंकि वह एक 'विरक्त पुरुष' होगा। जहाँ तक उसके समुदाय के भौतिक हित की बात है, अविवाहित तपस्वी के रूप में उसकी अवस्था दहन किए हुए पुरुष की तरह है। एक ओजमयी सामुदायिक जीवन का व्यय-भार वहन करने के लिए 'जाति' द्वारा एक निश्चित संख्या-शक्ति का पालन अनिवार्य है। इसकी आशा करना और अविवाहित जीवन की घोषणा करना ठीक वैसा ही है जैसे योग्य आहार के अभाव में क्षीण होते शरीर को रक्तस्राव द्वारा ठीक करना।

****

समुदाय के अवशिष्ट पुरुष पर अविवाहित जीवन की अनिवार्यता सैद्धांतिक और व्यावहारिक दृष्टि से असफल सिद्ध होती है। अत: उसे गृहस्थ (जो परिवार

का संचालन करता है) बनाए रखने में ही समुदाय का हित है। लेकिन समस्या है उसके लिए जाति के मध्य से एक पत्नी के प्रबंध की। प्रारंभ में यह संभव नहीं है, क्योंकि 'जाति' में 'एक पुरुष के लिए एक स्त्री' का अनुपात नियम होने से किसी के लिए पुनर्विवाह का अवसर नहीं। साथ ही, स्वयं में परिवेष्ठित जाति में विवाह-योग्य स्त्रियों की एक बड़ी संख्या के लिए विवाह-योग्य पुरुषों की खोज की जाती है। अतः उस परिस्थिति में अवशिष्ट पुरुष को समुदाय से संबद्ध रखने के लिए उसके निमित्त अल्प-आयु वर्ग से एक दुलहन का प्रबंध किया जा सकता है। अवशिष्ट पुरुष के लिए निश्चय ही यह सर्वश्रेष्ठ समाधान है। इसमें वह अपनी 'जाति' में बना रहता है। इसके साथ ही इस उपाय से बाहर की ओर निरंतर बहाव द्वारा समुदाय की संख्या संबंधी क्षति में तो सुरक्षा होती ही है, सगोत्रीयता और नैतिकता की सुरक्षा भी की जा सकती है।

***

स्त्री-पुरुष में संख्या समानता चार प्रकार से सुविधापूर्वक बनाई रखी जा सकती है—(1) मृत पति के साथ विधवा को जलाकर (2) स्त्री के लिए दाह से दयालुतर, अनिवार्य वैधव्य (3) विधुर के लिए अविवाहित जीवन प्रभावी बनाना (3) कम उम्र (जो अभी विवाह योग्य नहीं हुई) कन्या से उस (विधुर) का विवाह। यद्यपि, जैसाकि मैंने पहले ही कहा है, विधवा का दहन और विधुर का प्रभावी अविवाहित जीवन—दोनों ही—सगोत्रीयता की सुरक्षा के लिए प्रयत्नरत समुदाय के लिए शंकास्पद कार्य हैं। दोनों ही 'साधन' के रूप में फलीभूत होते हैं और ये (साधन) शक्ति के रूप में मुक्त होने अथवा गति प्राप्त करने पर 'परिणाम' को जन्म देते हैं। तब प्रश्न उठता है, साधन किस परिणाम को जन्म देते हैं? ये सगोत्रीयता को जन्म देते हैं, स्थिरता प्रदान करते हैं। 'जाति' की भिन्न-भिन्न परिभाषाओं के विश्लेषण के अनुसार 'जाति' और 'सगोत्रीयता' एक है। इस प्रकार इन साधनों का अस्तित्व 'जाति' के अनुरूप है और ये साधन 'जाति' के अंतर्गत हैं।

***

भारत में जातिवाद निश्चित रूप से अति प्राचीन संस्था है। प्रायः रीति-रिवाज और नैतिकताएँ अपना इतिहास सुनानेवाले, किंतु पृथ्वी में गढ़े हुए कोई पाषाण खंड अथवा अस्थि-पंजर नहीं हैं।

***

हिंदू समाज, एक बाहरी द्रष्टा के लिए, पत्नी संबंधी तीन पेचीदा और विलक्षण रीतियाँ प्रस्तुत करता है। सती होना अर्थात् मृत पति की चिता-अग्नि में विधवा दहन, अनिवार्य वैधव्य जिसमें विधवा को पुनर्विवाह की अनुमति नहीं तथा बालिका विवाह।

****

जहाँ तक मुझे ज्ञात है, इन रीतियों की उत्पत्ति की वैज्ञानिक व्याख्या आज तक नहीं की गई। यों हमारे पास दर्शन शास्त्रों का विपुल भंडार है, जो यह तो बताते हैं कि इन रीति-रिवाजों का 'आदर' क्यों किया गया, परंतु इनकी उत्पत्ति और अस्तित्व के विषय में मौन हैं। पति और पत्नी के मध्य शरीर और आत्मा की एकता तथा समाधि से परे भक्ति को प्रमाण मानकर सती प्रथा को मान्यता दी गई। यह उमा द्वारा अभिव्यक्त पत्नीत्व के आदर्श को साकार करता है—अपने स्वामी के प्रति निष्ठा ही स्त्री का सम्मान है। यही उसका अनंत स्वर्ग है। हे महेश्वर! उसने मर्मस्पर्शी मानवीय पुकार के साथ कहा, 'यदि आप मुझसे प्रसन्न नहीं तो मैं बैकुंठ की भी कामना नहीं करती।' अनिवार्य वैधव्य को मान्यता कैसे मिली, मैं नहीं जानता और न ही मेरी भेंट ऐसे किसी व्यक्ति से हुई है, जो इसकी प्रशंसा के गीत गाता हो। हाँ, ऐसे अनगिनत हैं, जो इससे चिपके हुए हैं। डॉ. केलकर ने बालिका-विवाह की प्रशंसा इस प्रकार प्रस्तुत की है—'यथार्थ में एक सच्चा पुरुष या स्त्री अपने से विवाहित स्त्री या पुरुष के अतिरिक्त किसी अन्य के प्रति आकर्षण न रखे। ऐसी पवित्रता विवाह के पश्चात् ही नहीं, पहले भी आवश्यक है; क्योंकि यह सदाचार का सत्य आदर्श है। कोई भी कुमारी पवित्र नहीं समझी जाएगी, यदि वह उस व्यक्ति—जिससे उसका विवाह होने वाला है—के अतिरिक्त अन्य किसी से प्रेमभाव रखती है। वह नहीं जानती कि उसका विवाह किससे होगा, इसीलिए उसे ब्याह से पूर्व किसी भी पुरुष से प्रेम नहीं करना चाहिए। यदि वह करती है तो यह पाप है। अतः भोग-विलास की भावना जाग्रत् होने से पहले ही कन्या को यह जान लेना श्रेयस्कर होगा कि वह किस पुरुष से प्रेम करे। इसलिए ही बालिका-ब्याह आवश्यक है।'

****

मेरी अपनी विवेचना के अनुसार, इन व्यवस्थाओं को चलन में आने के कारण मान्यता मिली। अठारहवीं सदी के 'व्यक्तिवाद' के उद्भव का तनिक भी ज्ञान रखनेवाला कोई भी व्यक्ति मेरे इस निरूपण की सराहना करेगा। सभी कालों में आंदोलन आवश्यक माना गया है। इसे उचित ठहराने और नैतिक समर्थन देने के

लिए दर्शनशास्त्री कालांतर में जनमे हैं। इन रीति-रिवाजों की इतनी अधिक प्रशंसा इस तथ्य को प्रमाणित करती है कि इन्हें अपने गुण-वर्णन और प्रचार की आवश्यकता थी। इस प्रश्न के संबंध में कि ये रीति-रिवाज क्यों पैदा हुए, मैं निवेदन करना चाहता हूँ कि 'जाति' के ढाँचे की संरचना और जनसाधारण में इसे लोकप्रिय बनाने के लिए 'इनकी' आवश्यकता पड़ी, ठीक उसी तरह जैसे कड़वी गोली पर मीठा मुलम्मा चढ़ाना। हम यह कह सकते हैं कि रीतियाँ सामान्य जन के नैतिक आचरण के लिए इतनी घिनौनी और उद्वेगकारी रही होंगी कि उन्हें अधिक रोचक बनाने की आवश्यकता पड़ी। इन रीतियों का स्वरूप यद्यपि 'साधन' का है, तथापि इन्हें 'आदर्श' के रूप में प्रदर्शित किया गया है। इतने पर भी ये हमें इनसे ज्ञान होनेवाले निष्कर्ष—परिणामों को समझने से भ्रमित नहीं कर सकते। कोई भी व्यक्ति सुरक्षापूर्वक कह सकता है कि 'साधनों' का आदर्शीकरण अनिवार्य है और विशेषकर ऐसे मामलों में, जहाँ रीति-रिवाजों को स्थायित्व प्रदान करना ध्येय हो। साधन को परिणाम कहने से, इसके मूल चरित्र पर आवरण पड़ने के अतिरिक्त, कोई हानि नहीं होती। साथ ही इन्हें इसके मौलिक स्वरूप 'साधन' से वंचित नहीं किया जा सकता। जिस प्रकार आप बिल्लियों को कुत्तों की संज्ञा देने का कानून पास कर सकते हैं, ठीक उसी प्रकार 'साधन' को 'परिणाम' भी कह सकते हैं। लेकिन इससे आप साधन की प्रकृति नहीं बदल सकते हैं। परिणामस्वरूप मैं यह मत स्थापित करने में न्यायसंगत हूँ कि साधन कहें या परिणाम—सती प्रथा, अनिवार्य वैधव्य और बालिका विवाह के पीछे 'जाति' में अवशिष्ट स्त्री और पुरुष की समस्या का समाधान और सगोत्रीयता बनाए रखना अभीष्ट रहा है। इन रीतियों के बिना 'सगोत्र विवाह नीति' की रक्षा नहीं की जा सकती और 'सगोत्र विवाह नीति' के अभाव में 'जाति' एक बेईमानी है।

***

भारत में जातिवाद की यांत्रिक रचना और असुरक्षा की विवेचना के उपरांत इसके जन्म के विषय में प्रश्न उठना स्वाभाविक है। मूल कारण का प्रश्न सर्वथा कष्टकर प्रश्न है और 'जाति' के उद्‍गम की गवेषणा में संभवतः इसलिए ही इसकी पूर्ण उपेक्षा की गई है। कुछ ने इसकी ओर से आना-कानी की है और कुछ ने इसे धोखे में डाल दिया है। कुछ तो इस प्रश्न से ही हैरान हैं कि जाति के उद्‍भव का कोई मूल कारण हो सकता है। ऐसे लोगों का प्रस्ताव है कि यदि हम जाति के 'उद्‍गम' या 'उद्‍भव' शब्द के प्रति अपना अनुराग संवरित नहीं कर पाते तो हमें

इसके बहुवचन अर्थात् 'भारत में जाति प्रणाली की मौलिकताओं का प्रयोग करना श्रेयस्कर होगा। जातिवाद के उद्गम के प्रश्न से मैं हैरान-परेशान नहीं हूँ, क्योंकि मैं 'सगोत्र विवाह नीति' को 'जाति' का विशिष्ट गुण स्थापित कर चुका हूँ।

***

व्यक्तियों से समाज बनता है, कहना अधमता है। समाज-निर्माण तो सदैव वर्गों (जातियों) से होता है। वर्ग-संघर्ष के सिद्धांत को स्वीकारना एक अतिशयोक्ति हो सकती है, परंतु समाज में अनेक वर्गों का अस्तित्व एक वास्तविकता है। उनका आधार भिन्न हो सकता है। वे आर्थिक, बौद्धिक अथवा सामाजिक हो सकते हैं; परंतु समाज में हर व्यक्ति एक वर्ग का सदस्य होता है। यह एक पूर्ण तथ्य है और प्रारंभिक हिंदू समाज इस नियम का अपवाद नहीं रहा होगा। यह वास्तव में हम जानते हैं, अपवाद था भी नहीं। इस सामान्यीकरण को ध्यान में रखने से हमें 'जातिवाद' के उद्गम के अध्ययन में सुविधा होगी, क्योंकि हमें यह निर्णीत करना है कि वह वर्ग क्या था, जो सर्वप्रथम 'जाति' में परिणत हुआ। वर्ग या जाति, जो भी कहा जाए, निकट के पड़ोसी हैं और दोनों में बित्ता भर अंतर है, जो उन्हें अलग करता है। 'जाति' स्वयं में परिवेष्ठित एक वर्ग (श्रेणी) है।

***

जाति की उत्पत्ति के अध्ययन से 'वर्ग क्या है, जिसने अपने चारों ओर घेरा बनाया' प्रश्न का उत्तर प्राप्त होगा। यह प्रश्न अधिक गवेषणा का होने की वजह से प्रासंगिक है। इसका उत्तर समस्त भारत में जातियों की उत्पत्ति और विकास के रहस्य को उजागर करने में उपयोगी होगा। दुर्भाग्य से, इस प्रश्न का सीधा उत्तर मेरे पास नहीं। मैं इसका उत्तर टेढ़े ढंग से ही दे सकता हूँ। मैंने कहा है कि विचाराधीन रीतियाँ हिंदू समाज में प्रचलित थीं। तथ्यों की सच्चाई के लिए कथन को नियमानुकूल बनाना है, क्योंकि रीतियों का चलन सार्वभौमिकता को सूचित करता है।

***

भारत की अ-ब्राह्मण जनसंख्या में जाति-संस्था का प्रसार कैसे हुआ? संपूर्ण भारत में जातिवाद के फैलाव के प्रश्न को, जातिवाद की उत्पत्ति की अपेक्षा, अधिक दुर्भाग्य का सामना करना पड़ा है। इसका कारण, मुझे लगता है प्रसार और उत्पत्ति के प्रश्नों को अलग-अलग न करना है। इसका कारण विद्वानों का यह सामान्य विश्वास है कि या तो किसी विधि-दाता ने अपने अधीनस्थ जनसंख्या पर जाति-प्रथा को 'पवित्र व्यवस्था' के रूप में लागू किया है अथवा यह प्रणाली

भारतीयों के लिए अनभिज्ञ किसी सामाजिक नियम से उभरी है।

***

प्रत्येक देश का एक विधि-दाता है, जिसका आविर्भाव आपातकाल में एक अवतार के रूप में पापाचारी मानवता का उद्धार करने, न्याय तथा नैतिकता प्रदान करने के निमित्त हुआ है।

***

पाश्चात्य शोधकारों ने कुछ दूसरे प्रकार की व्याख्याएँ प्रस्तुत करने का प्रयत्न किया है। उनके अनुसार, भारत में नाना जातियों की उत्पत्ति का कारण मुख्य मुद्दे हैं—(1) व्यवसाय (2) जातीय संगठनों का अतिजीवन आदि (3) नए विश्व की उत्पत्ति (4) वर्ण संकर संतति तथा (5) देशांतर गमन। यहाँ यह प्रश्न उठाया जा सकता है कि क्या ये मुद्दे दूसरे समाजों में भी हैं अथवा भारतवर्ष के लिए ही विशिष्ट हैं? यदि ये भारत के लिए विशेष नहीं और संसार के लिए सामान्य हैं, तब इन मुद्दों ने इस ग्रह के अन्य भागों में जाति-रचना क्यों नहीं की? इसका कारण क्या है? क्या इसकी वजह उन भागों का वेद-भूमि भारत से अधिक पावन होना है? या शोधवेत्ता ही भूल कर बैठे हैं? मुझे विश्वास है, अंतिम बात सच है।

***

नेस्फील्ड महाशय कहते हैं—'व्यवसाय और मात्र व्यवसाय ही वह आधार था, जिस पर भारत की समूची जाति प्रणाली का निर्माण हुआ।' उन्हें यह स्मरण कराया जा सकता है कि उनका यह कथन कि भारत में जातिवाद व्यावसायिक है, हमारी खोज को आगे बढ़ाने में सहायक नहीं। यह एक ओछा आविष्कार है। अभी तो नेस्फील्ड महाशय से हमें यह जानना है कि एक व्यावसायिक समुदाय जाति कैसे बन गया।

***

विचार के आरंभ में हमें यह याद रखना आवश्यक है कि अन्य समाजों की भाँति ही हिंदू समाज वर्गों में निर्मित था, जिसका प्राचीन स्वरूप चार वर्ग हैं—(1) ब्राह्मण या पुरोहित वर्ग (2) क्षत्रिय या सैनिक वर्ग (3) वैश्य या व्यापारी वर्ग और (4) शूद्र या शिल्पकार एवं नौकर वर्ग। यहाँ इस तथ्य की ओर विशेष ध्यान देना होगा कि यह एक ऐसी वर्ग व्यवस्था थी, जिसमें लोग वांछित योग्यता प्राप्त कर वर्ग बदल सकते थे और इसी वजह से वर्गों ने अपने सामाजिक कार्यकर्ता बदले। हिंदुओं के इतिहास-क्रम में किसी समय पुरोहित वर्ग स्वयं को शेष जनता से सामाजिक तौर पर अलग होकर 'बंद कपाट की नीति' से एक जाति बन बैठा। इसी तरह श्रम के

सामाजिक बँटवारे के नियम के आधार पर अन्य वर्ग भी कुछ बहुत बड़े और कुछ बहुत छोटे समुदायों में बँट गए। अविकसित प्राणी वर्ग—वैश्य और शूद्र—आज की अनेक जातियों की उत्पत्ति के स्रोत बने। सैनिक व्यवसाय का उप-विभाजन न हो पाने पर क्षत्रिय सैनिक और प्रशासनिक वर्ग में ही विभाजित हो पाए।

***

समाज का उप-विभाजन स्वाभाविक है; किंतु इस उप-विभाजन की अस्वाभाविक बात यह रही कि वर्ग प्रणाली अपना 'खुले द्वार' का चरित्र खोकर स्वत: में परिवेष्ठित इकाइयाँ—जिन्हें 'जाति' कहा जाता है, बन गईं। यहाँ प्रश्न उठता है—क्या उन्हें किसी ने द्वार बंद करने के लिए मजबूर किया था, जिसके फलस्वरूप वे सगोत्रीय विवाह-पद्धतिवाली जातियाँ बनीं अथवा उन्होंने अपने द्वार दूसरों के लिए खुद ही बंद कर दिए? इस प्रश्न का उत्तर दो प्रकार से है—कुछ ने अपने द्वार दूसरों के लिए बंद कर दिए और कुछ ने अपने लिए द्वार बंद पाए। इनमें एक व्याख्या मनोवैज्ञानिक और दूसरी यांत्रिक होने पर भी एक-दूसरे की पूरक हैं। साथ ही जाति प्रथा की समग्र व्याख्या के निमित्त दोनों ही आवश्यक भी।

***

व्यावसायिक, धार्मिक और अन्य उप-विभाग या वर्ग स्वयं परिवेष्ठित या सगोत्र विवाह-पद्धति की इकाइयाँ क्यों बन गए? मेरा उत्तर है—ब्राह्मण ऐसे थे। सगोत्रीय विवाह पद्धति या 'बंद कपाट की प्रथा' हिंदू समाज की एक रीति थी। ब्राह्मण (पुरोहित) वर्ग में इसका उद्‌भव होने के कारण सभी अब्राह्मण वर्गों ने खुले दिल से इसका अनुकरण किया। फल यह हुआ कि वे बारी-बारी से सगोत्रीय विवाह-पद्धतिवाली जातियाँ बनते गए। यह 'अनुक्रमण का संक्रमण' है, जिसमें सभी उप-विभागों की पृथकता की गति को प्रभावित कर जातियों में बदल दिया। अनुकरण की प्रवृत्ति की मानव मस्तिष्क में गहरी पैठ है, अत: भारत की विभिन्न जातियों की रचना के लिए इसे अपर्याप्त न समझा जाए। यह पैठ इतनी गहरी है कि वाप्टर बॉगहॉट लिखते हैं—'हमें अनुकरण को स्वैच्छिक अथवा अभिन्न नहीं समझना चाहिए। इसके विपरीत, इसकी पैठ मुख्यत: मस्तिष्क के दुर्बोध अंगों में होती है। कल्पनाओं के स्वत: उत्पन्न होने या इनके पहले ही ध्यान में आ जाने पर भी इसके भावों का अस्तित्व उस समय मालूम नहीं हो पाता। हमारी प्रवृत्ति में अनुकरण भाव की पैठ हमारा विश्वास है। इस पर विश्वास जमाने या उससे विश्वास हटाने के कारण स्रोत हमारी प्रवृत्ति में अदृश्य रूप में विद्यमान रहते हैं। जहाँ तक सहज विश्वास के अनुकरण

का प्रश्न है, इसमें किंचित्‌मात्र भी संदेह नहीं।'

***

गैशियल टार्डे ने नकल की प्रवृत्ति को अपने वैज्ञानिक अध्ययन का विषय बनाया और अनुकरण के तीन नियम प्रतिपादित किए। इनमें से एक नियम के अनुसार अनुकरण ऊपर से नीचे की ओर होता है। उनके अपने शब्दों में, 'कुलीनता सर्वदा और सर्वत्र, अवसर दिए जाने पर, अपने नायकों, राजाओं और महाराजाओं का अनुकरण करती है।' टार्डे महाशय के दूसरे नियम के अनुसार, उन्हीं के शब्दों में, प्रायः समीप की चीजों से अधिक उत्तम चीजों का अनुकरण किया जाता है। वस्तुतः आदर्श के उदाहरण का प्रभाव अपनी श्रेष्ठता तथा दूरी से विपरीत दिशा में समर्थ होता है। दूरी को यहाँ सामाजिक अर्थ में प्रयुक्त किया गया है। एक अजनबी व्यक्ति कितना ही दूर क्यों न हो, वह दृष्टि-बिंदु के समीप होता है, यदि हम नित्य उनके साथ संबंध स्थापित करते हैं तथा उसके अनुकरण में हमारी अपनी इच्छाओं की पूर्ति की हर सुविधा सुलभ है। कम-से-कम दूरी के और समीप के अनुकरण का यह नियम समाज के उच्चतर श्रेणी के विद्वानों द्वारा तय उदाहरण के धीमे किंतु उत्तरोत्तर प्रसार-चरित्र की व्याख्या करता है।

***

सर्वोत्तम ढंग मुझे यह लगता है कि हम यह पता लगाएँ कि अनुकरण के माध्यम से जाति रचना के आवश्यक गुण हिंदू समाज में हैं या नहीं। प्रामाणिक आधार पर अनुकरण की अवस्थाएँ हैं—(1) अनुकरण का स्रोत समुदाय में गौरव का विषय माना जाए तथा (2) समुदाय के सदस्य एतदर्थ प्रायः और दैनिक संबंध संपर्क करें। इसमें रंच मात्र भी संदेह नहीं कि ऐसी अवस्थाएँ भारत में थीं।

***

जातिवाद की रचना में अनुकरण के खेल-प्रदर्शन का दूसरा ढंग उन रीति-रिवाजों के प्रति अब्राह्मण वर्गों की मनोवृत्ति को समझना है, जिन्होंने जननोन्मुख दिनों में जाति प्रणाली के ढाँचे का तब तक समर्थन किया, जब तक कि इतिहास क्रम में यह हिंदू मस्तिष्क में गढ़ नहीं गया और बिना किसी आश्रय के आज तक वहाँ लटकता है। किसी पोखरे की सतह पर तैरती घास की तरह उसे कोई सहारा नहीं, विश्वास चाहिए। इन रीतियों का चलन दूरी के साथ बदलता रहता है और जातियों को अलग-अलग दरशाता है। ब्राह्मणों की निकटतम जातियों ने इन तीनों रीतियों का अनुकरण किया है और ये इनके दृढ़तापूर्ण पालन पर बल देती हैं। कुछ कम निकट की जातियों ने अनिवार्य वैधव्य और बालिका विवाह का केवल बालिका

विवाह में और अधिक दूरी की जातियों ने जातिवाद के सिद्धांत में विश्वास का अनुकरण किया। यह अपूर्ण अनुकरण—टार्डे महाशय के शब्दों में—कुछ अंशों में दूरी और कुछ अंशों में इन रीतियों का निर्दय चरित्र है। यह घटना टार्डे महाशय के नियम का पूर्ण दृष्टांत चित्र है। अत: इसमें संदेह नहीं कि भारत में जातियों की समग्र प्रक्रिया निम्नतर लोगों द्वारा उच्च वर्गीय व्यक्तियों के अनुकरण का प्रतिफल है।

****

अब हमको यह देखना है कि 'बंद द्वार की नीति' के फलस्वरूप अन्य लोगों को (द्वार के) बाहर कैसे कर दिया गया। मैं इसे जाति-रचना की यांत्रिक क्रिया कहता हूँ। यह आवश्यक है, अत: यांत्रिक है, निरूपण की इस मनोवैज्ञानिक दिशा ने पूर्ववर्ती शोधकारों को जाति को वर्ग-व्यवस्था के अंतर्गत एक युक्ति न मानकर 'स्वत: में एक इकाई' मानने की भूल करने के लिए भ्रमित कर दिया है। भूल और अदूरदर्शिता विषय की विवेचना और यथार्थ व्याख्या में हानिकारक सिद्ध हुई है।

****

आप इस तथ्य को अवश्य याद रखें कि 'जाति' एकवचन में वास्तविकता नहीं है। जातियों का अस्तित्व तो बहुवचन में है। 'जाति' नाम की कोई वस्तु नहीं, सदैव 'जातियाँ' रही हैं।

****

जातिवाद की व्यवहार संबंधी धार्मिक और सामाजिक संहिता के विपरीत कोई भी नवीन पद्धति जाति द्वारा सहन नहीं की जाती, क्योंकि इससे जाति-च्युत किए बिना तथा दूसरी जातियों में समाविष्ट होने के विकल्प के अभाव में अड़ियल सदस्यों को उनके अपने भाग्य पर छोड़ दिए जाने का भय रहता है। जाति के नियम दृढ़ होते हैं। वे अपराध की किस्मों में भेद की प्रतीक्षा भी नहीं करते। नवीन पद्धति कैसी ही क्यों न हो, सबको समान दंड भुगतना पड़ेगा। विचार की उत्तर दिशा एक नई जाति को जन्म देगी, क्योंकि पुरातनपंथी इसे सहन नहीं कर सकेंगे। हानिकारक विचारक का भाग्य-निर्णय अवैध प्रेम के दोषी व्यक्ति की तरह किया जाता है। पूर्ववर्ती वर्ग धार्मिक वृत्ति की जाति को और उत्तरवर्ती मिश्रित जाति को जन्म देता है। आचार-संहिता का उल्लंघन करनेवाले दोषी दु:साहसी व्यक्तियों के लिए जातियों में दया-भाव नहीं होता। एकमात्र दंड दोषी व्यक्तियों को जाति-च्युत करता है और उसका परिणाम है एक नई जाति का उदय।

□

विविध चिंतन

भारत की आजादी की इस जद्दोजहद में यदि स्वतंत्रता का कोई हेतु बनता है तो वह अछूतों का हेतु है।

~***~

व्यक्तिगत स्तर पर मैं स्पष्ट रूप से कहना चाहता हूँ कि मेरे लिए इस देश में किसी खास संस्कृति के लिए कोई स्थान नहीं है, फिर चाहे वह संस्कृति हिंदू संस्कृति हो, मुसलिम संस्कृति हो, कर्नाटक की संस्कृति हो या गुज़रात की संस्कृति हो। वे संस्कृतियाँ अस्तित्व में हैं, इसे नकारा नहीं जा सकता। हमारा सामान्य लक्ष्य इस भावना को जन्म देने में है कि हम सब भारतीय हैं। मैं उन लोगों की तरह नहीं सोचता, जो कहते हैं कि हम भारतीय पहले हैं और बाद में हिंदू हैं, या हम भारतीय पहले हैं और बाद में मुसलमान हैं। मैं इससे संतुष्ट नहीं हूँ, मैं साफ-साफ कहना चाहता हूँ कि मेरी इससे कतई संतुष्टि नहीं होती। मैं बिलकुल नहीं चाहता कि भारतीय होने की हमारी वफादारी किसी भी दूसरी प्रतियोगी-वफ़ादारी से तनिक भी प्रभावित हो, चाहे वह प्रतियोगी-वफादारी धर्म से उत्पन्न होती हो, संस्कृति से उत्पन्न होती हो या भाषा से उत्पन्न होती हो। मैं सारे लोगों से चाहता हूँ कि वे भारतीय पहले हों और बाद में भी भारतीय हों और भारतीय के सिवा कुछ न हों।

~***~

किसी सिद्धांत की असली परख इस बात से की जाती है कि वह बुरी से भी बुरी स्थिति में कितना टिक पाता है।

~***~

समाज को तीन कालों में से एक का चुनाव करना ही पड़ेगा। समाज चाहे तो

अपने 'अनुशासन' के लिए धर्म का चुनाव नहीं कर सकता। यदि धर्म 'अनुशासन' नहीं करता तो वह 'धर्म' ही नहीं है। इसका मतलब है कि समाज 'अराजकता' के पथ पर आगे बढ़ना ठीक समझता है। तीसरे, समाज 'धर्म' और 'मजिस्ट्रेट' दोनों का चुनाव कर सकता है। जितने अंश में समाज 'धर्म' का पालन करे उतने अंश में 'धर्म' और जहाँ 'धर्म' का पालन न करे, वहाँ 'मजिस्ट्रेट'। न अराजकता में स्वतंत्रता है और न डिक्टेटर राज्य में स्वतंत्रता है। केवल तीसरी व्यवस्था में ही स्वतंत्रता जीवित रहती है।

****

राजनीतिक और धार्मिक होने से ज्यादा मनुष्य एक सामाजिक पशु है। उसे धर्म की आवश्यकता नहीं भी रह सकती, ऐसे ही उसे राजनीति की जरूरत नहीं भी पड़ सकती। लेकिन उसे समाज की आवश्यकता होती ही है। वह बिना समाज के नहीं रह सकता।

****

समाज मनुष्य की सारी संस्थाओं में सबसे शक्तिशाली संस्था है। समाज सरकार से भी अधिक ताकतवर होता है। समाज की यह ताकत मनुष्य के भले के लिए और उसके बुरे के लिए भी लगाई जा सकती है। ज्यादातर लोग यह महसूस नहीं करते कि एक व्यक्ति के खिलाफ समाज सरकार की तुलना में ज्यादा कठोर और निर्दय हो सकता है। मनुष्य के उत्पीड़न के लिए समाज के पास जो साधन हैं, वे सरकार के साधनों से कहीं अधिक ज्यादा हैं और वे ज्यादा प्रभावी भी हैं। दंड संहिता में ऐसा कौन सा दंड है, जो परिमाण और कठोरता में सामाजिक बहिष्कार के सामने ठहर सकता है। हिंदू समाज का अछूत वर्ग इसी सामाजिक मार का मारा हुआ वर्ग है। एक तरफ अछूत को समाज में अपने जीवन-यापन की जिम्मेदारी खुद निभानी पड़ती है और दूसरी तरफ उसके जीवन-यापन के सारे साधन बंद कर दिए जाते हैं। इससे अछूत वर्ग के लोग समाज-व्यवस्था का एक लाभ भी प्राप्त नहीं कर सकते और उन्हें खुली समाज-व्यवस्था के सारे नुकसान उठाने पड़ते हैं।

****

कहीं भी मानव समाज एक अकेली पूर्णता नहीं है। यह हमेशा बहुवाची होता है। कार्यकलापों के संसार में एक छोर पर व्यक्ति पड़ता है और दूसरे छोर पर समाज इन दोनों के बीच में परिवार, मित्रता, सहकारी संस्था, व्यापारिक समूह, राजनीतिक दल तथा चोरों और लुटेरों के गुट आदि पड़ते हैं। ये छोटे समूह प्रायः आपस में गुँथे हुए और जुड़े हुए रहते हैं तथा कभी भी एक-दूसरे के प्रति इतने

निषेधात्मक हो जाते हैं जितनी कि जातियाँ।

***

वर्ग व्यवस्था में समूह मात्र समाज-रहित होते हैं, लेकिन जाति व्यवस्था में जातियाँ निश्चित रूप से समाज विरोधी होती हैं।

***

जाति-प्रथा वर्ण व्यवस्था के ठीक उलटी दिशा में चलती है। जाति-प्रथा ने वर्ण व्यवस्था को पूरी तरह उलट-पलट दिया है। वर्ण व्यवस्था जाति-प्रथा के मुकाबले में वर्ग व्यवस्था ज्यादा है। वर्ण व्यवस्था के ब्राह्मण, क्षत्रिय, वैश्य और शूद्र को पुन: दो भागों में बाँटा गया है। इनमें ब्राह्मण, क्षत्रिय और वैश्य को द्विज समाज और शूद्र को अद्विज समाज में रखा जाता है।

***

'कम्युनिटी' शब्द व्युत्पत्तिशास्त्र की दृष्टि से मूलत: धार्मिक कम्युनिटी के रूप में इस्तेमाल किया गया है। इसकी निष्पत्ति 'कम्युनियन' से निकलती है। 'कम्युनियन' एक धार्मिक शब्द है। यदि किसी व्यक्ति का धर्माधिकारी द्वारा धर्म-निकाला कर दिया जाता है तो वह उस समुदाय का सदस्य नहीं रह जाता। यह इस शब्द का मूल अर्थ है।

***

मनुष्य समुदायों में इसलिए रहते हैं, क्योंकि उनकी कुछ बातें आपस में एक समान होती हैं। जो उनमें एक समान बातें होती हैं वे उनके उद्देश्य, विश्वास, आकांक्षाओं, जानकारियों और समान समझदारी की है। समाज-शास्त्रियों की भाषा में वे समान मस्तिष्क के लोग होने चाहिए।

***

यह साफ है कि भारत जैसे देश को जहाँ समाज पूर्णतया सांप्रदायिक रूप से संगठित हुआ खड़ा है, प्रशासनिक उद्देश्यों के लिए चाहे किन्हीं भी क्षेत्रों में विभाजित किया जाए, यह जरूर होगा कि हर क्षेत्र में कोई एक समुदाय बहुसंख्यक समुदाय में हो जाएगा। मैं जानता हूँ कि इस देश में जब अल्पसंख्यक समुदाय से संबंधित व्यक्ति अपने समुदाय के हितों की रक्षा के लिए लड़ने को खड़ा होता है तो बहुसंख्यक समुदाय की तमाम भीड़ उसके विरोध में खड़ी हो जाती है, उसे सांप्रदायिक और भारत-विरोधी होने का खिताब देती है, तथा उसे इस देश को नष्ट करने के लिए ब्यूरोक्रेट के हाथ का खिलौना मानती है।

***

मानव इतिहास के संक्रांति कालों में नए रास्ते खोजे जाया करते हैं। यदि ऐसे समय नए रास्ते नहीं मिलते तो समाज पतन की ओर चला जाता है। तब समय खुद अपने नए रास्ते ढूँढ़ लिया करता है। लेकिन समय के लिए ऐसा नहीं है कि वह ठीक रास्ता ही चुनेगा। ठीक रास्ता चुनने का काम मनुष्य का है। इसलिए मनुष्य इतिहास के बनाने में एक महत्त्वपूर्ण घटक है और यदि बाहरी ताकतें, चाहे वे निर्वैयक्तिक हों या सामाजिक, पहली घटक हैं तो वे उसके बाद की घटक निश्चित रूप से नहीं हैं।

***

चाहे हम इस बात को मानें या न मानें, परंतु इस देश की राजनीति-व्यवस्था चातुर्वर्ण्य का प्रतिबिंब है। शूद्र इस देश के शासन में कोई हिस्सेदारी नहीं रख रहे हैं। वर्ण व्यवस्था में अस्पृश्य लोग सामाजिक रूप से शूद्र या अति शूद्र थे, लेकिन वर्तमान में उन्हें राजनीतिक रूप से भी शूद्र बनाया जा रहा है। वर्ण व्यवस्था के शूद्रों पर शासक वर्ग के सामाजिक, आर्थिक और धार्मिक प्रभुत्व में राजनीतिक प्रभुत्व भी जोड़ा जा रहा है।

***

मैं यह नहीं भूल सकता कि कुछ भारतीय लोग सांप्रदायिक दिमाग के लोग हैं। हम यह आशा जरूर कर सकते हैं कि एक ऐसा समय आएगा, जब भारत के लोग अपने प्रशासनिक मामलों के हल में सांप्रदायिक दृष्टिकोण से देखना छोड़ देंगे; लेकिन वह एक आशा हो और सच्चाई नहीं। इसके विपरीत, सच्चाई यह है कि यदि भारतीय लोगों को कानून के प्रशासन में स्वविवेक का अधिकार दिया जाता है तो वे वर्ग और वर्ग में, समुदाय और समुदाय में भेद जरूर करेंगे।

***

एक मुसलमान मजदूर यह सोचता है कि पूँजीपतियों से मजदूरों की लड़ाई में उसे एक मुसलमान मिल मालिक के विरोध में जाना पड़ जाएगा। वह इस बात का अच्छा जानकार है कि एक अमीर मुसलमान, एक मुसलमान भू-स्वामी और एक मुसलमान मिल मालिक को कोई नुकसान अंत में मुसलमान समाज का नुकसान होगा।

***

यह सोचना मुश्किल है कि भारत जैसे देश में, जहाँ ज्यादातर लोग सांप्रदायिक दृष्टि के हैं, सत्ता प्राप्त करके अधिकारी लोग उन समुदायों के साथ समानता का बरताव करेंगे, जो उनके समुदाय के नहीं हैं। इसी कारण भारत के प्रशासन में

अछूतों के साथ अधिकारियों द्वारा असमान व्यवहार उनका भाग्य है। विशेषकर सांप्रदायिक दंगों के समय अधिकारियों का यह सांप्रदायिक चरित्र और भी प्रकट हो जाता है।

***

मुझे ऐसा लगता है कि जब किसी शहर में सांप्रदायिक दंगे भड़कते हैं तो अल्पसंख्यक समुदाय के लिए यह जानना एक बड़ा रक्षा-कवच सिद्ध होता है कि किस समुदाय का कौन सा अफसर उनकी बस्ती में कानून और व्यवस्था को चला रहा है। हम सर्भ जानते हैं कि विभिन्न समुदायों के लोगों द्वारा पुलिस अधिकारियों पर पक्षपातपूर्ण रवैया अपनाने का आरोप लगाया जाता है कि वे इस या उस समुदाय का पक्ष लेते हैं। यह हो सकता है कि ऐसे आरोपों के लिए पर्याप्त कारण न हों, लेकिन ऐसे अनेक मामले होते हैं, जिनमें कानून और व्यवस्था को लागू करनेवाले अधिकारियों की पक्षधरता सिद्ध होती है।

***

जब इस देश के लोगों के दिलों में सांप्रदायिकता इतनी गहरी उतरी हुई है तो इसे समाप्त न कर सकने की स्थिति में इसे समाज-दार्शनिकों द्वारा ईमानदारी से स्वीकार कर लिया जाना चाहिए। फिर इस भावना को बहुसंख्यक समुदाय में और अल्पसंख्यक समुदाय में बाँटकर अलग-अलग नाम देना उपयुक्त नहीं है। यह बात ठीक नहीं है कि यदि बहुसंख्यक समुदाय अपनी मरजी से अल्पसंख्यक समुदाय पर अपने दिव्य अधिकार का प्रयोग करता है तो इसे राष्ट्रवाद कहा जाए और यदि अल्पसंख्यक समुदाय शासन व सत्ता में अपनी थोड़ी भागीदारी चाहे तो उसे सांप्रदायिकता का बुरा नाम दिया जाए। बहुसंख्यक समुदाय द्वारा सत्ता पर एकाधिकार रखने की सांप्रदायिक चेष्टा को राष्ट्रवाद के अच्छे शब्द से सुशोभित नहीं किया जा सकता।

***

मैं इस सदन के सम्मुख यह कहना चाहता हूँ कि हिंदू और मुसलमान, ऐसे होते हुए जैसे कि वे हैं, क्या यह बात ईमानदारी से कह सकते हैं कि वे एक-दूसरे के प्रति सांप्रदायिक नहीं हैं? इस सदन का एक भी सदस्य यह कहने का साहस नहीं जुटा पाएगा कि वह अपने दृष्टिकोण में गैर-सांप्रदायिक भावना का है।

***

मेरे कथन का विरोध करने से कोई लाभ नहीं है। सच यह है कि हर हिंदू और हर मुसलमान एक जाति या समुदाय में जन्म लेता है। हम सब एक सांप्रदायिक

माहौल में पलते और बड़े होते हैं। हम अपने-अपने समुदायों की इच्छाओं और चाहतों के भागीदार होते हैं। हम उन समुदायों की मुश्किलों और कमजोरियों को समझते व महसूस करते हैं। इसलिए मेरे मस्तिष्क में इस बात के लिए कोई संदेह नहीं है कि इस सदन का हर सदस्य—भीतर भी और बाहर भी—हर सवाल को, जाने में या अनजाने में, सांप्रदायिक रंग से देखता और सोचता है।

****

सांप्रदायिक प्रतिनिधित्व कोई बुरी चीज नहीं है। यह कोई जहर नहीं है, बल्कि देश के विभिन्न वर्गों की सुरक्षा और महफूजी के लिए सर्वोत्तम व्यवस्था है।

****

मैं पूरी तरह आश्वस्त हूँ कि संप्रदायवाद मेरी नीति होनी चाहिए। मैं इसके लिए तनिक भी शर्मिंदा नहीं हूँ।

****

मैं यह कहना चाहूँगा कि मेरे मत में भारत देश बिना संप्रदायवाद के अपना राजनीतिक विकास नहीं कर सकेगा। भारत के लिए बिना संप्रदायवाद में कोई स्वशासन नहीं हो सकता।

****

जाति ने जन-चेतना को मार दिया है। जाति ने जन-कल्याण को पूरी तरह नष्ट कर दिया है। जाति ने जन-जागरण को असंभव बना दिया है। एक वर्ग के लिए उसकी जनता उसकी अपनी जाति है। उसकी जिम्मेदारी केवल अपनी जाति के लिए है। उसकी वफादारी उसकी अपनी जाति तक सीमित है। उसके गुण जातीय गुण हो गए हैं। उसके शील उसके जातीय शील बनकर रह गए हैं। उसमें जरूरतमंदों के लिए कोई सहानुभूति नहीं है। जाति में योग्यता और गुण के लिए कोई स्थान नहीं है। गरीबों के लिए सहायता की भावना नहीं है। जाति में किसी के दुःखों की कोई चिंता नहीं है। यदि दान व सहायता की भावना है तो वह जाति से शुरू होती है और जाति में ही समाप्त हो जाती है।

****

अपनी जाति से बाहर के लोगों में खूबियाँ देखना और मानना सबकी क्षमता में नहीं है। गुणों की प्रशंसा की जा सकती है, परंतु वह व्यक्ति अपना जाति-भाई होना चाहिए। सारी नैतिकता एक आदिम नैतिकता के समान है कि मेरा जाति-भाई हो, चाहे वह सही हो या गलत; मेरा जाति भाई हो, चाहे वह अच्छा हो या बुरा।

हिंदू समाज में जाति-प्रथा के कारण इसका राजनीतिक ढाँचा भी इसके सामाजिक ढाँचे के अनुरूप अलग होना चाहिए।

***

कुछ लोग हैं, जो यह तर्क पेश करते हैं कि हिंदू समाज से जाति-प्रथा को मिटाया जा सकता है; लेकिन मैं उनकी इस बात को स्वीकार नहीं करता। जो ऐसे विचार सामने रखते हैं, वे शायद यह सोचते हैं कि जाति-प्रथा एक क्लब, नगरपालिका या देशीय परिषद् की तरह की कोई संस्था है। यह उनकी भारी गलती है। जाति-प्रथा धर्म का मामला है और धर्म किसी भी संस्था से बड़ी चीज होती है। यह संस्थागत मामला हो सकता है, लेकिन यह स्वयं वह नहीं है जिस संस्था से यह जुड़ा हुआ है। धर्म एक प्रभाव और शक्ति है, जो हर व्यक्ति के चरित्र को मोड़ती है, उसकी क्रियाओं और अतिक्रियाओं, चाहतों और गैर-चाहतों को निश्चित करती है।

***

अस्पृश्यों के जन्म और उनके पेशों का ब्योरा इस प्रकार है—

(1) ब्राह्मण, क्षत्रिय, वैश्य और शूद्र : इन चार वर्णों से बाहर जितने लोग हैं, वे चाहे म्लेच्छ भाषा बोलते हों या आर्य भाषा, वे सब दस्यु कहलाते हैं।

(2) ये गाँव की सीमा पार के वृक्ष समूहों के नीचे, श्मशानों में, पर्वतों और वनों में अपने अलग-अलग पेशों के अनुसार जीवन-निर्वाह करते हैं।

(3) लेकिन चांडाल और श्वपच, ये दोनों गाँव के बाहर बसें। ये बरतन आदि से वंचित हैं और इनका धन कुत्ते व गधे हैं।

(4) उनके कपड़े मुरदे के हों। वे टूटे-फूटे बरतनों में भोजन करें। उनके आभूषण लोहे के हों और वे सदा घुमक्कड़ जीवन व्यतीत करें।

(5) धर्मात्मा पुरुष इन लोगों के सम्मुख न पड़ें। इनका विवाह इन्हीं में परस्पर होता है और ये आपस में ही संबंध रखें।

(6) इनका भोजन दूसरों के अधीन है। इन्हें फूटे बरतन में अन्न दिया जाना चाहिए और ये लोग रात के समय गाँवों व शहरों में घूमने न आएँ।

(7) ये लोग राजा की आज्ञा से अपनी जाति के चिह्न सहित दिन में अपने कार्य के लिए जाएँ और जिस मृतक का कोई संबंधी न हो उस शव को ले जाएँ, यह शास्त्र का नियम है।

(8) ये लोग राजा की आज्ञा से शास्त्र-विधि के अनुसार वध योग्य पुरुषों का

वध करें और उन्हीं वध्य पुरुषों के वस्त्र, शय्या और आभूषण ग्रहण करें।

(9) जो निम्न जाति की स्त्री के साथ संबंध रखता है, वह वध्य है।

(10) यदि कोई अस्पृश्य जानबूझकर किसी सवर्ण को छूता है तो ऐसा अस्पृश्य वध्य है।

अस्पृश्यता आंतरिक तिरस्कार की बाहरी अभिव्यक्ति है।

***

अस्पृश्यता गुलामी से ज्यादा बुरी है—

(1) गुलामी कभी भी बाध्यकारी नहीं रही, लेकिन अस्पृश्यता हमेशा बाध्यकारी रही है। दासता में एक व्यक्ति दूसरे व्यक्ति को गुलाम बनाकर रखने के लिए प्रस्तुत रहता है। यदि वह ऐसा न करना चाहे तो उस पर कोई पाबंदी नहीं है, लेकिन एक अस्पृश्य के पास कोई विकल्प नहीं है।

(2) गुलामी के कानून ने गुलाम को उसकी मुक्ति की स्वीकृति दी है। ऐसा नियम नहीं था कि कोई एक व्यक्ति यदि एक बार गुलाम हो गया तो सदा गुलाम ही बना रहेगा; लेकिन अस्पृश्यता में इससे कोई बचाव नहीं है। मनुष्य एक बार अछूत है तो सदा के लिए अछूत हो जाता है।

(3) अस्पृश्यता अप्रत्यक्ष है, इसलिए यह गुलामी का सबसे निकृष्ट रूप है। किसी मनुष्य की आजादी का खुला और प्रत्यक्ष हनन गुलामी का चिह्न है। इसमें गुलाम को अपनी गुलामी की याद रहती है और गुलामी के प्रति जागरूक रहना आजादी की लड़ाई में सबसे पहला और महत्त्वपूर्ण कदम है; लेकिन यदि एक व्यक्ति की स्वतंत्रता का हनन अप्रत्यक्ष रूप से किया जाता है तो वह अपनी गुलामी के बारे में जानता नहीं। एक अछूत से यह कहना, "तुम स्वतंत्र हो। तुम एक नागरिक हो। तुम्हारे पास एक नागरिक के सारे अधिकार हैं। और उसके बंधन इस प्रकार कसना कि वह इनमें से किसी एक की भी प्राप्ति न कर सके, एक खतरनाक और निर्दयी धोखा है। यह अछूत को बिना उसकी गुलामी के प्रति जागरूक बनाकर उसकी गुलामी है। यह गुलामी है, यद्यपि यह अस्पृश्यता है। यह वास्तविक है, लेकिन अप्रत्यक्ष है। यह सतत है, क्योंकि यह अचेत है।"

***

यहूदी समाज के दु:ख अपने पैदा किए दु:ख हैं, लेकिन अछूत समाज के सामाजिक दु:ख ऐसे नहीं हैं। यहूदियों ने जन-समुदाय के सदस्य बनने के बजाय अजनबी रहना पसंद किया था।

***

अछूत समाज के अंग नहीं हैं। यदि वे अंग भी हैं तो एक अलग ही अंग हैं और किसी पूर्ण का कोई हिस्सा नहीं हैं।

***

दलित वर्ग अपने में एक पृथक् वर्ग है। वह मुसलमानों से स्पष्ट रूप से भिन्न है और पृथक् है। यद्यपि इस वर्ग के लोगों की गिनती हिंदुओं में की जाती है, तो भी, वे किसी भी तरह हिंदू जाति की कमी को पूरा करनेवाला उस जाति का एक हिस्सा नहीं है। इतना ही नहीं कि उनका पृथक् अस्तित्व है, बल्कि दोस्त लोगों ने उन्हें एक ऐसा दर्जा दिया है, जो भारत में किसी दूसरे जाति-समूह को प्राप्त दर्जे से स्पष्टतया भिन्न और पृथक् है। हिंदुस्तान में ऐसे जन-समूह अथवा ऐसी जातियाँ हैं, जिन्हें सीढ़ी के निचले दर्जे पर, निम्नतर स्तर पर रखा गया है; लेकिन दलित वर्ग को जो दर्जा प्राप्त है, वह इससे सर्वथा भिन्न है। उसका स्थान कुछ-कुछ डाँगर चरानेवाले खेतिहर और गुलाम के बीच का-सा है। उन्हें 'दासों' का दर्जा प्राप्त है, यह भी कहा जा सकता है। भेद इतना है कि 'दास' अस्पृश्य नहीं रहे। यह सारा दलित वर्ग अस्पृश्य है। जो सबसे बुरी बात है वह यह है कि यह जबरदस्ती की गुलामी और उनके 'अछूतपन' से पैदा हुआ सामाजिक पार्थक्य न केवल उन्हें सार्वजनिक जीवन में किसी भी प्रकार का हिस्सा लेने के अयोग्य बनाता है, बल्कि उनके लिए किसी भी ऐसे अवसर की प्राप्ति असंभव बना देता है; उन्हें किसी भी ऐसे नागरिक अधिकारों का अधिकारी नहीं रहने देता है, जिस पर मानव का सारा अस्तित्व ही निर्भर करता है।

***

अंग्रेजों के साथ दलित वर्ग का जो संबंध रहा है, वह कुछ अनोखे किस्म का है। अंग्रेज लोगों ने भी अपने आपको दलित वर्ग के हितों के संरक्षक के रूप में उपस्थित किया। दोनों पक्षों के बीच ऐसा नजदीकी का रिश्ता रहने के बावजूद भारत के अंग्रेजी सरकार के प्रति दलित वर्ग के दृष्टिकोण में यह परिवर्तन असंदिग्ध रूप से एक असाधारण महत्त्वपूर्ण घटना है; किंतु दृष्टिकोण के इस परिवर्तन के जो कारण हैं, वे ऐसे नहीं हैं कि जिनका पता न लगाया जा सके। हमने यह निर्णय इसलिए नहीं ग्रहण किया कि हम बहुसंख्यक लोगों के साथ एक ही नौका में पार

होने या डूब मरने के लिए तैयार हो गए हों। आप जानते हैं कि इन बहुसंख्यक लोगों में और जिस अल्प मत विशेष का मैं यहाँ प्रतिनिधित्व कर रहा हूँ, उनमें जो संबंध है वह बहुत कुछ 'तीन' 'छह' का ही संबंध है। यह निर्णय हमारा अपना स्वतंत्रतापूर्वक किया हुआ निर्णय है। हमने वर्तमान शासन-व्यवस्था को अपनी विशेष परिस्थिति के प्रकाश में समझने का प्रयास किया है और हमें लगा है कि सुशासन के हिसाब से वर्तमान सरकार में अनेक बड़ी-बड़ी अत्यंत महत्त्वपूर्ण कमियाँ हैं। अंग्रेजों के भारत में पदार्पण करने से पूर्व भारतीय समाज में हमारी जो स्थिति व परिस्थिति थी, जब हम उसकी तुलना अपनी वर्तमान स्थिति व परिस्थिति से करते हैं तो हमें लगता है कि हमने कुछ भी प्रगति नहीं की है। हमारी हालत में कुछ भी परिवर्तन नहीं हुआ है। हम बैठे-बैठे किसी अनुकूल समय के आगमन की प्रतीक्षा भर कर रहे हैं।

***

ब्रिटिश लोगों से पूर्व हम 'अछूतपन' की लानत के शिकार थे। क्या ब्रिटिश सरकार ने हमें इससे मुक्त कराने के लिए कुछ भी किया है? अंग्रेजों के भारत आगमन से पूर्व हम मंदिरों में नहीं जा सकते थे। क्या अब जा सकते हैं? अंग्रेजों से पूर्व हमें पुलिस में भरती नहीं किया जाता था। क्या अब भरती किया जाता है? हम इनमें से किसी भी प्रश्न का स्वीकारात्मक उत्तर नहीं दे सकते। हम यह प्रसन्नतापूर्वक स्वीकार करते हैं कि जिन अंग्रेजों ने इतने बड़े सुदीर्घ काल तक हमें अपने अधीन रखा, उन्होंने हमारी कुछ भलाई अवश्य की है। लेकिन यह स्पष्ट है कि हमारी स्थिति में कोई महत्त्वपूर्ण परिवर्तन नहीं आया है। जहाँ तक हमारी बात है, अंग्रेजों ने उस सामाजिक व्यवस्था को, जैसा भी उन्होंने उसे पाया वैसा ही स्वीकार कर लिया और उसे वैसा ही बनाए रखा। उनकी उपमा उस चीनी दर्जी से दी जा सकती है, जब उसे एक पुराना कोट नमूने के लिए दिया गया तो उसने जो नया कोट तैयार किया वह उसकी हू-ब-हू नकल थी, यहाँ तक कि पुराने कोट की ही तरह उसमें भी जगह-जगह पैबंद लगे हुए थे। अंग्रेजों को हम पर शासन करते हुए डेढ़ सौ साल हो गए, लेकिन हमारे साथ जो अन्याय हो रहा था, वह आज भी पूर्ववत् हो रहा है।

***

दलित वर्ग यह स्पष्ट देख रहा है कि भारत की अंग्रेजी सरकार के सम्मुख दो बड़ी बाधाएँ हैं। एक तो स्वाभाविक भीतरी बाधा है, जो शासकों की अपनी स्थिति, प्रवृत्ति और स्वार्थों की सहज संतान है, जो उन्हें हमारे समाज के नवजागरण

की ओर से अंधा बनाए रखती है तथा जो उन्हें हमारी आकांक्षाओं के प्रति सहानुभूति नहीं रखने देती और जो उन्हें हमारी शिक्षा की ओर से सर्वदा उपेक्षित ही नहीं, उसका विरोधी तक बना देती है। इस सरकार के मार्ग में जो दूसरी बाधा है, वह यह है कि इसे बाह्य प्रतिरोध का डर है। भारत की अंग्रेजी सरकार इस बात को समझती है कि उन सामाजिक बुराइयों को, जो भारतीय समाज को भीतर-ही-भीतर घुन की तरह खा रही हैं और जिन्होंने चिरकाल से दलित वर्ग का जीना दूभर बना रखा है, इसे दूर करने की आवश्यकता है।

***

भारत सरकार इस बात को समझती है कि जमींदार लोग जनता का खून चूस रहे हैं और पूँजीपति लोग श्रमिकों को न पूरी मजदूरी ही देते हैं और न उन्हें ढंग से जीने लायक सुविधाएँ ही देते हैं; लेकिन, इतना सब होने पर भी, यह कितने बड़े अनुताप की बात है कि भारत सरकार ने इनमें से किसी भी एक बुराई को दूर करने के लिए अपनी उँगली तक नहीं हिलाई है। ऐसा क्यों? क्या इसलिए कि इसे इन बुराइयों को दूर करने के लिए कानूनी अधिकार प्राप्त नहीं है? नहीं। इसका एक ही कारण है और वह यह है कि सरकार डरती है कि यदि उसने वर्तमान सामाजिक तथा आर्थिक जीवन-पद्धति में कुछ भी हेर-फेर करने का प्रयास किया तो निहित स्वार्थियों की ओर से उसका विरोध होगा। इस तरह की सरकार से किसी को भी क्या लाभ हो सकता है? जीवन की सारी प्रगति, जीवन का सारा सौष्ठव निरर्थक बन जाता है, जब हम ऐसी सरकार की अधीनता में रहते हैं, जो इस प्रकार की दो-दो कमियों, दो-दो बाधाओं के कारण निकम्मी बनी हुई हो। हमें ऐसी सरकार चाहिए, जिसमें शक्ति ऐसे लोगों के हाथ में हो, जो एकमात्र देश-हित को ही सर्वोपरि स्थान देते हों। हमें ऐसी सरकार चाहिए, जिसमें शक्ति ऐसे लोगों के हाथ में हो, जो जब यह जान लें कि यही करना न्याय है और यही करना समय की माँग है तो फिर समाज के सामाजिक व आर्थिक ढाँचे में परिवर्तन करने से न घबराए, भले ही उन्हें यह अंदाज लग जाए कि ऐसा करने से लोग सरकारी आदेशों का उल्लंघन कर उसका विरोध करने तक उतारू हो सकते हैं। अंग्रेजी सरकार से इस कर्तव्य-पूर्ति की कभी आशा नहीं की जा सकती। इस देश में ऐसी सरकार ही कुछ भी करने में समर्थ हो सकती है, जिसे जनता की सरकार कहा जा सके, जनता द्वारा संचालित सरकार कहा जा सके, जनता का हित करनेवाली सरकार कहा जा सके।

****

दलित वर्ग अनिवार्य तौर पर इस परिणाम पर पहुँचा है कि वर्तमान नौकरशाही भले ही हमारी कितनी भी शुभचिंतक होने का दावा करे, वह हमारी विशेष शिकायतों को दूर करने में असमर्थ सिद्ध होगी। हम अनुभव करने लगे हैं कि हम स्वयं ही अपनी शिकायतों को दूर कर सकते हैं और हम भी अपनी शिकायतों को तब तक दूर नहीं कर सकते, जब तक कि हमारे हाथ में राजनीतिक शक्ति नहीं आ जाती। और यह भी स्पष्ट है कि जब तक अंग्रेजी सरकार सत्तासीन है, तब तक राजनीतिक सत्ता का एक हिस्सा किसी भी तरह हमारे हाथ में नहीं आ सकता। केवल 'स्वराज्य' मिलने पर ही यह संभव है कि हमारे हाथ में कुछ राजनीतिक सत्ता आ सके, जिसके बिना हम अपने लोगों के उद्धार के कार्य में एक कदम भी आगे नहीं बढ़ सकते।

****

दलित वर्ग को यह अनुभव होता है कि यदि भारत के नए विधान की राजनीतिक मशीन विशेष प्रकार से निर्मित नहीं होती तो उसके हाथ में नाममात्र के लिए भी राजनीतिक शक्ति आनेवाली नहीं है। उस राजनीतिक मशीन की रचना करते समय भारतीय सामाजिक जीवन की कुछ खास ठोस बातों की ओर से आँखें नहीं मूँदी जा सकतीं। हमें यह स्वीकार कर लेना होगा कि भारतीय समाज जाति-पाँति का एक पीढ़ी-दर-पीढ़ी सिलसिला है, जिसमें हर नीची सीढ़ी पर का वर्ग अपने से ऊपर की सीढ़ी के वर्ग को सम्मान की दृष्टि से देखता है और हर ऊपर की सीढ़ी का वर्ग अपने से नीचे की सीढ़ी के वर्ग को घृणा की दृष्टि से देखता है। यह एक ऐसी सामाजिक पद्धति है, जिसमें उस समता व भाईचारे की भावना के विकास की तनिक भी गुंजाइश नहीं, जो किसी भी प्रजातांत्रिक समाज-व्यवस्था के लिए नितांत आवश्यक है।

****

दलित वर्ग इस बात को जोर देकर कहना चाहता है कि जब जिस किसी राजनीतिक विधान की स्वीकृति अथवा उसका निर्माण हो, उस समय इस बात का ध्यान अवश्य रखा जाए कि वह विधान उस समाज की मानसिक स्थिति के अनुरूप हो, जिसके लिए उसकी रचना होने जा रही है। अन्यथा इसकी पूरी संभावना है कि आप कोई ऐसा विधान बना डालें, जो चाहे कितना सुडौल हो, लेकिन जिसकी ऊपर से छँटनी हो गई हो और जिस समाज के लिए उसकी रचना हुई हो, वह उसके किसी काम का सिद्ध न हो।

~***~

हमको प्रायः याद दिलाया जाता है कि दलित वर्ग की समस्या एक सामाजिक समस्या है और इसका हल 'राजनीति' में न होकर अन्यत्र है। हम इस मत से पूर्णतया असहमत हैं। हमारा मत है कि जब तक दलित वर्ग के अपने हाथों में राजनीतिक सत्ता नहीं आती, तब तक उनकी समस्या कभी हल नहीं हो सकती। यदि यह कथन सत्य है, और जहाँ तक मैं समझता हूँ, इसको विरोधी मत स्थापित नहीं किया जा सकता, तो मेरा निवेदन है कि दलित वर्ग की समस्या मुख्यतः राजनीतिक समस्या है और इसे यही मानकर सुलझाने का प्रयास करना चाहिए।

~***~

हम जानते हैं कि अंग्रेजों के हाथों से अब राजनीतिक सत्ता उनके हाथों में हस्तांतरित होने जा रही है जिनका हमारे आर्थिक, सामाजिक और धार्मिक जीवन पर एकाधिकार-सा है। यह हो जाए, हमें इसमें कुछ आपत्ति नहीं। यद्यपि इसमें भी कुछ संदेह नहीं कि हममें से बहुतों को 'स्वराज्य' के साथ-साथ उन अत्याचारों की, उन जुल्मों की, उन अन्यायों की भी याद आती है, जो भूतकाल में हम पर ढाए गए हैं और जिनके फिर भविष्य में भी ढाए जाने का डर हमें बराबर बना है। इसमें जो अनिवार्य खतरा हमें मोल लेना पड़ रहा है, हम उसे लेने के लिए तैयार हैं; क्योंकि हमें आशा और विश्वास है कि अपने देशवासियों के साथ-साथ हमारे हाथों में भी देश के शासन की बागडोर आएगी। हम इस बात के लिए केवल एक ही शर्त पर तैयार हो सकते हैं कि अब हमारी समस्या के निर्णय को और अधिक स्थगित न किया जाए। इस भरोसे से कि समय पर ही इस प्रश्न को सुलझा लिया जाएगा, दलित वर्ग एक अरसे तक प्रतीक्षा करता रहा है।

~***~

जब-जब भी अंग्रेजी सरकार ने अपनी प्रतिनिधि सरकार के अधिकारों को कुछ व्यापक स्वरूप दिया, ऐसे हर अवसर पर दलित वर्ग की उपेक्षा की गई, उनकी राजकीय सत्ता की माँग की ओर तनिक भी ध्यान नहीं दिया गया। मैं इस लापरवाही के विरुद्ध पूरे जोर से विरोध करता हूँ और घोषणा करता हूँ कि अब हम इसे और अधिक नहीं सहन करेंगे। हमारी समस्या का हल राजनीतिक सत्ता हस्तांतरित करने के समझौते का एक आवश्यक हिस्सा माना जाए। हम नहीं चाहते कि हमें भावी शासकों की दया और सहानुभूति की मृग-मरीचिका दिखाकर उनके हाथों में सौंप दिया जाए।

जिन कारणों से दलित वर्ग को अपनी बात इतने जोर से कहनी पड़ रही है वे कारण अस्पष्ट नहीं हैं। हममें से हर कोई इस बात को जानता है कि जिस आदमी का किसी चीज पर अधिकार होता है, वह उस आदमी की अपेक्षा जिसका उस चीज पर अधिकार नहीं होता, कहीं अधिक शक्तिशाली होता है। हममें से हर कोई इस बात को भी अच्छी तरह जानता है कि जिनके हाथ में अधिकार होता है, वे स्वेच्छा से प्राय: इन अधिकारों का त्याग नहीं करते। इसलिए हमें भरोसा नहीं है कि एक बार जब उन लोगों के हाथ में राजनीतिक सत्ता चली जाएगी तो फिर बाद में हमारी समस्या का कोई समाधान हो सकेगा, क्योंकि ऐसा करना उनके लिए घाटे की ही बात होगी। तब हमारे लिए एक ही उपाय बचा रहेगा कि जिनको आज हम सत्ता दिलाने में सहायक सिद्ध हो रहे हैं, उन्हीं को फिर शासनाधिकार से च्युत करने के लिए उन्हीं के विरुद्ध 'क्रांति' की बात सोचें। हम अत्यधिक विश्वास करके अपने ही हाथों अपने विनाश के बीज बोने के बजाय अत्यधिक शंकालु माने जाकर घृणा के पात्र बनना अधिक पसंद करते हैं। इसलिए मैं समझता हूँ कि यह न्यायसंगत है और यह हमारे लिए ठीक भी है कि हम इस बात पर जोर दें कि हमारी समस्या के समाधान की सबसे अच्छी 'गारंटी' यही है कि जिस राजनीतिक मशीनों पर दोस्त लोग अपना एकाधिकार चाहते हैं—ऐसा एकाधिकार, जिसमें कोई भी उनके एकाधिकार में हस्तक्षेप न कर सके—उस राजनीतिक मशीन पर हमारा भी कुछ अधिकार हो।

***

दलित वर्ग के साथ कभी किसी ने मित्रता का व्यवहार नहीं किया। सरकार ने केवल अपने आपको चिरंजीवी बनाने में उनका उपयोग किया है। हिंदू उनको अपना कहते हैं, ताकि वे उन्हें किसी प्रकार के भी अधिकार प्राप्त करने से रोक सकें अथवा ठीक कहा जाए, वे उनके अधिकार हड़प सकें। मुसलमान उनका पृथक् अस्तित्व स्वीकार करने से इनकार करते हैं; क्योंकि उन्हें डर है कि यदि दलित वर्ग को भी कुछ अधिकार मिल गए तो उन्हें जो आज विशेष अधिकार प्राप्त हैं, उनमें कुछ-न-कुछ कमी आ जाएगी। सरकार की ओर से निराश कर दिए गए, हिंदुओं द्वारा दबाए गए, मुसलमानों द्वारा अस्वीकार किए गए दलित वर्ग की दुर्व्यवस्था वर्णनातीत है। आज तक कभी कोई इतना 'असहाय' न हुआ होगा।

***

भारत एक विचित्र देश है। उसके 'राष्ट्रीयतावादी' तथा 'देशभक्त' भी अनोखे प्राणी हैं। वह जानना है कि उसके देश की स्त्रियाँ तथा पुरुष नागरिकता के अधिकारों

से वंचित हैं, लेकिन तब भी उसे ऐसी चुभन नहीं होती कि इसे मिटाने के लिए वह कुछ करे। वह देखता है कि उसके देशवासियों का एक पूरा 'वर्ग' कहीं किसी भी प्रकार की सरकारी नौकरी नहीं कर सकता। लेकिन इससे उसकी न्यायप्रियता की भावना को किसी भी प्रकार की चोट नहीं पहुँचती। आदमी और समाज को हानि पहुँचानेवाली सैकड़ों कुरीतियों को वह देखता है, लेकिन उसे उन्हें मिटाने की कोई चिंता नहीं है। हमारा 'देशभक्त' अपने लिए और अपने वर्ग के लिए अधिक-से-अधिक शक्ति चाहता है—और कुछ भी नहीं। मुझे खुशी है कि मैं उन 'देशभक्तों' में से एक नहीं हूँ। मैं उस वर्ग का हूँ, जो जनतंत्रवाद का पक्षपाती है, जो एकाधिकार को उसकी सभी शक्लों में समाप्त हुआ देखना चाहता है। हमारा आदर्श है कि हम देखें कि बाहर से सभी आदमी समान समझे जाएँ—राजनीतिक क्षेत्र में, आर्थिक क्षेत्र में तथा सामाजिक क्षेत्र में; क्योंकि जनता का प्रतिनिधित्व करनेवाली सरकार हमारे उस उद्देश्य तक पहुँचने का एक साधन है और क्योंकि दलित वर्ग ऐसी सरकार को इतना महत्त्व देता है, इसलिए मैंने आपसे इतना आग्रह किया है कि आपकी घोषणा की सार्थकता के लिए यह अनिवार्य है कि इस शर्त की अवश्यंभावी पूर्ति हो।

***

मैं विश्वास नहीं करता कि सम्मिलित चुनाव मात्र से 'दलित वर्ग' को हिंदू समाज में खपा देने की समस्या का अंतिम इलाज हो जाएगा। चुनाव संबंधी कोई भी व्यवस्था बड़ी सामाजिक समस्या का हल नहीं हो सकती। उसके लिए 'राजनीतिक समझौता' मात्र पर्याप्त नहीं है। मैं आशा करता हूँ कि आपके लिए संभव होगा कि आप इसे 'राजनीतिक समझौते' से आगे बढ़कर ऐसा कुछ कर सकें, जिससे दलित वर्ग के लिए न केवल हिंदू समाज का एक हिस्सा बने रहना संभव हो जाए, बल्कि उसे समाज में एक सम्मान और समानता का दर्जा प्राप्त हो जाए।

***

जब तक दलित वर्ग अज्ञान के गड्ढे में गिरा था, उसमें आत्मसम्मान की भावना नहीं जागी थी, तब तक उनके लिए यह संभव था कि हिंदू कानूनों ने उनके लिए समाज में जो दर्जा तय कर दिया—वे उसे स्वीकार कर लें; लेकिन ज्यों-ज्यों वे शिक्षित होंगे त्यों-त्यों उन कानूनों को सहन नहीं करेंगे और उनके हिंदू समाज से दूर चले जाने का खतरा उत्तरोत्तर बढ़ता चला जाएगा।

□

आर्य

जिस चातुर्वर्ण्य प्रणाली को हिंदू आर्य पूर्वजों की अनोखी खोज मानते हैं, वह कोई अनोखी खोज नहीं है। इसके बारे में कोई दावा मौलिक नहीं है। वैयक्तिक आर्य कालीन अमरता की कोई धारणा प्राचीन आर्य ग्रंथों में या प्लेटो की रचनाओं में नहीं मिलती।

~***~

पहले अफगानी लोग बौद्ध थे।

~***~

ईसा-पूर्व ब्रिटेन में बौद्ध धर्म प्रचलित था। (डॉ. डोनाल्ड ए. मेकीनरिक, 1928)

~***~

भारत का इतिहास इस बात से शुरू होता है कि आर्यों ने भारत पर आक्रमण किया, यहाँ अपना घर बनाया और अपनी संस्कृति का विकास किया। आर्यों की संस्कृति, धर्म और उनकी सामाजिक प्रणाली के बारे में हम जितना जानते हैं, उनके राजनीतिक इतिहास के बारे में उतना ही कम जानते हैं। अनार्यों की तुलना में आर्यों की श्रेष्ठता के दावे में कोई दम नहीं है। आर्यों ने इतिहास में अपनी राजनीतिक उपलब्धियों के नगण्य निशान ही छोड़े हैं। भारत के राजनीतिक इतिहास की शुरुआत एक अनार्य समुदाय नाग के उत्थान के साथ होती है। नाग समुदाय को पराजित कर पाने में आर्यों को सफलता नहीं मिल पाई थी और उन्हें संधि करते हुए उस समुदाय को बराबरी का दर्जा देना पड़ा था। प्राचीन काल में राजनीतिक क्षेत्र में भारत ने जो गौरव और यश अर्जित किया, उसका पूरा श्रेय अनार्य समुदाय नाग को जाता है। उन लोगों ने ही विश्व के मानचित्र पर भारत की महानता को

साबित कर दिखाया था।

***

मगध साम्राज्य के संस्थापक का नाम शिशुनाग था, जो अनार्य जाति नाग से संबंधित था।

***

आर्यों के सामाजिक जीवन की सूचनाएँ एकत्रित करने के लिए वेद उपयोगी स्रोत हो सकते हैं। आर्यों और अनार्यों के बीच जातिगत नहीं, बल्कि सांस्कृतिक अंतर था। आर्य चतुर्वर्ण में विश्वास रखते थे। अनार्य इस बात का विरोध करते थे। अपने धर्म के लक्ष्य के रूप में आर्य यज्ञ करना अनिवार्य समझते थे। अनार्य यज्ञ का विरोध जताते थे।

***

जो शूद्र नियमों का पालन करते थे, उन्हें हर महीने सिर का मुंडन करवाना पड़ता था। वैश्यों को भी शुद्धि की यही विधि अपनानी पड़ती थी।

***

जिन लोगों ने आर्य संस्कृति को स्वीकार कर लिया था, वे 'आर्य' कहलाने लगे थे।

***

एशिया माइनर के बोघजकोई में बरामद पुरातात्त्विक अवशेष ई.पू. 1400 के हैं, जिनमें इंद्र, वरुण और नासत्य जैसे देवों के नामों का उल्लेख मिलता है। ऋग्वेद में भी इन देवों का उल्लेख मिलता है। उसी युग के मिट्टी के पात्र मिस्र के तेलीह अमरना में बरामद हुए जिन पर उत्तर-पश्चिम मेसोपोटामिया के राजकुमारों के नाम मिलते हैं, जो नाम आर्य नाम लगते हैं।

□

हिंदू धर्म की मान्यताएँ

बकल ने अपनी 'सभ्यता का इतिहास' पुस्तक में लिखा है—यह स्पष्ट ही है कि जब तक संदेह ने जन्म नहीं लिया, प्रगति का होना असंभव था। क्योंकि हमने स्पष्ट तौर पर देख लिया है कि सभ्यता की प्रगति मानव बुद्धि द्वारा प्राप्त किए गए ज्ञान पर ही निर्भर करती है। वह उस बात पर भी निर्भर करती है कि उस ज्ञान का प्रचार किस हद तक हुआ है। लेकिन जो आदमी अपनी ही जानकारी से संतुष्ट हैं, वे उसे बढ़ाने का प्रयास कभी नहीं करेंगे। जो आदमी समझते हैं कि उनके मत शत-प्रतिशत ठीक हैं, वे कभी उन बातों का परीक्षण करने का प्रयास नहीं करेंगे, जिन पर वे मत आश्रित हैं। पूर्वजों के मतों से सर्वथा विपरीत मत प्रकट किए जा रहे हैं। जब तक उनकी ऐसी ही मन:स्थिति है तब तक यह असंभव है कि वे किसी भी ऐसे सत्य को स्वीकार कर लें, जो उनकी पूर्व की मान्यताओं से बेमेल हो।

~***~

सामाजिक प्रगति का कोई भी कदम तभी उठाया जा सकता है जब हमारे पास कुछ नया ज्ञान हो। और इस तरह का ज्ञान प्राप्त हो, इसके लिए यह आवश्यक है कि उससे पहले जिज्ञासा का आग्रह हो, संदेह की उत्सुकता हो। बिना संदेह के खोज नहीं की जाती और बिना खोज किए ज्ञान की प्राप्ति नहीं होती। ज्ञान कोई ऐसा बेजान तत्त्व नहीं है, जो हमारे पास अनायास चला आए। उसे प्राप्त करने के लिए प्रयास करना अनिवार्य है। यह बहुत परिश्रम करने से और बहुत परित्याग करने से प्राप्त होता है। और ऐसा मानना बेकार है कि आदमी उन विषयों का ज्ञान प्राप्त करने के लिए श्रम करेगा और परित्याग करेगा, जिनके बारे में वह पहले ही आश्वस्त है कि वह सबकुछ जानता है। जिन्हें अंधकार का अनुभव नहीं होता, वे

कभी प्रकाश की खोज नहीं करेंगे। यदि किसी बात के बारे में हम निश्चित हैं तो हम उस विषय में कुछ भी नई खोज नहीं करेंगे, क्योंकि खोज करना ही बेकार है और कभी-कभी खतरनाक भी हो सकता है खोज करना। आरंभ करने से पहले संदेह को दखल देना ही चाहिए। इसलिए जो संदेह करता है उसी से प्रगति आरंभ होती है या प्रगति की संभावना है।

***

भारत जमातों का देश है। यहाँ पारसी, मुसलमान और हिंदू रहते हैं। इन जमातों का आधार भिन्न-भिन्न नस्लें नहीं हैं। यह स्पष्ट है कि ये धार्मिक विभाजन है। किंतु ऐसा मान बैठना भी किसी बात की गहराई में जाना नहीं है। जो बात महत्त्व की है वह यह है कि जो पारसी है, वह पारसी क्यों है? जो ईसाई है, वह ईसाई क्यों है? जो मुसलमान है, वह मुसलमान क्यों है? जो हिंदू है, वह हिंदू क्यों है? जहाँ तक पारसी, ईसाई, मुसलमान का सवाल है, उनके लिए प्रश्न का उत्तर देना कठिन नहीं। किसी पारसी से पूछिए कि वह पारसी क्यों है? उसे इस प्रश्न का उत्तर देने में कुछ कठिनाई न होगी। वह ईसाई है, क्योंकि वह ईसा मसीह में विश्वास करता है। एक मुसलमान से भी यही प्रश्न पूछिए, उसे भी इस प्रश्न का उत्तर देने में तनिक भी हिचकिचाहट नहीं होगी। वह कहेगा कि वह इसलाम में विश्वास करता है, इसलिए वह मुसलिम है। अब यही प्रश्न एक हिंदू से पूछकर देखिए। यदि वह कहता है कि वह हिंदू इसलिए है कि वह उसी देवता की पूजा करता है, जिसकी बाकी सारे हिंदू करते हैं, तो इसका उत्तर सच्चा और सही नहीं हो सकता।

***

सभी हिंदू किसी भी एक देवता को नहीं मानते हैं और किसी भी एक देवता की पूजा नहीं करते। कुछ हिंदू एक-देववादी हैं, कुछ अनेक देवताओं को नहीं मानते। कुछ विष्णु की पूजा करते हैं, कुछ शिव की पूजा करते हैं, कुछ कृष्ण की पूजा करते हैं; कुछ किसी भी पुरुष देवता को नहीं मानते। वे किसी-न-किसी स्त्री देवी की पूजा करते हैं। देवियों को माननेवाले भी किसी एक ही देवी की पूजा नहीं करते। वे भिन्न-भिन्न देवियों को पूजते हैं। कुछ काली की पूजा करते हैं, कुछ पार्वती की पूजा करते हैं, कुछ लक्ष्मी की पूजा करते हैं। अनेक देववादियों को लें, वे सभी देवताओं को मानते, पूजते हैं। वे विष्णु को भी पूजेंगे, शिव को भी पूजेंगे; राम को भी पूजेंगे, कृष्ण को भी पूजेंगे। वे काली, पार्वती और लक्ष्मी को एक साथ पूजेंगे। एक हिंदू शिवरात्रि का व्रत रखेगा, क्योंकि वह शिव का पवित्र दिन है। वह

एकादशी व्रत रखेगा, क्योंकि वह विष्णु का पवित्र दिन है। वह बेल का गाछ लगाएगा, क्योंकि वह शिव का प्रिय वृक्ष है। वह तुलसी का पौधा लगाएगा, क्योंकि वह विष्णु का प्रिय पौधा है।

***

हिंदुओं में जो अनेक देववादी हैं, वे केवल हिंदू देवी-देवताओं की पूजा नहीं करते। वे किसी के भी देवी-देवताओं की पूजा करते हैं। किसी भी हिंदू को एक मुसलिम पीर तक को पूजने में हिचकिचाहट नहीं होती और किसी ईसाई देवी को भी। हजारों हिंदू अनेक मुसलिम पीरों के पास जाते हैं और चढ़ावे चढ़ाते हैं। यथार्थ में कुछ ऐसे स्थान हैं, जहाँ मुसलिम पीरों के परंपरागत पूजक ब्राह्मण हैं और मुसलिम पीरों का वेश पहनते हैं। बंबई के पास ही मंत मौली नाम की ईसाई देवी पर हजारों हिंदू चढ़ावा चढ़ाने जाते हैं। ईसाई या मुसलिम देवी-देवताओं की पूजा समय विशेष पर ही होती है। कुछ हिंदू लोग दूसरे धर्मों के साथ इसकी भी अपेक्षा निकट-वफादारी का संबंध निभाते हैं।

***

मलकानों-मुसलमानों की चर्चा करते हुए भी ब्लंट ने लिखा है कि वे सभी पहले के हिंदू हैं और आगरा जिले के तथा आस-पास के मथुरा, एटा, मैनपुरी आदि जिलों के रहनेवाले हैं। वे पहले के राजपूत हैं, जाट हैं और बनिए हैं। वे अपना परिचय अपने आपको मुसलमान कहकर नहीं देते। वे बहुधा अपनी पुरानी जाति का नामोल्लेख करते हैं और मुश्किल से 'मलकाना' शब्द तक को जानते हैं। उनके नाम हिंदुओं के से नाम हैं। वे अकसर हिंदू मंदिरों तक में आते-जाते हैं। आपस में भेंट-मुलाकात होने पर वे 'राम-राम' कहते हैं। शादी-विवाह वे केवल आपस में करते हैं। यूँ वे कभी-कभी किसी मसजिद में भी चले जाते हैं। वे सुन्नत करवाते हैं और अपने मुर्दों को गाड़ते हैं। यदि घनिष्ठ संबंध हो तो वे मुसलमानों के यहाँ खाना-पीना करते हैं।

***

गुजरात में ऐसे कई जन-समूह हैं, जैसे मटिया कुनबी, ये लोग अपने शादी-विवाह आदि संस्कार विधियाँ ब्राह्मणों के हाथ से कराते हैं, लेकिन वे पीरान संत इमाम शाह के अनुयायी हैं और अपने मुर्दों को मुसलमानों की तरह दफनाते हैं। शेखदास अपनी बिरादरी में होनेवाले शादी-विवाह के अवसरों पर पुरोहिताई के लिए ब्राह्मण पुरोहित के साथ मुसलमान मुल्लाजी को भी बुलाते हैं। मोमिन लोग सुन्नत करते-कराते हैं, अपने मुर्दों को जलाते हैं और गुजराती भाषा में छपे 'कुरान'

भी पढ़ते हैं। अन्य सभी बातों में हिंदू संस्कार विधि का आश्रय लेते हैं।

***

यदि कोई हिंदू कहे कि मैं इसलिए हिंदू हूँ, क्योंकि मैं दूसरे हिंदुओं के जो रीति-रिवाज हैं उन्हीं का अनुकरण करता हूँ, तो उसका उत्तर भी सही नहीं है। सभी हिंदुओं के रीति-रिवाज समान हैं ही नहीं। उत्तर भारत में नजदीकी रिश्तेदार परस्पर शादी नहीं कर सकते; किंतु दक्षिण में चचेरे, ममेरे भाई या बहन से शादी कर लेना विहित है। कभी-कभी इनसे भी अधिक निकट के रिश्तेदारों में विवाह संबंध हो जाते हैं। सामान्य रूप से स्त्रियों की शुद्धता को ऊँचा स्थान दिया जाता है; किंतु कुछ लोग इसे भी बहुत महत्त्व नहीं देते, कम-से-कम लड़कियों के विवाहित हो जाने तक। कुछ ऐसे भी लोग हैं, जो अपनी एक-न-एक लड़की को देवदासी का जीवन बिताने के लिए छोड़ देते हैं। देश के कुछ भागों में स्त्रियाँ स्वतंत्र विचरण करती हैं, दूसरे भाग में उन्हें परदों में रहना पड़ता है। कहीं-कहीं स्त्रियाँ साड़ियाँ पहनती हैं, कहीं-कहीं पाजामे।

***

फिर यदि कोई हिंदू कहे कि वह इसलिए हिंदू है, क्योंकि वह जाति-पाँति की प्रथा को मानता है, तो उसका यह उत्तर भी पूरी तरह संतोषजनक नहीं। यह सत्य है कि किसी भी हिंदू को इस बात की परवाह नहीं होती कि उसका पड़ोसी किन बातों में विश्वास रखता है और किनमें नहीं। लेकिन उसे यह जानने की बड़ी फिक्र रहती है कि उसका पड़ोसी उसके हाथ का पानी पी सकता है या नहीं! उसके हाथ का खाना खा सकता है या नहीं! इसका मतलब यह हुआ कि जाति-पाँति मानना हिंदू धर्म का अनिवार्य अंग है। जो आदमी किसी भी सम्माननीय जाति से संबंधित नहीं है, वह हिंदू नहीं ही है। यह सबकुछ सही है तो भी यह बात भुलाई ही नहीं जा सकती कि अकेला जाति-पाँति में विश्वास करना ही पर्याप्त नहीं है। बहुत से मुसलमान और ईसाई, भले ही वे खान-पान को लेकर न करते हों, भी शादी-विवाह के मामले में जाति प्रथा मानते हैं। लेकिन अकेले इसी एक वजह से भी वे हिंदू नहीं कहला सकते। दोनों बातें एक साथ होनी चाहिए। उसे हिंदू भी होना चाहिए और साथ-साथ जाति-प्रथा में विश्वास करनेवाला भी होना चाहिए।

***

ऐसा शायद ही कोई हिंदू होगा, जो वेदों को अपने धर्म का पवित्रतम ग्रंथ न

मानता हो। किसी भी वैदिक ब्राह्मण से पूछा जाए तो उसका उत्तर होगा कि वेद सनातन हैं। सनातन का क्या अर्थ है? सनातन धर्म की सर्वश्रेष्ठ व्याख्या 'मनुस्मृति' के प्रथम परिच्छेद के 22वें, 23वें श्लोक पर कुल्लुक भट्ट की टीका में है। 'सनातन' शब्द के संबंध में कुल्लुक भट्ट का कहना है—'सनातन' शब्द का मतलब है सदा से पूर्व-स्थित। जो वेद पूर्व-कल्प में विद्यमान थे, वे सर्वज्ञ ब्रह्म की स्मृति में सुरक्षित रहे। ब्रह्म और परमेश्वर एक थे। ये ही वेद इस कल्प के आरंभ में उसने अग्नि, वायु, सूर्य से प्राप्त किए। यह मान्यता, जो वेदों पर आधारित है, किसी तरह अमान्य नहीं हो सकती। वेदों का ही वचन है, 'ऋग्वेद का मूल अग्नि है, यजुर्वेद का वायु और सामवेद का सूर्य'।

***

कुल्लुक भट्ट की व्याख्या को हृदयंगम कर सकने के लिए यह आवश्यक है कि हम यह समझें कि 'कल्प' से क्या अभिप्राय है? 'कल्प' वैदिक ब्राह्मणों द्वारा स्वीकार किया गया समय का हिसाब-किताब है। ब्राह्मणों ने समय को पाँच भागों में बाँटा है—(1) वर्ष (2) युग (3) महायुग (4) मन्वंतर और (5) कल्प। वर्ष का अर्थ सामान्य है समय। यह 'काल' का पर्याय है। युग का ठीक-ठीक क्या अभिप्राय है, इसमें एक मत नहीं है। महायुग का मतलब है चार युगों का समूह (1) कृत (सत्), (2) त्रेता (3) द्वापर तथा (4) कलि। चारों युग साइकिल के पहियों की तरह एक-दूसरे का अनुसरण करते हैं। जब एक युग समाप्त हो जाता है तो दिए हुए क्रम के अनुसार अगला युग उसका अनुसरण करता है। जब चक्र पूरा हो जाता है, एक महायुग संपूर्ण होता है। तब दूसरे महायुग का आरंभ होता है। प्रत्येक महायुग का पहला युग होता है कृत युग और अंतिम युग होता है कलियुग।

***

महायुग और कल्प के बीच समय के संबंध की दृष्टि से कुछ भी संदिग्ध नहीं है। इकहत्तर महायुगों का एक कल्प होता है। लेकिन महायुग और मन्वंतर के बीच का समय की दृष्टि से जो संबंध है उसमें कुछ अनिश्चयात्मकता है। एक मन्वंतर के इकहत्तर महायुग होते हैं और कुछ और। ब्राह्मण ग्रंथों में यह कहीं नहीं दिया गया है कि इस कुछ और का क्या अभिप्राय है? इसी कारण से मन्वंतर और कल्प का समय की दृष्टि से परस्पर का संबंध अस्पष्ट है। हमारे सामने इस समय जो विचारणीय प्रश्न है, उसकी दृष्टि से इस बात का कुछ भी विशेष महत्त्व नहीं। इस समय हमारे लिए 'कल्प' का ही चिंतन करना पर्याप्त है। कल्प के संबंध में जो

मान्यता है, उसका विश्व की रचना और विश्व के विनाश की मान्यता की कल्पना से निकट का संबंध है। विश्व की रचना सृष्टि कहलाती है और विश्व का विनाश प्रलय। सृष्टि और प्रलय के बीच का समय 'कल्प' कहलाता है। इस प्रकार वेदों के मूल की कल्पना 'कल्प' की कल्पना के साथ निकट संबंध रखती है।

***

जब कल्प का आरंभ होता है, तभी सृष्टि का आरंभ होता है। जिस समय सृष्टि का आरंभ होता है, तभी वेदों का एक नया संस्करण अस्तित्व में आता है। कुल्लुक यह कहना चाहता है कि यद्यपि एक दृष्टि से प्रत्येक नए कल्प का अपना एक नया वेद होता है, लेकिन वह वही पुराना वेद होता है, जो पहले था और उसे ब्रह्मा अपनी स्मरण-शक्ति के बल पर नए सिरे से उपस्थित करते हैं। इसी अर्थ में उसका कहना है कि वेद सनातन हैं, अर्थात् सदा से पूर्व-स्थित।

***

कुल्लुक भट्ट का कहना है कि वेद स्मृति से उपस्थित किए गए। प्रश्न यह नहीं है कि किसने उपस्थित किए? प्रश्न यह है कि किसके द्वारा रचित हैं? यदि हम यह स्वीकार भी कर लें कि प्रत्येक कल्प के आरंभ में वेद नए सिरे से उपस्थित किए गए, तब भी यह प्रश्न बना ही है कि प्रथम कल्प के आरंभ में किसने प्रथम बार वेदों की रचना की। वेद यूँ ही अभाव में से अस्तित्व में नहीं आए होंगे। उनका आदि तो होगा ही, भले ही उनका अंत न हो।

***

वेदों के मूल की खोज हम स्वयं वेदों से आरंभ कर सकते हैं। ऋग्वेद वेदों के मूल के बारे में एक सिद्धांत का प्रतिपादन करता है। यह प्रसिद्ध सूक्त में समाविष्ट है। इसके अनुसार पुरुष नाम के जीव ने एक यज्ञ किया था और इसी यज्ञ ने ऋक्, यजु और साम नाम के लिए तीन वेदों को जन्म दिया।

***

सामवेद और यजुर्वेद में वेदों के मूल के बारे में एक शब्द भी नहीं। अथर्ववेद ही एक ऐसा दूसरा वेद है, जिसे वेदों के मूल के बारे में कुछ कहना है। वेदों के मूल के संबंध में इसमें कई मान्यताएँ हैं। एक मान्यता है—काल ने ऋग्वेद के मंत्रों को जन्म दिया। यजुर्वेद के मंत्रों को जन्म दिया। यजुर्वेद काल से उत्पन्न हुआ।

***

वेदों के मूल को समझाने का प्रयत्न करनेवाले ब्राह्मण ग्रंथ केवल चार हैं—

(1) शतपथ ब्राह्मण, (2) तैतिरीय ब्राह्मण, (3) ऐत्रेय ब्राह्मण और (4) कौशीतकी ब्राह्मण। शतपथ ब्राह्मण में विविध व्याख्याएँ दी गई हैं। एक व्याख्या के अनुसार, वेदों की उत्पत्ति प्रजापति से हुई है। इसके अनुसार—'एक समय प्रजापति ही एकमात्र विश्व था। उसकी आकांक्षा हुई—मैं होऊँ, मेरा विस्तार हो।' उसने भक्ति की, उसने तपस्या की। जब उसने इस प्रकार परिश्रम किया था और तपस्या की थी, तीन लोक उत्पन्न हुए—पृथ्वी, वायु और आकाश। उसने इन तीनों को उष्णता प्रदान की। इस प्रकार गरम होने पर उन तीनों लोकों में से तीन प्रकार के प्रकाश उत्पन्न हुए। अग्नि से ऋग्वेद, वायु से यजुर्वेद, सूर्य से सामवेद। उसने तीनों वेदों को उष्णता प्रदान की। गरम होने पर उन तीनों वेदों में तीन प्रकाश-सार उत्पन्न हुए। ऋग्वेद से भूः, यजुर्वेद से भुवः और सामवेद से स्वः।

****

प्रजापति से वेदों की उत्पत्ति के प्रकरण को लेकर शतपथ ब्राह्मण की ही एक दूसरी व्याख्या भी है। इसका अभिप्राय है कि प्रजापति ने वेदों को जल से उत्पन्न किया। शतपथ ब्राह्मण के शब्द हैं—उस पुरुष प्रजापति ने इच्छा की, मैं एक से अनेक हो जाऊँ, मेरा विस्तार हो जाए। उसने श्रद्धा की। उसने तपस्या की। ऐसा करके उसने सर्वप्रथम तीनों वैदिक विद्याओं को उत्पन्न किया। यह उसके लिए एक आधार हो गया। इसलिए लोग कहते हैं कि पवित्र ज्ञान ही विश्व का आधार है। इसके बाद से वेद ज्ञान ही आदमी के लिए खड़े रहने का आधार है, क्योंकि पवित्र ज्ञान ही उसका आश्रय है। इसी आधार पर स्थित होकर उसने तपस्या की। उसने वाक् से जल को उत्पन्न किया। वाक् उसकी वाणी थी। वह उत्पन्न की गई थी। क्योंकि वह व्याप्त थी, इसलिए वह 'अप्' कहलाई। क्योंकि उसने सभी कुछ ढक लिया, इसलिए पानी 'वर' कहलाया। उसने इच्छा की कि मैं इन जलों से विस्तृत हो जाऊँ। इस त्रिविध वैदिक विद्या के साथ जलों में प्रविष्ट हो गया। वहाँ से एक अंडा उत्पन्न हुआ। उसने उस अंडे को प्रेरित किया और कहा—'होने दो, होने दो, फिर होने दो।' उसी से सर्वप्रथम पवित्र विद्या की उत्पत्ति हुई, त्रिविध-वैदिक विद्या। इसलिए लोग कहते हैं कि विश्व में पवित्र ज्ञान की उत्पत्ति हुई है। इसके अतिरिक्त उस पुरुष के अग्रभाग से उत्पन्न होनेवाली पवित्र विद्या ही थी, यह उसके मुख स्वरूप प्रकट हुई। इसलिए जो आदमी वेदविद् होता है, उसके बारे में कहा जाता है कि वह अग्नि के समान है, क्योंकि पवित्र विद्या अग्नि में ही उत्पन्न है।

****

शतपथ ब्राह्मण में एक तीसरी व्याख्या दी गई है— 'मैं तुम्हें समुद्र रूपी आसन पर विराजमान करूँगा।

***

'चित्त समुद्र है। मानस समुद्र से वाणी रूपी फावड़े की सहायता से देवताओं ने त्रिविध वैदिक विद्याओं को खोदकर निकाला। इसलिए यह मंत्र उच्चरित हुआ—उस प्रकाश-स्वरूप देवता को आज इस बात की जानकारी होनी चाहिए कि उन्होंने उस भेंट को कहाँ रखा, जिसे देवताओं ने तेज फावड़े से खोद निकाला। मन समुद्र है। वाणी तेज फावड़ा है। त्रिविध वैदिक विद्या भेंट है।

***

तैतिरीय ब्राह्मण में तीन मत अभिव्यक्त किए गए हैं। उसका यह भी मत है कि वेदों के मूल में प्रजापति है। उसका यह भी कहना है कि प्रजापति ने सोम को उत्पन्न किया और इसके बाद तीन वेदों की उत्पत्ति हुई। इसी ब्राह्मण ग्रंथ में एक और अभिमत है, जिसका प्रजापति से कोई संबंध नहीं।

***

वाक् अविनाशी है। उसका जन्म सबसे पूर्व हुआ है। वह वेदों की जननी है। वह अमृतत्व का केंद्र-बिंदु है। हम पर प्रसन्न होकर वह यज्ञ में पधारी। हमारी संरक्षक देवी हमारे आह्वान को सुनने के लिए प्रस्तुत रहे। उस वाक् को ही वे ज्ञानी ऋषि, जिन्होंने वेद मंत्रों की रचना की है और देवतागण तपस्या व भक्ति द्वारा तलाश करते रहे हैं। इन सबके ऊपर तैतिरीय-ब्राह्मण की एक अपनी तीसरी मौलिक व्याख्या है। इसके अनुसार वेद प्रजापति की दाढ़ी से उत्पन्न हुए।

***

उपनिषदों ने वेदों के मूल के बारे में अपना मत अभिव्यक्त किया है। छांदोग्य उपनिषद् में वेदों के मूल की जो व्याख्या की है, वह वही है जो शतपथ ब्राह्मण की है अर्थात् ऋग्वेद अग्नि से उत्पन्न हुआ, यजुर्वेद वाणी से और सामवेद सूर्य से।

***

बृहदारण्यक उपनिषद् में दो तरह से समझाया है। एक जगह उसका कहना है कि जैसे गीली लकड़ी जलाने से तरह-तरह का धुआँ निकलता है, उसी प्रकार से उस महान् पुरुष की साँस ही सभी कुछ है—ऋग्वेद, यजुर्वेद, सामवेद, अथर्वेगीरस, इतिहास, पुराण, विद्या, उपनिषद्, श्लोक, सूत्र, नाना प्रकार की टीकाएँ—ये सब उसी पुरुष परमेश्वर की साँसें हैं। इसी का दूसरी जगह कथन है—प्रजापति (यमराज) ने वाक् को उत्पन्न किया और उसके माध्यम से आत्मा सहित सभी वस्तुओं को

उत्पन्न किया। वेद का भी उसमें समावेश है।

***

'मनुस्मृति' में वेदों के मूल के बारे में दो मत मिलते हैं। एक जगह कहा गया है कि वेदों की उत्पत्ति ब्रह्मा से हुई। आरंभ में ब्रह्मा ने वेदों के शब्दों से बहुत से नामों, कार्यों तथा सभी प्राणियों की स्थितियों की रचना की। उसी ने सूक्ष्म क्रियारत जीवित देवताओं की रचना की, साध्यों की रचना की, अनंत यज्ञों की रचना की और यज्ञों का विधान करने के लिए उसने अग्नि, वायु और सूर्य से ऋक्, यजु और सामवेद नामक त्रिविध वेदों की रचना की।

***

दूसरी जगह ऐसा प्रतीत होता है कि 'मनुस्मृति' ने वेदों के ब्रह्ममूलक होने के सिद्धांत को स्वीकार कर लिया है। इस उद्धरण को देखें—'प्रजापति ने तीन वेदों, अ, उ तथा म वर्णों को भी'''।' साथ ही भूः भुवः तथा स्वः शब्दों को। इसी सर्वोपरि प्रजा ने तीनों वेदों में से प्रत्येक में से मूल के तीन हिस्सों में से एक को दुहा। यह 'सावित्री' या 'गायत्री' कहलाता है और 'तत्' से आरंभ होता है।

***

कुछ वेदों के मूल के बारे में पुराणों में भी वर्णन आया है। विष्णु पुराण की स्थापना है, "ब्रह्मा ने अपने पूर्वाभिमुख मुँह से गायत्री, ऋग्वेद के मंत्र, त्रिवृत्त, सोमरथंतर, यज्ञप्रथा अग्निस्रोतम् की रचना की। दक्षिणाभिमुख से उन्होंने यजुर्वेद के मंत्रों, त्रिष्टुम छंदस् तथा अतिरत्र की रचना की। उत्तराभिमुख मुँह से उन्होंने एकविंश विज छंदों की रचना की।"

***

भागवत पुराण का कथन है— "एक बार चतुर्मुख ब्रह्मा से वेदों की उत्पत्ति हुई। उस समय वे यही सोच रहे थे कि मैं पूर्ववत् समस्त विश्व को कैसे उत्पन्न करूँगा? उन्होंने अपने पूर्वाभिमुख मुँह से तथा दूसरे मुँहों से ऋक्, यजु, साम एवं अथर्व की रचना की। उसके साथ-साथ स्तुतियों की रचना की, प्रार्थनाओं की रचना की तथा प्रायश्चित्तों की रचना की।"

***

उनकी आँखों में प्रवेश करके—उनसे तब एक चार गुना जीव उत्पन्न हुआ, जो पुरुष जाति का था। वह ब्रह्मा की तरह प्रकाशमान था। वह वाणी से अगोचर था, अनंत था, स्थिर था, शारीरिक इंद्रियों या गुणों से रहित था, अति प्रकाशमान था, चंद्रमा की किरणों की तरह स्वच्छ था, तेजयुक्त था तथा वर्णों से मुक्त था।

परम पुरुष ने अपनी आँखों से ऋग्वेद व यजुर्वेद की रचना की, अपनी जीभ के सिरे से सामवेद की रचना की और अपने सिर से अथर्ववेद की रचना की। इन वेदों के अस्तित्व में आते ही उन्हें क्षेत्रों की प्राप्ति हो जाती है। इसलिए वे अपने-अपने वेद स्वरूप को प्राप्त कर सकते हैं, क्योंकि उन्हें क्षेत्रों की प्राप्ति हो जाती है। तब ये वेद 'पूर्व स्थित ब्रह्मा' को उत्पन्न करते हैं, एक दिव्य पुरुष अपने मानस गुण लिये हुए।

***

पुराण प्रजापति को भी वेदों का मूल स्वीकार करते हैं। उनका कहना है कि जब परं पुरुष को विश्व की रचना की चिंता थी, प्रजापति ने अपने मुख से 'ॐ' शब्द का उच्चारण किया।

***

हरिवंश पुराण की स्थापना है—जब वह इस प्रकार विचारमग्न था, उसके मुँह से 'ॐ' शब्द निकला। वह शब्द पृथ्वी, हवा और आकाश में गूँज गया। जब वह देवाधिदेव बार-बार इस शब्द का उच्चारण करने में रत थे, उनके हृदय से वषट्कार का निर्गमन हुआ।

***

इस सिद्धांत का आधार क्या है? इस सिद्धांत का आधार यह दृष्टि है कि वेद अपौरुषेय हैं। जब वैदिक ब्राह्मण कहते हैं कि वेद अपौरुषेय हैं तो उनका मतलब इतना ही होता है कि वेद किसी भी पुरुष की रचना नहीं हैं। क्योंकि वे मनुष्यों की रचना नहीं हैं, इसलिए उनमें वैसा कोई भी दोष नहीं है, वैसी कोई भी कमी नहीं है, वैसी कोई भी कमजोरी नहीं, जो किसी भी मनुष्य की रचना में हो सकती है। इसीलिए वेद निर्भ्रांत हैं।

***

वेदों की प्रामाणिकता के प्रश्न को लेकर गौतम धर्म-सूत्र ने प्रामाणिकता के विषय में निम्नलिखित मर्यादा तय की है—

(1) वेद हैं पवित्र ज्ञान के भंडार।

(2) जो वेद के ज्ञाता हैं, उनका आचरण तथा परंपरा।

(3) यदि परस्पर विरोध हो तो स्वेच्छानुसार किसी भी आचरण का अनुसरण किया जा सकता है।

***

वशिष्ठ धर्म-सूत्र का दृष्टिकोण है—

(1) पवित्र ज्ञान का निर्णय वेद और उसके ज्ञात ऋषियों के आचरण से होता है।

(2) इन दोनों स्रोतों से यदि मार्गदर्शन उपलब्ध न हो तो शिष्ट जनों का आचरण प्रमाणभूत माना जाता है।

बौधायन के मत में इस प्रकार बताया गया है—

(1) प्रत्येक वेद में पवित्र ज्ञान उपलब्ध है।

(2) हम उसी के अनुसार निर्णय देंगे।

(3) इसके बाद स्मृतियों में दिए गए ज्ञान का दर्जा है।

(4) शिष्टों के आचरण का दर्जा तीसरा है।

(5) यदि तब भी कोई विषय अनिर्णीत रहे तो कम-से-कम दस आदमियों की सभा उसका निर्णय करे।

***

आपस्तंब धर्म-सूत्र का दृष्टिकोण इस प्रकार है—

(1) अब हम उन पुण्य कर्मों का उल्लेख करेंगे, जो दैनिक जीवन के अंग हैं।

(2) इन कर्तव्यों के विषय में उनकी सहमति प्रमाण है, जो स्मृतियों के जानकार हैं।

(3) और उन जानकारों के लिए एकमात्र वेद ही प्रमाण हैं।

***

न केवल वशिष्ठ धर्म सूत्र ने, बल्कि बौद्धायन धर्मसूत्र ने भी इसका खयाल रखा है कि शिष्ट की परिभाषा की जाए। वशिष्ठ धर्म सूत्र का कथन है—'जो आदमी तृष्णा-मुक्त है, वही शिष्ट है।' बौद्धायन शिष्ट के गुणों के विषय में अधिक ब्योरा देता है। उसका मत है—'शिष्ट वे हैं जो ईर्ष्या से मुक्त हैं, अहंकार से मुक्त हैं, दस दिनों के लिए पर्याप्त अनाज से संतुष्ट हैं, लालसा से मुक्त हैं, ढोंग से मुक्त हैं, अभिमान से मुक्त हैं, लालच से मुक्त हैं, असमंजस से मुक्त हैं और मुक्त हैं क्रोध से।'

***

जिस सभा को बौद्धायन ने प्रमाण माना है, उसके विषय में एक खास बात कही है—'अब वे इस श्लोक को उद्धृत करते हैं, जिसका भावार्थ है—चारों वेदों में से कम-से-कम एक वेद के जानकार चार व्यक्ति, एक मीमांसक, वेदांग आदि का एक जानकार, जो वेदमंत्रों का उच्चारण करता है, वह एक, तीन भिन्न-भिन्न परंपराओं का प्रतिनिधित्व करनेवाले तीन ब्राह्मण—ये सब मिलकर कम-से-कम

दस सदस्यों की एक-एक सभा का निर्माण करते हैं। पाँच, तीन या परिशुद्ध आदमी भी पवित्र धर्म के बारे में निर्णय दे सकता है।

***

धर्मसूत्रों की समालोचना से यह स्पष्ट होता है कि किसी विवादास्पद विषय पर विचार करना और कुछ निर्णय करना होता था तो सत्यासत्य का निर्णय करने के लिए चार बातों को प्रमाण माना जाता था—(1) वेद, (2) स्मृति, (3) शिष्ट पुरुषों का आचरण और (4) कम-से-कम दस व्यक्तियों की सभा की सहमति।

***

जिस सभा का निर्णय कानून माना जाता था, उसका सदस्य चुने जाने के लिए वेदों का ज्ञान एक आवश्यक योग्यता मात्र थी।

***

क्या सचमुच कोई वेदों का रचयिता नहीं है? क्या वे वास्तव में अपौरुषेय हैं? इस विषय में सबसे बड़ी गवाही, सबसे बड़ी प्रामाणिकता अनुक्रमणिकाओं की प्रामाणिकता है। अनुक्रमणिकाएँ कहती हैं, वे और कुछ नहीं हैं, वे केवल प्राचीन वैदिक साहित्य के कुछ हिस्सों की व्यवस्थित अनुक्रमणिकाएँ हैं। हर वेद की एक अनुक्रमणिका है। कभी-कभी किसी वेद की एकाधिक अनुक्रमणिकाएँ भी संभव हैं। ऋग्वेद की सात अनुक्रमणिकाएँ विद्यमान मानी जाती हैं। शौनक की पाँच, कात्यायन की एक और एक अज्ञात लेखक की एक। यजुर्वेद की तीन अनुक्रमणिकाएँ विद्यमान हैं। तीनों शाखाओं के लिए एक-एक ऐत्रेयी, चर्यनियस तथा मध्यंदिन। सामवेद की दो अनुक्रमणिकाएँ हैं। एक आर्षेय ब्राह्मण कहलाती है और दूसरी कहलाती है परिशिष्ट। अथर्ववेद के लिए एक ही अनुक्रमणिका विद्यमान मानी जाती है। यह वृहत् सर्वानुक्रमणी नाम से जानी जाती है।

***

प्रो. मैक्समूलर का कहना है कि ऋग्वेद की कात्यायन की अनुक्रमणिका सर्वाधिक प्रामाणिक अनुक्रमणिका है। इसका महत्त्व इस बात में है कि यह प्रत्येक मंत्र का पहला शब्द देती है, मंत्रों की संख्या देती है, रचयिता ऋषि का और उसके परिवार का नाम देती है, देवताओं का नाम देती है, छंदों का उल्लेख करती है। सर्वानुक्रमणिका का निर्देश करने से जो बात प्रकट होती है कि ऋग्वेद की ऋचाओं के रचयिता ऋषि हैं। इसलिए अनुक्रमणिका के अनुसार ऋग्वेद को एक मानव रचना मानना ही पड़ेगा। दूसरे, वेदों के बारे में भी हमें परिणाम पर पहुँचना पड़ेगा।

ये अनुक्रमणिकाएँ यथार्थ हैं, यह बात स्वयं ऋग्वेद में दिए गए अनेक अनुच्छेदों से प्रमाणित होती है, जिनमें ऋषियों ने अपने आपको उन ऋचाओं का रचयिता कहा है।

****

स्वयं ऋषियों ने वेदों को मानवी रचनाएँ और ऐतिहासिक महत्त्व की कृतियाँ स्वीकार किया है। ऋग्वेद की ऋचाएँ प्राचीन ऋषियों और आधुनिक ऋषियों में भेद करती हैं; जैसे—पहले के ऋषियों ने भी तुम्हारा आह्वान किया था। अब यह मेरी भी स्तुति सुनो—'हे इंदु! हमारी नवीनतम स्तुतियों से प्रसन्न होकर हमारे लिए धन, खाद्य सामग्री और संतति की व्यवस्था करो।'

****

न्याय दर्शन के संस्थापक गौतम ऋषि का कहना था—सूत्रों की और आयुर्वेद के ग्रंथों की प्रामाणिकता की तरह वेदों की प्रामाणिकता भी उनके रचयिताओं की योग्यता पर निर्भर करती है, क्योंकि वेदों का रचयिता योग्य था, प्रामाणिक था, सत्यवादी था, इसलिए यह बात अनुमानित होती है कि वेद प्रामाणिक ग्रंथ हैं।

****

वैशेषिक दर्शन वेदों को प्रामाणिक मानता है, किंतु उसके प्रामाणिकता के आधार भिन्न हैं—

(1) वेद किसी चेतन मानस की उपज हैं।

(2) वे ईश्वरकृत हैं, इसलिए वे प्रामाणिक हैं।

****

कपिल मुनि द्वारा स्थापित सांख्य दर्शन का कहना था कि वेदों को अनादि नहीं माना जा सकता; क्योंकि वेदों के अनेक अंश स्वयं अपने बारे में कहते हैं कि वे उत्पन्न किए गए हैं। सांख्य मत के अनुसार सूर्य के समान वेद स्वयं प्रकाशमान हैं। उनमें एक ऐसी दिव्य शक्ति है, जिससे वे स्वयं भी प्रकाशित होते हैं तथा दूसरी सभी वस्तुओं को भी प्रकाशित करते हैं—भूत और वर्तमान की, बड़ी और छोटी, नजदीक की और दूर की।

****

वेदांत दर्शन दो भिन्न-भिन्न मतों का समर्थन करनेवाला मालूम होता है। इसका मंतव्य है कि वेदों का मूल ब्रह्मा है या वह मूल कारण है। वह ब्रह्म पुल्लिंग नहीं है, साकार ब्रह्मा, बल्कि ब्राह्मन् है, न स्त्रीलिंग न पुल्लिंग—परम सत्ता। यह

वेदों के अनादि होने की बात को भी मानता है और आत्माश्रित रचयिता के अस्तित्व को स्वीकार करता है।

***

ब्राह्मण बहुत आगे बढ़े और कहा कि वेद ईश्वरकृत भी नहीं हैं। पूर्व मीमांसा के संस्थापक जैमिनी का यही मत है।

***

पूर्व मीमांसा के संस्थापक जैमिनी पहले नैयायिकों के तर्क का परीक्षण करते हैं, जो वेदों को परमेश्वर की रचना मानते हैं और पूर्व पक्ष के रूप में नैयायिकों का मत उपस्थित करते हैं। मीमांसकों का तर्क है—'वेद ऐसे अशरीरी परमेश्वर द्वारा कभी भी उच्चरित नहीं हुए होंगे, जिसके मुँह में तालु नहीं है, या भाषण दे सकने लायक कोई भी अंग नहीं है। इसलिए यह कल्पना ही नहीं की जा सकती कि जिन वर्णों से वेद की निर्मिति हुई है, उन वर्णों का कभी उसने उच्चारण किया होगा। नैयायिकों का मानना है कि यह तर्क युक्तियुक्त नहीं है क्योंकि यद्यपि ईश्वर अशरीरी है, लेकिन वह लीला की तरह अपने श्रद्धा-संपन्न लोगों के प्रति करुणार्द्र होकर शरीर धारण कर सकता है। इसलिए यह कहना कि वेदों का कोई व्यक्ति रूप से रचयिता नहीं था, सार्थक नहीं है।'

***

जैमिनी अपने इस मत की स्थापना के लिए कि वेद ईश्वर-वचन नहीं हैं, वे इससे भी बढ़कर कुछ हैं, अपने निजी तर्क देते हैं। जैमिनी की स्थापना है—इससे पहले के सूत्रों में यह कहा गया है, शब्दों और उनके अर्थों का संबंध नित्य है। अब यह सिद्ध करना है कि शब्दों और उनके अर्थों की नित्यता शब्दों की नित्यता पर निर्भर करती है। उसके लिए पहले उन नैयायिकों का खंडन करना होगा, जिनकी स्थापना है कि शब्द नित्य नहीं हैं।

□

कांग्रेस और गांधी ने अस्पृश्यों के लिए क्या किया?

जो हिंदू महात्मा गांधी के अंध समर्थक हैं उन्हें मेरी बात पसंद नहीं आ सकती है और संभव है कि उन्हें गुस्सा भी आ जाए। चूँकि इस अध्ययन का निष्कर्ष 'गांधी से सावधान' है, इसीलिए वैसी ही प्रतिक्रिया की अपेक्षा की जा सकती है।

~***~

मेरा मानना है कि जिस तरह आज सभी महात्मा गांधी को संशय की नजरों से देख रहे हैं, वह एक नेता के लिए बहुत बड़ी त्रासदी कहलाएगी।

~***~

गोलमेज सम्मेलन के दूसरे सत्र में अस्पृश्यों की माँग के प्रति अपने रवैए के जरिए महात्मा गांधी ने अपने चरित्र की कमजोरी का परिचय दे दिया था।

~***~

महात्मा गांधी पहले मुसलमानों की चौदह माँगें मानने के लिए तैयार नहीं थे, मगर जैसे ही उन्हें पता चला कि मुसलमान अस्पृश्यों का समर्थन करने के लिए तैयार हैं, वे फौरन उनकी चौदह माँगों को मानने के लिए तैयार हो गए। इसके लिए उन्होंने मुसलमानों के सामने अस्पृश्यों को दिया जानेवाला समर्थन वापस लेने की शर्त रख दी।

~***~

वे समझ गए कि अगर वे अस्पृश्यों के वजूद को स्वीकार कर लेंगे तो इस तरह उनका हित प्रभावित होगा। उन्होंने सत्य और निष्ठा को न्योछावर करना ही

उचित समझा और सुरक्षित मार्ग पर चलने का निश्चय किया। उन्होंने इस बात से इनकार कर दिया कि भारत में अस्पृश्यता की कोई समस्या मौजूद है और इस तरह उन्होंने अस्पृश्यों की राजनीतिक माँग पर प्रहार किया। इस तरह उन्होंने जता दिया कि अपने स्वार्थ के लिए वे किस तरह अस्पृश्यों के खिलाफ सोची-समझी साजिश को अंजाम दे सकते हैं।

***

सत्ता के भूखे कांग्रेसी ईर्ष्यालु भी हैं, जिन्हें डर सता रहा है कि कहीं उन्हें अपनी मेहनत के फल से वंचित न होना पड़े।

***

यह कांग्रेस की नीति रही है कि कैबिनेट में अस्पृश्यों के अधिकारों को किसी सूरत में प्रतिनिधित्व करने की इजाजत न दी जाए और महात्मा गांधी ऐसी नीति का समर्थन करते रहे हैं।

***

सेंट्रल प्रोविंस की कांग्रेस मिनिस्ट्री में डॉ. खरे प्रधानमंत्री थे। पुरानी कैबिनेट की तुलना में डॉ. खरे की नई कैबिनेट की उल्लेखनीय बात यह थी कि उसमें एक अस्पृश्य श्री अग्निभोज को शामिल किया गया, जो सेंट्रल प्रोविंस के सदस्य थे, कांग्रेस पार्टी से ताल्लुक रखते थे और उच्च शिक्षित होने के कारण मंत्री बनने की योग्यता रखते थे।

***

डॉ. खरे ने खुलेआम बताया कि किस तरह महात्मा गांधी ने कैबिनेट में एक अस्पृश्य को शामिल करने के निर्णय को अनुशासनहीनता का उदाहरण बताया था। महात्मा गांधी ने डॉ. खरे से कहा था कि इस तरह अस्पृश्यों की आकांक्षाओं और महत्त्वाकांक्षाओं को प्रोत्साहित करना उचित नहीं था और यह ऐसा गलत निर्णय था जिसके लिए वे कभी डॉ. खरे को माफ़ नहीं करनेवाले थे। डॉ. खरे कई मंचों पर इस बयान को दोहरा चुके हैं। महात्मा गांधी ने कभी इस बयान का खंडन नहीं किया।

***

कार्यकारिणी में वे ऐसा कोई सवाल पूछने नहीं देते, जो उन्हें पसंद नहीं हो। वे अपनी मरजी से कोई प्रस्ताव पारित नहीं कर सकते। वे कोई विधेयक अपनी मरजी से नहीं ला सकते। वे अपनी अंतरात्मा की आवाज पर मतदान नहीं कर सकते। वे वहाँ मूक मवेशियों की भूमिका निभाते रहते हैं।

***

अगर आप स्वयं को भद्र पुरुष की जगह केवल हिंदू कहलाना पसंद करते हैं तो अपने दरवाजे को बंद ही रखें, क्योंकि मैं आपके घर में कदम रखना पसंद नहीं करूँगा।

***

जब सवर्ण हिंदुओं को अपना रवैया बदलने के लिए मजबूर किया जाएगा, तभी वंचित वर्ग को सही अर्थों में न्याय मिल पाएगा। इसके लिए आपको परंपरागत आचार-संहिता के विरुद्ध सीधी काररवाई करनी होगी। ऐसा करने पर वह सोचने के लिए मजबूर हो जाएगा और फिर बदलाव के लिए तैयार भी हो सकेगा।

***

इसीलिए मैं अस्पृश्यता-निवारण लीग से आग्रह करता हूँ कि वह वंचित तबके को नागरिक अधिकार दिलाने के लिए सीधी कार्रवाई का तरीका अपनाए।

***

मालिक के तौर पर सवर्ण हिंदू का स्थान सुरक्षित है। वहीं दलित वर्ग के श्रमिक उसके दास के रूप में कार्य करते रहेंगे। उम्र या योग्यता से कोई फर्क नहीं पड़ने वाला है।

***

मैं सही वस्तु-स्थिति के बारे में बताना जरूरी समझता हूँ। मैं नहीं चाहता कि अस्पृश्यों को लेकर किसी तरह की भ्रामक स्थिति पैदा हो। अस्पृश्यों ने 'स्वतंत्रता की लड़ाई' में भाग लेना इसलिए जरूरी नहीं समझा कि वे ब्रिटिश साम्राज्यवाद के पुर्जे बनकर जीना चाहते हैं, बल्कि उन्हें डर है कि भारत को आजादी मिलने पर हिंदुओं का वर्चस्व स्थापित हो जाएगा और इस तरह अस्पृश्यों के जीवन की सारी संभावनाएँ समाप्त हो जाएँगी। उन्हें फिर दयनीय जीवन गुजारने के लिए मजबूर कर दिया जाएगा।

***

जब हिंदू राज कायम होगा तब ब्राह्मण और बनिए राज करेंगे और छोटी जातियों के लोग पुलिसकर्मी की भूमिका निभाएँगे और ये तमाम लोग परंपरागत रूप से अस्पृश्यों के शत्रु साबित होंगे।

***

स्वतंत्रता और भ्रातृत्व के लिए वर्गगत असमानता खतरे की तरह है। यह विषमता खत्म हो जाएगी, ऐसा विश्वास नहीं किया जा सकता। न ही यह माना जा

सकता है कि हिंदू इसे मिटाने की कोशिश कर सकते हैं। ऐसा नामुमकिन है। यह असमानता संयोगवश या दुर्घटनावश नहीं है।

****

'हिंदुस्तान में रहनेवाला प्रत्येक व्यक्ति हिंदू है।' इस परिभाषा के तहत दावा किया जा सकता है कि अछूत भी हिंदू हैं।

****

एक साझे धर्म का मतलब साझी सहभागिता भी है।

****

धार्मिक भिन्नता से सामाजिक गठजोड़ नहीं टूट सकता।

****

जब फौज में अछूतों के लिए शिक्षा का प्रावधान था तो उससे उन्हें एक अभूतपूर्व लाभ पहुँचा। इस शिक्षा के जरिए उन्हें नई दृष्टि मिली, नए मूल्य मिले। वे समझने लगे कि जिस हीनता के माहौल में वे जीवन गुजारते रहे हैं, वह उनकी नियति नहीं है, बल्कि कतिपय लोगों ने उन्हें इस तरह की नारकीय जिंदगी जीने के लिए मजबूर कर रखा है।

****

भारत में जाति-प्रथा विशिष्ट और अलग-अलग है। उद्देश्यों और लक्ष्यों के बीच किसी तरह का तालमेल या संशोधन दिखाई नहीं देता। जातियाँ अपने स्वार्थों को ही सर्वोपरि मानती हैं और उसी के आधार पर अन्य जातियों के साथ बरताव करती हैं। ये जातियाँ भले ही आपस में मेल-जोल बढ़ाती हैं, मगर ऐसा करने से उनके चरित्र में कोई परिवर्तन नहीं आता। सहयोग का उनका रवैया यांत्रिक होता है।

****

भारत में जातियाँ महज गैर-सामाजिक ही नहीं हैं, बल्कि समाज-विरोधी भी हैं। अछूतों के प्रति लोगों के रवैए से इस सच्चाई का अंदाजा लगाया जा सकता है। कई तथ्य हैं, जो अछूतों के प्रति उनकी क्रूरता को उजागर करते हैं।

****

इतिहास गवाह है कि जब भी नैतिकता और अर्थव्यवस्था में टकराव होता है तो जीत हमेशा अर्थव्यवस्था की होती है।

****

भारत में समाचार प्रसारित करनेवाली प्रमुख एजेंसी एसोसिएटेड प्रेस ऑफ

इंडिया के तमाम कर्मचारी मद्रासी ब्राह्मण हैं—वैसे भी पूरे भारत में मीडिया उन्हीं के हाथों में है। ये लोग स्वाभाविक रूप से कांग्रेस के समर्थक हैं और कांग्रेस के विरुद्ध किसी भी समाचार को ये लोग आम जनता तक पहुँचने नहीं देते। अस्पृश्यों के लिए हालात इसलिए भी प्रतिकूल बने हुए हैं।

***

नेहरू इस बात को लेकर काफी सजग रहते हैं कि वे एक ब्राह्मण हैं। नेहरू की बहन ने कहा कि उन्हें ब्राह्मण वंश में पैदा होने के नाते गर्व क्यों नहीं करना चाहिए।

***

क्या इस बात में तनिक भी संदेह है कि ब्राह्मण ही भारत के शासक वर्ग का निर्माण करते हैं? क्या इस बात में तनिक भी संदेह है कि आजादी के लिए कांग्रेस की लड़ाई शासक वर्ग की आजादी के लिए है? क्या इस बात में तनिक भी संदेह है कि कांग्रेस ही शासक वर्ग है और शासक वर्ग ही कांग्रेस है? क्या इस बात में तनिक भी संदेह है कि जब वर्ष 1937 में प्रांतीय स्वायत्तता के रूप में स्वराज आया तो कांग्रेस ने बेहयाई के साथ सत्ता और स्वायत्तता के शीर्ष पर शासक वर्ग को बैठा दिया?

***

यह अफसोस की बात है कि वे शोषक और शोषित के बीच फर्क नहीं कर पा रहे हैं। शोषक को आजादी इसलिए चाहिए, ताकि वह दमन कर सके। वहीं शोषित वर्ग शोषक वर्ग के दमन से आजाद होना चाहता है। भारत में आजादी लाने की हड़बड़ी में वे इस बात को नहीं समझ पा रहे हैं कि कांग्रेस का पक्ष लेकर वे भारत में लोकतंत्र को महफूज नहीं बना रहे हैं और शोषकों को शोषण जारी रखने की खुली छूट दे रहे हैं। क्या उन्हें यह बताना जरूरी है कि कांग्रेस का समर्थन करने का अर्थ है शोषणकारियों को दमन जारी रखने की खुली छूट देना?

***

हम महात्मा गांधी की निष्ठा पर कैसे भरोसा कर सकते हैं, जो गुरुवयूर मंदिर का द्वार अछूतों के लिए नहीं खोले जाने पर अनशन करने का संकल्प करता है, मगर द्वार नहीं खुलने पर भी अनशन शुरू नहीं करता? वे ऐसे व्यक्ति को कैसे स्वीकार कर सकते हैं, जो मंदिर में प्रवेश के लिए एक विधेयक प्रस्तुत करता है और फिर विधेयक को वापस ले लेता है? वे ऐसे व्यक्ति की निष्ठा पर कैसे भरोसा

कर सकते हैं, जो अछूतों के मंदिर में प्रवेश सुनिश्चित करने की जगह सिर्फ इतना कहकर चुप रह जाता है कि वह उस मंदिर में नहीं जाएगा, जिसमें अछूतों का प्रवेश वर्जित हो? वे ऐसे व्यक्ति की निष्ठा पर कैसे भरोसा कर सकते हैं, जो हर चीज के लिए अनशन कर सकता है, मगर अछूतों के हक के लिए अनशन नहीं कर सकता?

****

गुरुवयूर मंदिर का द्वार अभी भी अछूतों के लिए बंद ही है, मगर गांधीजी ने अनशन शुरू करने का अपना वादा पूरा नहीं किया है।

****

अछूतों के पक्ष में किए गए वादों को वापस लेने की माँग में गांधीजी ने आमरण अनशन शुरू कर दिया।

****

महात्मा गांधी एक राजनीतिक खिलाड़ी हैं। उन्हें अपने खेल में जुटा रहना चाहिए।

****

कोई भी व्यक्ति अपना सम्मान खोकर आभार व्यक्त नहीं कर सकता। कोई भी स्त्री अपनी इज्जत गँवाकर कृतज्ञता नहीं व्यक्त कर सकती। कोई भी देश आजादी गँवाकर कृतज्ञता नहीं व्यक्त कर सकता। आयरिश नेता डेनियल ओ कोनेल ने यह बात कही थी।

****

गांधीवाद क्या है? यह किसका पक्ष लेता है? आर्थिक संकट को लेकर इसके पास क्या समाधान है? सामाजिक समस्याओं को लेकर इसके पास क्या समाधान है?

****

यह बात स्पष्ट रूप से बता देना जरूरी है कि कुछ गांधीवादी जिस गांधीवाद की अवधारणा प्रस्तुत कर रहे हैं, वह पूरी तरह कपोल-कल्पना है।

****

महात्मा गांधी कहते हैं—

(1) पश्चिमी सभ्यता की रचना शैतान ने की है।

(2) परंपरागत पेशा जाति प्रथा की आत्मा है, जो एक अहम सिद्धांत है। इसमें परिवर्तन करने से अव्यवस्था पैदा हो सकती है।

(3) वर्ण-व्यवस्था के अंतर्गत किसी भी व्यक्ति को अपना पेशा चुनने की आजादी नहीं मिली हुई है। उसका पेशा आनुवंशिक रूप से निर्धारित होता है।

****

संक्षेप में, गांधीवाद जब प्रकृति की तरफ लौटने की बात कहता है तो इसका अर्थ आदिम युग में लौटना होता है, गरीबी में लौटना होता है, अज्ञान के अँधेरे में जनता का लौटना होता है।

****

जिस तरह की वर्ग संरचना है, उसके बारे में व्यावहारिक रूप से कहा जाए तो एक तरफ निरंकुशता, प्रदर्शनप्रियता, अहंकार, आक्रामकता, लालच, स्वार्थ है तो दूसरी तरफ असुरक्षा, दरिद्रता, तिरस्कार, गुलामी और दीनता है। ऐसी स्थिति में स्वाभिमान, स्वतंत्रता, मर्यादा और गरिमा की कल्पना नहीं की जा सकती।

****

गांधीवाद लोकतांत्रिक नहीं है, चूँकि यह जाति और वर्ण-व्यवस्था का समर्थन करता है। वे कहते हैं, "इसमें कोई बुराई नहीं कि एक वर्ण का व्यक्ति दूसरे वर्णों के लिए आरक्षित विज्ञान और कला में पारंगत हो जाता है। मगर जहाँ तक आजीविका के लिए पेशे के चयन का सवाल है, उसे अपने वर्ण के अनुसार पेशे को चुनना होगा। इसका अर्थ है कि उसे परंपरागत रूप में अपने पूर्वजों के पेशे को ही अपनाना पड़ेगा।"

****

महात्मा गांधी जानते हैं कि उनके अपने प्रांत गुजरात में किसी भी जाति ने कभी कोई सैन्य टुकड़ी नहीं बनाई। न तो वर्तमान विश्व युद्ध में ऐसा हुआ, न ही पिछले विश्व युद्ध में ऐसा हो पाया, जब गांधी गुजरात में भ्रमण कर ब्रिटिश साम्राज्यवाद के एजेंट के रूप में फौजियों की भरती का अभियान चला रहे थे।

****

कई लोग जाति व्यवस्था का समर्थन करते रहे हैं, मगर पहली बार मैं किसी को एक असामान्य तर्क का प्रयोग करते हुए देख रहा हूँ। ऐसी हालत में तो कोई पोंगापंथी भी कह उठेगा कि हमें गांधी से बचाओ। इससे पता चलता है कि महात्मा गांधी किस कदर हिंदू आदर्शों से जुड़े हुए हैं।

****

कौन कह सकता है कि स्वैच्छिक प्रणाली की तुलना में अपरिहार्य प्रणाली

अधिक कारगर साबित हुई है?

~***~

महात्मा गांधी ने कहा, ''मैं जाति के नियमों को लागू करने के लिए आया हूँ।''

~***~

जिस धर्म से वे बँधे हुए हैं वह उन्हें सम्मानजनक स्थान देने की जगह अस्पृश्य मानता है। उन्हें स्वाभाविक रूप से जीने का अवसर नहीं देता। आर्थिक रूप से अस्पृश्य लोगों को पूरी तरह सवर्ण हिंदुओं की दया पर निर्भर रहना पड़ता है और उनके पास जीविका का कोई स्वतंत्र मार्ग नहीं है। अस्पृश्यों के लिए हिंदुओं के पूर्वाग्रह के कारण सारे दरवाजे बंद हैं और हिंदू समाज ऐसे सभी दरवाजों को बंद कर देना चाहता है, जिनके जरिए दलित वर्ग का उत्थान संभव है।

~***~

इस तरह की परिस्थितियों में सुनियोजित दमन के कुचक्र से अस्पृश्यों को बचाने के लिए उनके पास राजनीतिक शक्ति का होना जरूरी है। इस तरह वे संकट की घड़ी में कम-से-कम अपना बचाव तो कर सकते हैं।

~***~

महात्मा आते रहे हैं और महात्मा जाते रहे हैं, लेकिन जो पहले अछूत थे वे आज भी अछूत ही बने हुए हैं।

~***~

अछूतों को पृथक् तत्त्व माना जाता है।

~***~

यह प्रस्ताव दलित वर्ग के लिए किस तरह लाभदायक साबित हो सकता है? संक्षेप में कहा जाए तो दलितों के हाथ-पैर बाँधकर उन्हें हिंदुओं के हवाले किया जा रहा है। उनके पास देने के लिए कुछ भी नहीं है।

~***~

उच्च जाति के हिंदुओं को असेंबली के लिए दलित वर्ग के प्रतिनिधियों को नामांकित करने का अधिकार दिया गया है। दलित वर्ग के ऐसे प्रतिनिधियों को उच्च जाति के हिंदू अपने हथियार की तरह इस्तेमाल करना पसंद करेंगे।

~***~

रोमन लोगों के पास दास थे, स्पार्टन लोगों के पास हेलोट थे, ब्रिटिश लोगों के पास वीलेन थे, अमेरिकियों के पास नीग्रो थे, जर्मन लोगों के पास यहूदी थे,

उसी तरह हिंदुओं के पास अपने अछूत हैं।

~***~

पहले अंग्रेजों ने सत्ता हथियाई, अब कांग्रेस सत्ता हथिया रही है।

~***~

एक कहावत है कि आदमी की कमीज की बजाय उसकी चमड़ी उसके अधिक नजदीक होती है।

~***~

भारत गाँवों के देश के रूप में जाना जाता है और ग्रामीण व्यवस्था में अछूतों को अलग-थलग रहने के लिए मजबूर करना आसान होता है। गाँव में अछूतों के लिए भेदभाव से बच पाना संभव नहीं होता। चूँकि ग्रामीण व्यवस्था अस्पृश्यता पैदा करती है, इसीलिए इस व्यवस्था को ध्वस्त करना आवश्यक है। अछूत अगर अलग-थलग हैं तो भौगोलिक और क्षेत्रीय रूप से भी उन्हें पृथक् पहचान मिलनी चाहिए। अछूतों के लिए अलग गाँव बसाए जाएँ, जहाँ ऊँच-नीच, स्पृश्य-अस्पृश्य का कोई भेदभाव न रह पाए।

~***~

वर्ग प्रणाली से समूह का जन्म होता है। इस बात से इनकार नहीं किया जा सकता कि हिंदुओं की सामाजिक संरचना भिन्न है, इसीलिए इसकी राजनीतिक संरचना भी भिन्न होनी चाहिए।

~***~

कई लोग युक्ति दरशाते हैं कि हिंदू समाज से जाति-व्यवस्था को समाप्त किया जा सकता है। मैं इस बात से सहमत नहीं हूँ। जो लोग इस तरह का तर्क देते हैं उन्हें लगता है कि जाति क्लब, नगर निकाय या देशी परिषद् जैसी ही किसी संस्था का नाम है। यह मानना भूल है। जाति धर्म है और धर्म एक संस्था है।

~***~

अछूतों से यह कैसे कहा जा सकता है कि वे अपने भाग्य का निर्णय करने का अधिकार ऐसे लोगों को सौंप दें, जो उनकी आशाओं-आकांक्षाओं का हमेशा विरोध करते रहे हैं, जो अछूतों के बीच मौजूद जीवनी शक्ति को मान्यता देने से इनकार करते रहे हैं, जो अपने पूर्वाग्रहों को बदल पाने में असमर्थ रहे हैं, जो अछूतों के साथ न्यायपूर्ण बरताव करने से इनकार करते रहे हैं और जिन्हें अछूतों के साथ अमानवीय बरताव करते हुए शर्म नहीं आती, चूँकि ऐसे बरताव को वे

धर्मसम्मत समझते हैं। ऐसे लोगों से बचाव के लिए अछूतों के पास राजनीतिक शक्ति का होना जरूरी है। सुरक्षा के लिए ऐसी माँग कर क्या अछूत कुछ गलत कर रहे हैं?

***

कुछ लोग अछूतों के लिए राजनीतिक अधिकार की माँग का विरोध कर रहे हैं और इसके लिए बेतुके तर्कों का सहारा ले रहे हैं। देश में जो मीडिया है वह जान-बूझकर अस्पृश्यों के हितों को नजरअंदाज करता रहा है। जब वे उपेक्षा करने में नाकाम रहते हैं तो किसी नेता को खरीद लेते हैं। जब कोई नेता बिकने के लिए तैयार नहीं होता तो वे उसकी तीखी आलोचना शुरू कर देते हैं, उसकी छवि को बिगाड़ना शुरू कर देते हैं, उसका भयादोहन करना शुरू कर देते हैं और उसे दबाने-खामोश कराने के लिए हर तरह के हथकंडों का इस्तेमाल करते हैं। जो भी नेता अस्पृश्यों के हक की लड़ाई लड़ रहा है, उसे राष्ट्र-विरोधी कहकर फटकारा जाता है।

***

हरिजन सेवक संघ गांधीजी के कई हथकंडों में से एक है, जिनके सहारे वे सफलता हासिल करते रहे हैं।

***

हिंदुओं की उच्च जातियों की संतानों ने अछूतों के साथ किए गए अपने पूर्वजों के दुर्व्यवहार को सुधारने के लिए क्या कदम उठाया है? आपको राजनीतिक आंदोलन में शामिल होने के लिए हजारों हिंदू युवक मिल जाएँगे, मगर जाति-प्रथा या अस्पृश्यता मिटाने के लिए आपको एक भी युवक नहीं मिल पाएगा।

कांग्रेसी इस बात को स्वीकार नहीं करेंगे कि अस्पृश्यों ने अपने खिलाफ कांग्रेसी अभियान को देखते हुए चुप रहकर देश-प्रेम का परिचय दिया है। अस्पृश्यों ने खामोशी का और टकराव टालने का रास्ता चुना है; मगर इससे गलतफहमी पैदा हुई है कि कांग्रेस अस्पृश्यों के साथ सभी समुदाय के लोगों का प्रतिनिधित्व करती है।

***

संसदीय लोकतंत्र से सावधान रहने की जरूरत है। यह वैसा उपयोगी नहीं है जैसा बाहर से प्रतीत होता है।

***

जहाँ सामाजिक और आर्थिक जनतंत्र नहीं होता, वहाँ राजनीतिक जनतंत्र

लागू नहीं हो सकता। कुछ लोग इस कथन पर सवालिया निशान लगा सकते हैं। ऐसे लोगों से मैं पूछना चाहता हूँ कि इटली, जर्मनी और रूस में संसदीय जनतंत्र आसानी से कैसे नष्ट हो गया? इंग्लैंड और अमेरिका में यह आसानी से नष्ट क्यों नहीं हुआ? इसका एक ही अहम कारण मुझे दिखाई देता है। इंग्लैंड और अमेरिका में बाकी तीन देशों की तुलना में आर्थिक और सामाजिक जनतंत्र मजबूत रहा है। संसदीय जनतंत्र को मजबूती प्रदान करने के लिए आर्थिक और सामाजिक जनतंत्र का मजबूत होना जरूरी है। जनतंत्र समता का ही दूसरा नाम है। संसदीय जनतंत्र लोगों में स्वतंत्रता के प्रति लगाव का भाव पैदा करता है।

****

प्रत्येक समुदाय का जो बौद्धिक वर्ग होता है उसके ऊपर अपने समुदाय के हितों की सुरक्षा की जवाबदेही होती है।

****

सन् 1942 में लॉर्ड लिनलिथगो ने 52 महत्त्वपूर्ण भारतीय प्रतिनिधियों को आमंत्रित किया। उनमें अनुसूचित जाति के प्रतिनिधि भी शामिल थे। वल्लभभाई पटेल को यह नागवार गुजरा कि वायसराय ने ऐसे छोटे लोगों को बैठक में भाग लेने के लिए आमंत्रित किया था।

****

अगर कोई व्यक्ति अपने पिता का देहांत होने के बाद गंगा नदी के किनारे पुरोहित की देखरेख में हिंदू धर्म के परंपरागत अनुष्ठानों को संपन्न करवाता है तो वह खुद को धर्मनिरपेक्ष नहीं कह सकता। पं. जवाहरलाल ने सन् 1931 में ऐसा ही किया, जब उनके पिता का देहांत हो गया।

****

कांग्रेस शिकायत करती रही है कि अंग्रेज भारत के लोगों का पूरी तरह नि:शस्त्रीकरण कर भारत पर शासन चलाते रहे हैं। मगर वह भूल जाती है कि उच्च वर्ग ने शूद्रों और अस्पृश्यों का नि:शस्त्रीकरण कर शासन चलाने की नीति अपनाई थी।

****

सिर्फ दक्षता को ही कसौटी नहीं माना जा सकता। अगर इसे कसौटी मान लिया जाए तो फ्रांसीसियों का काम जर्मन चला सकते हैं, तुर्कों का काम रूसी चला सकते हैं, चीनियों का काम जापानी चला सकते हैं। जो लोग अच्छे प्रशासन के

लिए दक्षता को ही कसौटी मानते हैं उन्हें फ्रांसीसियों, तुर्कों और चीनियों से इसके बारे में राय पूछनी चाहिए।

***

भारत का शासक वर्ग दक्षता को ही प्रशासन के लिए सबसे अनिवार्य गुण मानता है, चूँकि इस तर्क के जरिए वह अपने एकाधिकार को कायम रखना चाहता है।

***

भारत में शासक वर्ग और शोषित वर्ग के बीच गहरी खाई है।

***

आरक्षण की माँग का डिकी ने संकेत किया है, जिसका समर्थन मार्क्स ने किया और जिसे रूस ने लागू किया।

***

इससे जहाँ वंचित वर्ग के हितों की सुरक्षा सुनिश्चित होगी वहीं आनेवाले समय में शासक वर्ग से दूरी कम करने में भी मदद मिलेगी।

***

अस्पृश्यों के लिए मीडिया में कोई स्थान नहीं है। कांग्रेसी मीडिया का दरवाजा उनके लिए बंद है, जो उन्हें तनिक भी प्रचार देने के लिए तैयार नहीं हैं। अस्पृश्यों का अपना अखबार नहीं हो सकता, चूँकि विज्ञापन के बगैर अखबार चला पाना मुमकिन नहीं है।

□

धर्मांतरण क्यों?

धर्मांतरण का प्रश्न मेरी दृष्टि में बहुत ही महत्त्वपूर्ण बात है। इतना ही नहीं, हम दलित, शोषित, पीड़ित जनता का अगला भविष्य भी मेरी दृष्टि से धर्मांतरण की समस्या पर ही अवलंबित होने के कारण मुझे बहुत अधिक महत्त्वपूर्ण लगता है। इस प्रश्न की गंभीरता और महत्ता को आप सभी लोग अच्छी तरह समझ गए होंगे, यह कहने में अब मुझे कुछ भी संकोच नहीं होगा।

***

जब से हमने धर्मांतरण की घोषणा की है उस दिन से आज तक कई जगह पर छोटी-बड़ी मात्रा में सभा-गोष्ठियों का आयोजन कर अपने लोगों ने इस विषय पर अपना मत प्रदर्शित किया है। यह सभी लोग जानते भी हैं और सबसे सुना भी है, किंतु सब लोगों ने मिलकर, किसी एक जगह इकट्ठा होकर विचार-विमर्श द्वारा धर्मांतरण के प्रश्न की निर्णयात्मक और निश्चयात्मक चर्चा करने का अवसर आज तक हम सब लोगों को नहीं दिया था। ऐसे महत्त्वपूर्ण अवसर की आप सब लोगों से अधिक मुझे आवश्यकता थी। आप सभी लोगों को यह बात माननी पड़ेगी कि धर्मांतरण का आंदोलन सफल बनाने के लिए प्रारंभिक तैयारी की बड़ी आवश्यकता है।

***

धर्मांतरण कोई खेल नहीं, न तो यह मौज का या मनोरंजन का ही विषय है। यह तो संपूर्ण दलित समाज की जिंदगी और मौत का प्रश्न है; लेकिन जैसे नाविक को अपना जहाज एक बंदरगाह पर ले जाने के लिए जितनी मात्रा में प्रारंभिक तैयारी करनी पड़ती है, उतनी ही प्रारंभिक तैयारी धर्मांतरण के लिए भी करनी पड़ेगी। बगैर इस प्रकार की तैयारी किए एक नया किंतु मानवता से परिपूर्ण जीवन

प्राप्त करना कदापि संभव नहीं। नौका में कितने यात्री आने वाले हैं, इसका अंदाज किए बिना नाविक सफर की प्रारंभिक तैयारी नहीं कर सकते। बस, उसी प्रकार की मेरी भी स्थिति है। मेरे लिए धर्मांतरण की प्रारंभिक तैयारी के लिए तब तक जुट जाना संभव नहीं जब तक इस बात का सही अंदाजा न हो कि कितने लोग मेरे साथ धर्मांतरण करेंगे। जब तक हम लोगों द्वारा कहीं-न-कहीं इस प्रश्न पर सोच-विचार के लिए परिषद् का आयोजन नहीं होता तब तक जनमत जानने का अवसर प्राप्त ही नहीं हुआ होता।

***

महारों की ही परिषद् क्यों बुलाई गई और तमाम अछूत वर्ग की परिषद् क्यों नहीं बुलाई गई, इसके कई कारण हैं। इसमें पहला कारण यह है कि इस परिषद् में न तो हमें किसी भी प्रकार की माफी माँगनी है और न ही हिंदू लोगों से कुछ भी सामाजिक हक माँगने हैं। हमें अपने तमाम दलित भाइयों की जिंदगी और मनुष्यता के लिए क्या करना जरूरी है? हमें अपने जीवन-क्रम की रूपरेखा किस प्रकार से तय करनी चाहिए? केवल इतना ही प्रश्न इस परिषद् के सामने है। हमें लगता है कि यह प्रश्न हर दलित जाति और जनजातियों को स्वतंत्र रूप से सोचकर ही हल करना चाहिए। जिन कारणों के लिए मुझे सभी अछूत जातियों की संयुक्त सभा आयोजित करने की आवश्यकता महसूस नहीं हुई, उसमें यह एक कारण है। केवल महारों की ही परिषद् करने में दूसरा भी एक कारण है। धर्मांतरण की घोषणा के बाद आज करीब दस महीने का समय बीत चुका है। इस कालावधि में जन-जागृति का, समाज-प्रबोधन का कार्य बहुत कुछ हुआ है और अब मैं समझता हूँ कि जनमत आजमाने की आवश्यकता है। मुझे लगता है कि सबसे सरल और आसान उपाय वही है।

***

धर्मांतरण के प्रश्न को कृति में परिवर्तित करने के लिए प्रयास करना आवश्यक है। यह प्रयास करने के पहले वास्तव में दलित जनमत की तैयारी कहाँ तक है, इस बात का पूरा विश्वास कर लेना एकदम आवश्यक है। मेरी मान्यता है कि दलित समाज की प्रत्येक जाति को अपनी-अपनी जाति की सभाएँ आयोजित करके दलित जनता का विश्वास प्राप्त करना चाहिए। उसमें जो वास्तविकता उभरकर आएगी, इस प्रकार का स्वरूप आम अछूत वर्ग की परिषद् में न तो दिखाई देगा और न ही तमाम वर्ग के प्रति कुछ निश्चित निष्कर्ष पर पहुँच पाएँगे। चूँकि यदि हम तमाम अछूतों की परिषद् का नाम देकर भी परिषद् का आयोजन करते, तब भी

यह तमाम अछूतों की प्रतिनिधि स्वरूप की ही सभा न हो सकती थी, जिसमें तमाम दलित वर्ग के जनमत का पूरा विश्वास करना असंभव था। इसलिए केवल महार वर्ग की ही परिषद् आयोजित की गई है।

~***~

इस परिषद् में अन्य दलित अछूत जातियों को सम्मिलित न करने पर भी उनकी कुछ हानि होनेवाली नहीं है। यदि उनको धर्मांतरण करने की आवश्यकता महसूस न हो तो उनको इस सभा में सम्मिलित न करने के कारण उनको कुछ भी बुरा लगने का प्रश्न ही नहीं है और यदि उनको धर्मांतरण करने की आवश्यकता महसूस हो तो इस परिषद् में उनको सम्मिलित नहीं किया गया, इसलिए उनके धर्मांतरण करने में किसी भी प्रकार की कठिनाई नहीं होगी। जैसे महार लोगों की परिषद् आयोजित की गई है उसी प्रकार दलित अछूत वर्ग की हर जाति को अपनी-अपनी जाति की परिषदें आयोजित करके इस प्रश्न पर अपना जनमत जाग्रत् करने का और अपनी भावनाओं को पूरी तरह से व्यक्त करने का पूरा अवसर प्राप्त होगा। इसलिए मैं इन्हें सलाह देना चाहूँगा कि इस प्रकार की परिषद् उन्हें भी आयोजित करनी चाहिए। इस काम के लिए उन्हें मेरी ओर से जो भी सहायता चाहिए, मैं अवश्य करूँगा। यहाँ तक तो इस परिवार की केवल प्रारंभिक चर्चा ही हुई है।

~***~

धर्मांतरण का प्रश्न जितना महत्त्वपूर्ण है उतना ही गहन और गंभीर भी है। यह प्रश्न आम आदमी की बौद्धिक बातों की समझ से कोसों दूर है। उसी प्रकार आम आदमी को धर्मांतरण जैसे प्रश्न को समझा देना भी चाहिए, यह उतनी आसान बात नहीं है। इसके बावजूद जब तक आप सभी लोगों को धर्मांतरण का प्रश्न अच्छी तरह नहीं समझाया जा सकता, तब तक धर्मांतरण को अमल में लाना बहुत ही मुश्किल है, इस बात को मैं अच्छी तरह से जानता हूँ और इसलिए मैं जितनी सरलता से इस प्रश्न का विश्लेषण करूँगा, जितने आसान तरीके से और खुले रूप में आप सब लोगों के सामने रखूँगा, उतना ही आप अच्छी तरह समझ पाएँगे और मैं इसकी पूरी कोशिश करूँगा।

~***~

धर्मांतरण के प्रश्न पर दो दृष्टियों से विचार होना चाहिए। सामाजिक दृष्टि से विचार करना चाहिए, धार्मिक दृष्टि से भी विचार करना चाहिए। उसी प्रकार धर्मांतरण पर तात्त्विक दृष्टि से भी विचार करना चाहिए। किंतु किसी भी दृष्टि से

धर्मांतरण का विचार करने पर भी सबसे पहले अस्पृश्यता क्या बला है, उसका वास्तविक स्वरूप क्या है, इसको वास्तविक रूप में समझ लेना बहुत आवश्यक है। इस बात को बिना समझे आपको मेरी धर्मांतरण की घोषणा का अर्थ समझना संभव नहीं है। अस्पृश्यता क्या है और उसका सही स्वरूप क्या है, इसकी जानकारी आप लोगों को होनी चाहिए।

***

सरकारी पाठशालाओं में अपने बच्चों को पढ़ाई के लिए दाखिल करने का हक जताने के कारण, सार्वजनिक और सरकारी कुओं पर, तालाब और पनघट पर पानी भरने का हक जताने के कारण, बारात घोड़े पर ले जाने का हक जताने के कारण, जुलूस और मोर्चे निकालने का, सभा और सम्मेलन कराने का हक जताने के कारण सवर्ण हिंदुओं द्वारा अछूत दलित वर्ग के साथ मारपीट करने की घटनाएँ सुबह से शाम तक और हमेशा ही होते रहने के कारण ये दृश्य सबकी आँखों के सामने हैं।

***

किंतु मारपीट करने के दूसरे कई कारण हैं, जिनका उल्लेख करने पर भारत के बाहर के लोगों को भी दुःख होगा। अछूतों ने कीमती और अच्छे कपड़े पहने, इस कारण सवर्ण हिंदुओं द्वारा उनसे मारपीट किए जाने के असंख्य उदाहरण दिए जा सकते हैं। अछूतों द्वारा पानी भरने के लिए ताँबे और पीतल के बरतन उपयोग करने के कारण सवर्ण हिंदुओं द्वारा उनसे मारपीट किए जाने की असंख्य घटनाएँ दर्ज की जा सकती हैं।

***

अछूतों ने भूमि खरीदी, इस कारण सवर्ण हिंदुओं द्वारा उनके घर-मकान जलाए जाने की असंख्य घटनाएँ बताई जा सकती हैं। अछूतों के जनेऊ पहनने के कारण सवर्ण हिंदुओं द्वारा उन्हें मारने-पीटने की असंख्य घटनाएँ हैं। मरे हुए जानवर न उठाने के कारण सवर्ण हिंदुओं द्वारा उन्हें मारने-पीटने के अनगिनत उदाहरण दिए जा सकते हैं।

***

अछूतों द्वारा अपने पाँवों में जूते और जुराब पहनकर गाँव के भीतर से जाने के कारण, रास्ते में मिले सवर्ण हिंदुओं को झुककर सलाम नहीं करने के कारण, शौच के लिए ताँबे का लोटा ले जाने के कारण उन पर अमानवीय अत्याचार किए जाने की दर्द भरी अनगिनत कहानियाँ हैं। अछूतों द्वारा अपने समाज के पंचों के

भोज में रोटी परोसने के कारण सवर्ण हिंदुओं द्वारा उनके भोज को ही बरबाद करने की एकदम नई घटना है।

***

जुल्म और अत्याचार सवर्ण हिंदुओं द्वारा अस्पृश्य समाज पर सदियों से हो रहे हैं और अब भी इस प्रकार के जुल्म होने की अनगिनत घटनाएँ हैं। सवर्ण हिंदुओं द्वारा अमानवीय व्यवहार करने की कई घटनाएँ आपके सुनने में नहीं बल्कि देखने और स्वयं भोगने में भी आई होंगी। जहाँ सवर्ण हिंदुओं द्वारा अछूतों को मारना-पीटना संभव नहीं होता, वहाँ उन्होंने आपके विरुद्ध बहिष्कार के हथियार का कैसे उपयोग किया, इसका भी आप लोगों को अच्छा-खासा अनुभव होगा।

***

सवर्ण हिंदू पूरी कोशिश करते हैं कि अछूतों को पूरी मजदूरी न मिलने पाए। आपके पशुओं को जंगल में से होकर नहीं जाने दिया जाता। गाँव में अछूतों को नहीं जाने दिया जाता। इस तरह सवर्ण हिंदुओं द्वारा आपके लोगों को परेशान करने की घटनाएँ आप में से अधिकांश लोगों को मालूम होंगी ही; किंतु इस तरह के अन्याय अपने ही समाज पर क्यों होते हैं? इसके मूल में ऐसे कौन से कारण हैं? ऐसा कौन सा रहस्य छिपा हुआ है? इस बात को आप में से बहुत ही कम लोगों ने समझ पाया होगा; किंतु मेरी दृष्टि में इस बात को समझ लेना, इसको जानना बहुत आवश्यक है।

***

उपर्युक्त वर्ग कलह की जो घटनाएँ दी गई हैं, उनसे व्यक्ति की अच्छाई और बुराई का कुछ भी संबंध नहीं है। इन दोनों घटनाओं का आपसी संबंध नहीं है। छुआछूत का प्रश्न वर्ग कलह का प्रश्न है। वह कलह सवर्ण और अछूत—इन दो समाजों का है। एक समाज पर दूसरे समर्थ समाज द्वारा हो रहे अतिक्रमण का यह प्रश्न नहीं है बल्कि एक वर्ग द्वारा दूसरे वर्ग पर जबरदस्ती किए जा रहे अतिक्रमण, जुल्म, शोषण तथा उत्पीड़न का प्रश्न है।

***

यह वर्ग कलह सामाजिक दर्जे से संबंधित वर्ग कलह है। तथाकथित एक ऊँचे वर्ग को दूसरे दलित वर्ग से व्यवहार करते समय अपना व्यवहार कैसे रखना चाहिए, इससे संबंधित यह वर्ग कलह है। इस वर्ग कलह की जो उपर्युक्त घटनाएँ दी गई हैं, उनसे जो एक समस्या स्पष्ट रूप से सिद्ध हो रही है, वह यह है कि आप

लोग परंपरागत ऊपरी वर्ग से व्यवहार करते समय बराबरी के, समानता के रिश्ते से व्यवहार करने का आग्रह करते हैं, इसलिए यह वर्ग कलह उत्पन्न होता है। यदि ऐसा नहीं होता तो रोटी की मेजबानी देने के कारण, अच्छे और कीमती कपड़े पहनने के कारण, जनेऊ पहनने के कारण, ताँबे और पीतल के बरतन में पानी भरने के कारण, घोड़े पर बारात ले जाने के कारण, वह ऊँचे वर्ग के किसी भी व्यक्ति को हानि नहीं करता है, अपना ही रुपया-पैसा खर्च करता है। तब भी ऊँचे वर्ग को उसके इस व्यवहार के प्रति जलन, द्वेष और दर्द क्यों होना चाहिए?

***

इस जलन और क्रोध का एक ही कारण है और वह यह कि इस तरह का समता का व्यवहार सवर्णों की मानहानि का कारण है। वे मानते हैं कि आप निम्न वर्ग के हैं, अपवित्र हैं। निम्न दर्जे में ही आप रहें, तब तो आप लोगों को सुख व शांति से रहने देंगे, अन्यथा नहीं। आपने अपने परंपरागत स्तर से उठने की कोशिश की तो आपके विरुद्ध कलह की आग भड़कने लगती है। सवर्ण हिंदुओं के दिल में आपके प्रति आग के शोले भड़क उठते हैं।

***

एक बात अब निर्विवाद रूप से सिद्ध हो चुकी है, वह यह कि अस्पृश्यता नैमित्तिक न होकर नित्य की चीज है और सदियों से चलती आ रही है। यही बात स्पष्ट रूप से कहनी हो तो यह कहा जा सकता है कि यह सवर्ण और अछूत वर्गों का नित्य कलह है और यह हमेशा के लिए रहनेवाला है। जिस धर्म के कारण आप लोगों को इतना निम्न दर्जा दिया गया है, वह धर्म ऊँचे वर्ग के कथनानुसार सनातन है। उसमें किसी भी प्रकार से समयानुकूल परिवर्तन नहीं हो सकता है। आज आप लोग जिस निम्न दर्जे के हैं, उसी प्रकार आपको हमेशा के लिए ही निम्न दर्जे का रहना चाहिए।

***

इसका अर्थ यह है कि सवर्ण और अछूत इनमें यह कलह, यह वर्ग संघर्ष स्थायी रूप से रहनेवाला है। इस वर्ग कलह, वर्ग संघर्ष से आप लोगों का कैसे बचाव हो, यही मुख्य प्रश्न है और इस प्रश्न का विचार किए बिना आप लोगों के पास अन्य रास्ता नहीं है, यह मेरी स्पष्ट धारणा है। जिन्हें स्वाभिमान की जिंदगी से प्रेम हो, जिनको मानवता और मनुष्यता से प्रेम हो; जिन्हें समता, बंधुता, स्वतंत्रता और न्याय का जीवन आवश्यक लगता हो, उनके लिए इस प्रश्न पर विचार किए बिना कोई अन्य रास्ता नहीं है। किस तरह से इस वर्ग कलह से अपना बचाव किया

जा सकता है, इसका विचार करना अति आवश्यक है। इस वर्ग कलह से अपना बचाव कैसे होगा? इस प्रश्न का निर्णय करना मैं समझता हूँ, मेरे लिए कोई असंभव बात नहीं।

***

यहाँ एकत्र हुए आप सभी लोगों को इस एक बात को स्वीकार करना पड़ेगा, और वह यह कि किसी भी कलह में, संघर्ष में, जिनके हाथ में सामर्थ्य होता है, उन्हीं के हाथ में जय होती है। जिनमें सामर्थ्य नहीं है उन्हें अपनी विजय की आशा रखना बेकार बात है। यह बात आप सबको अनुभव से सिद्ध हो चुकी है। इसके समर्थन में और अन्य आधार खोजने की बात बेकार है। इसलिए तमाम दलित वर्ग को अब अपने हाथों में सामर्थ्य और शक्ति को इकट्‌ठा करना बहुत ही जरूरी है। अब उन्हें अपनी शक्ति का, अपने संगठन-सामर्थ्य का प्रदर्शन करने की नितांत आवश्यकता है।

***

इसके बाद का प्रश्न जिस पर आप लोगों को गंभीरता से सोचना है, वह यह है कि इस वर्ग कलह से मुक्ति पाने के लिए आपके पास आवश्यक सामर्थ्य है या नहीं? मनुष्य समाज के पास तीन प्रकार की सामर्थ्य होती है—एक है जन-बल अर्थात् संगठन की शक्ति। दूसरी है धन-बल, और तीसरी है मानसिक बल अर्थात् आत्मविश्वास और बुद्धि। मानव समाज के पास यह तीन तरह का बल होता है। इसमें से किस प्रकार का बल आपके पास उपलब्ध है? आपको अपने बारे में क्या लग रहा है?

***

आप लोग मनुष्य बल की दृष्टि से अल्पसंख्यक हैं यह बात एकदम स्पष्ट है। बंबई क्षेत्र में अछूत वर्ग की जनसंख्या 2/8 के बराबर है, और वह भी संगठित नहीं है। उनमें आपसी जाति-भेद के कारण संघ शक्ति का पूरी तरह से अभाव है। फिर संगठित भी नहीं। इतना ही नहीं, बल्कि इकट्‌ठे भी नहीं रहते हैं। ये सब लोग गाँव-देहात में बिखरे हुए हैं। इस कारण जो भी थोड़ी-बहुत जनसंख्या है, उसका भी अत्याचार और अन्याय से पीड़ित अछूत वर्ग की बस्ती के लिए किसी भी प्रकार का उपयोग नहीं हो सकता।

***

धन-बल की दृष्टि से देखा जाए तो आप लोगों का वही हाल है। आपके पास थोड़ा-बहुत जन-बल तो है, किंतु आपके पास धन-बल कुछ भी नहीं है। आप लोगों का न कोई उद्योग-धंधा है और न ही कोई व्यापार है। नौकरी-पेशा भी

आप लोग नहीं हैं और न कोई जमीन-जायदाद ही है। आप लोग न तो अच्छा खाना खा सकते हैं और न ही अच्छी तरह से रहने के लिए आपके पास कोई मकान है। न ही अच्छा कपड़ा-लत्ता पहनने को मिलता है और न लेने दिया जाता है।

***

आपके पास धन-बल की बात तो एक तरफ रही, यदि आप पर जुल्म और अन्याय हुआ तो कोर्ट से न्याय और संरक्षण माँगने की हैसियत आपके पास कहाँ है? आपमें से हजारों-लाखों लोग कोर्ट का खर्चा बरदाश्त करने की हैसियत नहीं होने के कारण सवर्ण हिंदुओं द्वारा होनेवाले अतिक्रमण, रोष और अत्याचार को मुर्दों की तरह बरदाश्त कर रहे हैं। मानसिक बल की तो उससे भी बुरी हालत है।

***

सैकड़ों सालों से सवर्ण हिंदुओं द्वारा हो रहे अत्याचार, धिक्कार मुर्दों की तरह सहन करने के कारण प्रतिकार करने की शक्ति और हिम्मत पूरी तरह से नष्ट हो चुकी है। आप लोगों के आत्मविश्वास, उत्साह और महत्त्वाकांक्षी होने की चेतना का मूलोच्छेद कर दिया गया है। आप सभी लोग निराशा, नाउम्मीद, विपन्न और निस्तेज बना दिए गए हैं, बन चुके हैं। निराशा का वातावरण चारों ओर फैला हुआ है। किसी के दिल-दिमाग में 'हम भी कुछ कर सकते हैं', इस तरह का विचार पैदा ही नहीं हो रहा है और न कोई सोचता है और न किसी में इस तरह के विचार आ सकते हैं।

***

मैंने वास्तविकता का यह जो विश्लेषण किया है, यदि यह सत्य है तो उससे जो सिद्धांत निकलेगा, उसे भी आप लोगों को स्वीकार करना होगा और वह यह है कि आप अपने ही व्यक्तिगत सामर्थ्य पर निर्भर रहेंगे तो आपके लिए इस अन्याय का प्रतिकार करना संभव नहीं है। आप लोगों के सामर्थ्यहीन होने के कारण ही आप लोगों पर अत्याचार और अन्याय होता है, इसमें मुझे किसी भी प्रकार का संदेह नहीं है।

□

भारत-विभाजन

आक्रामक साम्राज्यवाद के खिलाफ आजादी की लड़ाई रह रहे राष्ट्रवादी एक आक्रामक बहुसंख्यक के राष्ट्रवाद से आजादी की माँग कर रहे अल्पसंख्यकों के अधिकारों को नकारने के लिए ब्रिटिश साम्राज्यवादियों से मदद की माँग नहीं कर सकते। इस मुद्दे को हिंदुओं और मुसलमानों को आपस में बात कर सुलझाना होगा। ब्रिटिश उनके इस मुद्दे का फैसला नहीं कर सकते। यह पहला अहम बिंदु है, जिस पर गौर करने की जरूरत है।

~***~

यह मानना बहुत बड़ी गलती होगी कि अगर अभी पाकिस्तान की माँग को दबा दिया गया तो वह फिर कभी सिर नहीं उठाएगी। मैं यकीन के साथ कह सकता हूँ कि माँग को दबाने से पाकिस्तान का प्रेत पीछा करना नहीं छोड़ेगा।

~***~

हिंदुओं की युक्तियों में कई त्रुटियाँ हैं। जो पहली दलील दी गई है कि दोनों पक्षों के बीच समानता है, वह एक-दूसरे के साथ समन्वय करने के किसी जागरूक प्रयत्न का नतीजा नहीं है। अगर कोई समानता दिखाई देती भी है तो उसके पीछे कुछ यांत्रिक कारण हैं। ऐसा अपूर्ण धर्मांतरण के कारण भी है। भारत जैसे देश में निचली जाति के हिंदुओं का धर्मांतरण किया गया। इस तरह का धर्मांतरण कभी प्रभावशाली साबित नहीं हो पाया। शायद विद्रोह के डर से पूरी आबादी को मुसलिम बनाने का कभी प्रयास नहीं किया गया या फिर उस समय जिन मौलवियों ने धर्मांतरण की प्रक्रिया को संपन्न करवाया, वे असरदार तरीके से अपनी भूमिका का निर्वाह कर पाने में सफल नहीं हो पाए। यही वजह है कि मुसलमानों का एक बड़ा तबका अपने हिंदू मूल का उल्लेख करता रहता है और अपने धार्मिक व सामाजिक जीवन

में इस बात का परिचय भी देता रहता है। सदियों से एक ही माहौल में हिंदू और मुसलमान रहते आए हैं। समान परिवेश में रहने से प्रतिक्रियाओं में भी समानता होना स्वाभाविक है। कुछ समानता के पीछे अकबर कालीन धार्मिक समरसता के प्रयत्नों को माना जा सकता है।

***

क्या कोई ऐसी ऐतिहासिक घटना या विरासत है, जिसे लेकर हिंदू और मुसलमान समान रूप से गर्व महसूस करते हों? यही अहम सवाल है, जिसका जवाब हिंदुओं को देना चाहिए, अगर वे हिंदू और मुसलमानों को मिलाकर एक देश बनाना चाहते हैं। अब तक दोनों का रिश्ता सशस्त्र सैन्य टुकड़ियों जैसा रहा है, जो आपस में लड़ती रहती हैं। किसी साझी उपलब्धि में दोनों पक्षों की साझी प्रतियोगिता दिखाई नहीं देती। उनका अतीत शत्रुतापूर्ण रहा है। दोनों के बीच राजनीति के साथ-साथ धार्मिक क्षेत्र में भी टकराव होता रहा है।

***

इतिहास बताता है कि राष्ट्रीयता का सिद्धांत लोगों की इच्छा की संप्रभुता के जनतांत्रिक सिद्धांत से जुड़ा होता है। इसका मतलब है कि किसी राष्ट्र के निर्माण की माँग करने के लिए किसी तरह की शिकायत जरूरी नहीं होती। इसे तर्कसंगत ठहराने के लिए लोगों की इच्छा ही पर्याप्त होती है।

***

लेकिन, अपने दावे के लिए अगर उन्हें शिकायतें दर्ज करनी पड़ें, तो मुसलमानों के पास ढेर सारी शिकायतें हैं। उन्हें एक वाक्य में व्यक्त किया जा सकता है—हिंदू बहुसंख्यकों के जुल्म से बचा पाने में संवैधानिक सुरक्षा नाकाम साबित हुई है।

***

अंग्रेजों ने भारत में भारतीयों के हित में काफी अच्छे कदम उठाए हैं। उन्होंने सड़कें बनाईं, नहरों का निर्माण किया, रेल सेवा का विकास किया, एक व्यवस्था का प्रचलन शुरू किया, तार संदेश पहुँचाना शुरू किया, मुद्रा में सुधार किया, माप-तौल का सही निर्धारण किया। भूगोल, अंतरिक्ष विज्ञान और औषधि-शास्त्र के ज्ञान का प्रसार किया। अंदरूनी कलह को शांत कर कुछ हद तक उनकी भौतिक उन्नति की। क्या अच्छे प्रशासन के बदले भारतीयों से कहा जाएगा कि कृतज्ञतावश वे अंग्रेजों का विरोध करना छोड़ दें और अपने आंदोलन को वापस ले लें?

***

जो मनुष्य अपने अस्तित्व को लेकर जागरूक होता है वह पशु की तरह

सिर्फ अपना पेट भरने की चिंता नहीं करता। ऐसे मनुष्य के पास अपना स्वाभिमान होता है, जिसे धन-दौलत से भी नष्ट नहीं किया जा सकता।

***

यह कहने से कोई फायदा नहीं कि कांग्रेस एक हिंदू संगठन नहीं है। यह हिंदुओं का ऐसा संगठन है, जो अपने धर्म के नजरिए को उजागर करता है और जो अपने धर्म की आकांक्षाओं का समर्थन करता है। कांग्रेस और हिंदू महासभा के बीच यही अंतर है कि महासभा का रवैया अनुदार है, वहीं कांग्रेस नम्र बरताव करती है। इन बातों को छोड़कर कांग्रेस और हिंदू महासभा के बीच कोई अंतर नहीं है।

***

भारत पर जब अंग्रेजों ने कब्जा कर लिया तो दो समुदायों की स्थिति में एक किस्म का राजनीतिक बदलाव आया। छह सौ सालों से मुसलमान हिंदुओं पर राज करते आ रहे थे। अंग्रेजों का राज स्थापित होते ही मुसलमानों को हिंदुओं के बराबर का ओहदा दिया गया। मालिक से नौकर बनना शर्म की बात थी। अगर अब उन्हें हिंदुओं का आज्ञा-पालक बनना पड़े तो यह और भी अपमानजनक बात होगी। मुसलमान पृथक् राष्ट्र बनाकर इस तरह के अपमान से बचना चाहते हैं।

***

भारत पर हमला करते समय अफगान तातार को नष्ट कर देना चाहते थे और मुगल तातार के साथ-साथ अफगानों को नष्ट कर देना चाहते थे। उनमें इसलामी भ्रातृत्व की भावना के चलते किसी तरह का मैत्रीपूर्ण संबंध नहीं था। वे एक-दूसरे के कट्टर प्रतिद्वंद्वी थे और अकसर एक-दूसरे के खिलाफ खूनी जंग छेड़ते रहते थे। लेकिन गौर करने की बात है कि हिंदुओं की आस्था को ध्वस्त करने के उद्देश्य को लेकर उनके बीच गजब की एकता थी।

***

इसमें कोई संदेह नहीं कि मुसलमान हमलावर हिंदुओं के खिलाफ नफरत की भावना लेकर भारत में आए थे। मगर वे केवल नफरत की भावना साथ लेकर कुछ मंदिरों को नष्ट करके ही वापस नहीं लौट गए थे। वे केवल नकारात्मक नतीजे ही नहीं चाहते थे। उन्होंने इसलाम के बीज बिखेरकर सकारात्मक कदम भी उठाया था। उस पौधे की वृद्धि अनोखी कही जा सकती है।

***

इस बात को स्वीकार करना पड़ेगा कि पाकिस्तान का निर्माण हो जाने के बाद भी भारत में सांप्रदायिकता का मसला मौजूद ही रहेगा। बँटवारे के समय अगर आबादी का आदान-प्रदान नहीं किया गया तो पाकिस्तान बन जाने के बावजूद भारत में अल्पसंख्यक बनाम बहुसंख्यक की समस्या बनी ही रहेगी।

***

दुर्भाग्य की बात है कि सवर्ण हिंदू नेता के रूप में भी फिसड्डी ही साबित होते हैं। वे दूसरों के साथ जीवन की सुखद चीजों को बाँटना पसंद नहीं करते। वे धन और शिक्षा पर एकाधिकार को कायम रखना चाहते हैं। धन और शिक्षा की सहायता से ही उन्होंने शासन पर वर्चस्व बनाए रखा है। वे अपने इस एकाधिकार को भविष्य में भी बनाए रखना चाहते हैं और यही उनके जीवन का मकसद भी होता है। अपना वर्चस्व कायम रखने के लिए वे दलित वर्ग को धन, शिक्षा और सत्ता से बाहर रखना चाहते हैं। ऐसा करने के लिए वे हर तरह के हथकंडों का इस्तेमाल करते हैं।

***

बंगाल और पंजाब के हिंदुओं को विचार करना चाहिए कि वे किस विकल्प को चुनना पसंद करेंगे। मुझे लगता है कि वह समय आ गया है कि बंगाल और पंजाब के सवर्ण हिंदुओं को बता देना चाहिए कि अगर रोजगार के समृद्ध क्षेत्र से वंचित हो जाने के डर से वे पाकिस्तान के निर्माण का विरोध कर रहे हैं तो बहुत बड़ी गलती कर रहे हैं। स्थान और सत्ता पर उनका अब तक जो एकाधिकार था, अब वह कायम नहीं रहनेवाला है। वे राष्ट्रवाद के नाम पर अस्पृश्यों को धोखा दे सकते हैं, मगर मुसलिम प्रांतों में वे मुसलमानों की आँखों में धूल नहीं झोंक सकते; जैसा लिंकन ने कहा था कि 'आप हमेशा लोगों को बेवकूफ नहीं बना सकते।'

***

जहाँ कानून का राज होता है वहाँ हत्यारों को सजा मिलती है। गण्यमान्य मुसलमानों ने कभी मुसलिम अपराधियों की निंदा नहीं की है; बल्कि वे धार्मिक शहीद के रूप में वे उन्हें महिमा-मंडित करते रहे हैं और उन्हें सजा से बचाने के लिए आंदोलन चलाते रहे हैं। उनके इस रवैये के उदाहरण के रूप में लाहौर के वकील बरकत अली का उल्लेख किया जा सकता है, जिन्होंने अब्दुल कयूम को बचाने के लिए मुकदमा लड़ा था। बरकत अली ने तर्क दिया कि कयूम नाथूरमल के समान हत्यारा नहीं था। चूँकि उसने जो गुनाह किया, वैसे गुनाह को 'कुरान'

की मंजूरी मिली हुई है। मुसलमानों के ऐसे रवैये को आसानी से समझा जा सकता है। केवल महात्मा गांधी का रवैया समझ में नहीं आता। वे इस मसले पर खामोश बने हुए हैं।

***

तुर्की और चेकोस्लाविया के घटनाक्रम का गहन विश्लेषण करने से स्पष्ट होता है कि स्थानीय स्वायत्तत या धर्म का बंधन ही राष्ट्रवाद को मजबूत बनाने के लिए पर्याप्त नहीं है।

***

बाल विवाह जैसे मामलों में क्या मुसलमानों की स्थिति हिंदुओं से बेहतर मानी जा सकती है?

***

मुसलिम स्त्रियों को पर्याप्त अधिकार मिले हुए हैं और काफी हद तक स्वतंत्रता भी मिली हुई है।

***

वे कहते हैं कि मुसलिम कानून में शादी की कोई उम्र निर्धारित नहीं की गई है और लड़कियों की शादी के लिए उनकी सहमति को जरूरी माना गया है। अगर बचपन में किसी लड़की की शादी हो जाती है तो वयस्क होने पर अपने पति को अस्वीकार करने का अधिकार उसके पास होता है।

***

मुसलमानों की शादी एक अनुबंध की तरह है। अनुबंध होने के कारण पति को अपनी पत्नी को तलाक देने का अधिकार होता है। वहीं स्त्रियों के हितों की सुरक्षा के लिए भी व्यापक प्रावधान रखा गया है।

□

पत्रों में व्यक्त विचार

आपको याद रखना चाहिए, जो शेक्सपियर ने कहा है, 'इनसानों के जीवन में उतार-चढ़ाव आता है। उससे लड़कर यदि इस बाढ़ को पार कर लिया तो किस्मत बन जाती है और यदि नजरअंदाज कर दिया तो उनके जीवन का सफर उथला हो जाता है और कष्टों से घिर जाता है।' यह जमादारों और सूबेदारों के बच्चों पर खासतौर से लागू नहीं होता, जिनको रेखांकित करके हम उनकी गिनती कर सकें। फिर भी उसका दोष मैं उनके अभिभावकों को देना चाहता हूँ, न कि उन बच्चों को, जैसे कि वे भी अभागे हैं। हमें इस धारणा को छोड़ देना चाहिए कि अभिभावक बच्चे को 'जन्म' देते हैं, न कि 'कर्म'। वे अपने बच्चों का भाग्य भी बदल सकते हैं और यदि हम केवल इस सिद्धांत का अनुसरण करते हैं, तब हमें निश्चित रूप से मान लेना चाहिए कि हम जल्द ही बेहतर दिन देखेंगे और हमारी प्रगति और बढ़ जाएगी।

~***~

आप मुझसे सहमत होंगे, यदि मैं यह कहूँ कि भारत जैसा देश सामाजिक, आर्थिक और राजनीतिक रूप से बुरी तरह जकड़ा हुआ है। अब समय की सर्वश्रेष्ठ माँग यह है कि ईमानदार नेतागण इसके पुनर्निर्माण का पूर्ण उत्तरदायित्व अपने ऊपर लें। ऐसे नेता की झलक मुझे सर्वश्री गोखले और पी.एम. मेहता में दिखाई दी थी, जिन्होंने अपने उद्‌देश्य-प्राप्ति हेतु उत्साह दरशाया एवं बलिदान दिया, दोनों ही अत्यंत धन्यवाद के पात्र हैं। लोगों के साथ उनकी पहचान और उनमें निर्भरता इतनी प्रगाढ़ थी कि उनकी अकस्मात् मृत्यु कुछ क्षणों के लिए प्रकृति का छल लगी थी। उनके प्रति आभार यह है कि उन्होंने हमारी बहुत सी समस्याओं को सुलझाने के लिए संघर्ष किया। उनमें से कुछ को सुलझा लिया गया और शेष के लिए हम लोगों को

सामना करने के लिए छोड़ दिया। किंतु उनके द्वारा किए गए कार्यों के प्रति अपनी भावनाओं को हम लोगों ने आनेवाले समय में किए जानेवाले काम के लिए पूरी तरह से अपने को अलग करके रख लिया—फिर भी हम लोग उनके लिए आभार प्रकट करते हैं।

***

यह स्वाभाविक है कि उनकी स्मृतियों को ताजा रखने के लिए उनके स्मारक बनाने चाहिए, जिन्होंने हमारे लिए हमारी आज की समस्याओं का सामना करते हुए निस्स्वार्थ भाव से कड़ी मेहनत की है।

***

भारत के समाचार-पत्र, जो मेरे पढ़ने में आए हैं, वे बताते हैं कि मिस्टर गोखले का स्मारक इंडियन सोसायटी के सेवकों की शाखाएँ विभिन्न स्थानों पर स्थापित करना चाहती हैं। जबकि सर पी.एम. मेहता का स्मारक बंबई म्युनिसिपल ऑफिस के सामने एक पुतले के रूप में खड़ा हुआ है।

***

मैं व्यक्तिगत रूप से कहना चाहता हूँ कि सर पी.एम. मेहता का यह स्मारक उनके सम्मान व गरिमा के मुकाबले बहुत ही मामूली और छोटा है। मैं उनके इस स्मारक को अच्छा तभी मानता जब यह आनेवाली पीढ़ियों के लिए स्थायी रूप से भी उपयोगी होता।

***

इन दो उद्देश्यों को साथ में मिलाते हुए मैं नम्रतापूर्वक यह विचार करने का प्रस्ताव करता हूँ कि सर पी.एम. मेहता का स्मारक बंबई में एक सार्वजनिक पुस्तकालय के रूप में होना चाहिए और उसका नाम हो 'सर फिरोजशाह मेहता लाइब्रेरी'।

***

बड़ा दुर्भाग्य है कि अभी तक हम पुस्तकालय का मूल्य नहीं समझ पाए हैं, जबकि यह समाज की उन्नति और प्रगति की संस्था होती है। उसकी अच्छाइयों को विस्तार से बताने के लिए यह जगह नहीं है। बंबई के प्रबुद्ध जनसाधारण एक अच्छे पुस्तकालय के लिए लंबे समय से तरस रहे हैं। इसमें निराश होने जैसी कोई बात नहीं, समय से पहले सुधार करना अच्छा रहता है।

***

बंबई में कुछ निजी पुस्तकालय हैं, जो कुछ लोगों द्वारा व्यक्तिगत रूप से

चलाए जा रहे हैं। यदि वे अव्यवस्थित संस्थाएँ सर पी.एम. मेहता स्मारक फंड से उनके नाम से एक ही इमारत में लामबंद हो जाएँ तो बंबई शहर इन दोनों उद्‍देश्यों को प्राप्त कर लेगा। जहाँ तक किताबें खरीदने के लिए फंड और आधुनिक प्रबंधन का सवाल है, मुझे विश्वास है कि यहाँ के लोग उदारता से दान दे देंगे।

***

मैं अमेरिका के बड़े विश्वविद्यालयों में से एक का विद्यार्थी होने के नाते यह विश्वास करता हूँ कि पुस्तकालय एक ऐसी जगह होती है, जहाँ लोगों के बौद्धिक एवं सामाजिक विकास और प्रतिभा में वृद्धि होती है। मुझे बड़े दुःख के साथ यह कहना पड़ रहा है कि बंबई शहर में इसकी कमी है। अतः इस सुअवसर का लाभ उठाते हुए मैं बंबई की जनता से प्रार्थना कर रहा हूँ कि आधुनिक भारत के इस महानायक का सुव्यवस्थित एवं स्थायी स्मारक बनाने का पुण्य कार्य करें।

***

'मूकनायक' समाचार-पत्र की व्यवस्था ही लीजिए। मेरी कितनी हानि हुई, मैंने कितने कष्ट झेले व उत्तरदायित्व लिया। इतना अन्य किसी ने नहीं किया। तब भी घोलप के 'अपना नया वर्ष' नामक अग्र लेख में मेरे नाम का उल्लेख भी नहीं है। उन्होंने स्वयं की और भटकर की तारीफ की है, परंतु इस बात का मुझे लेशमात्र का भी खेद नहीं है। 'मूकनायक' चल रहा है, इसमें ही मुझे संतोष है।

***

ऐसा आचरण यदि रखेंगे तो विवाद के लिए कोई कारण उपजेगा, ऐसा मुझे नहीं लगता है। जाति-द्वेष या अन्य किसी कारण से कुछ लोग आपको अपमानित करते हैं तो इसका मुझे बहुत दुःख है। ईर्ष्या करनेवाले लोगों से निभाना बहुत कठिन होता है, यह मैं जानता हूँ; फिर भी ईर्ष्या का कारण न बनें, ताकि बातें जिस स्तर पर आज पहुँची हैं, उस स्तर तक न पहुँचें।

□

संविधान विषयक विचार

मुझे नहीं लगता कि संविधान तैयार करने के लिए किसी संविधान सभा की कोई आवश्यकता है। भारत के संविधान के विषय में काफी कुछ भारत सरकार अधिनियम, 1935 में पहले से ही लिखा जा चुका है। ऐसा लगता है कि संविधान सभा की नियुक्ति मात्र उपलब्ध संविधान की नकल करने के लिए ही की गई है। जरूरी यह है कि हम भारत सरकार अधिनियम, 1935 के उन अनुच्छेदों को हटाएँ, जो हमारे प्रभुता-संपन्न राष्ट्र के लिए गैर-जरूरी हैं।

~***~

प्रारूप समिति को संविधान तैयार करने का दायित्व संविधान सभा के निर्देश पर दिया गया था। हमें विभिन्न समितियों; जैसे—केंद्रीय सत्ता समिति, संविधान समिति, प्रांतीय संविधान समिति तथा मूलभूत अधिकारों, अल्पसंख्यकों, जनजातीय क्षेत्रों हेतु सलाहकार समितियों द्वारा प्रस्तुत रिपोर्टों के आधार पर संविधान सभा के आदेश पर संविधान की रचना करनी थी। मैं आशा करता हूँ कि प्रारूप समिति ने उसे दिए गए निर्देशों के अनुसार ईमानदारीपूर्वक अपने दायित्व का निर्वहण किया है।

~***~

मैं यहाँ संविधान सभा में कोई बड़ी महत्त्वाकांक्षा लेकर नहीं आया था। मेरा यहाँ आने का एकमात्र उद्‍देश्य अनुसूचित जातियों के हितों के संरक्षण की बात उठाना था। मुझे कल्पना में भी यह विश्वास नहीं था कि मुझे सर्वाधिक महत्त्वपूर्ण दायित्व सौंपे जाने के लिए बुलाया गया है। यह देखकर मुझे और भी आश्चर्य हुआ कि संविधान सभा ने मुझे प्रारूप समिति में चुना। इससे भी अधिक आश्चर्य मुझे तब हुआ, जब प्रारूप समिति ने मुझे अपना अध्यक्ष चुन लिया।

~***~

कानून मंत्री बनाए जाने का प्रस्ताव मेरे लिए बड़े आश्चर्य की बात है। मैं विपक्षी गुट में था और अगस्त 1946 में अंतरिम सरकार के गठन के समय से ही सरकार की

असमता पर प्रहार कर रहा था। मुझे यह अनुमान अभी भी लगाना है कि ऐसा क्या हो गया, जिससे प्रधानमंत्रीजी के मन में मेरे प्रति परिवर्तित दृष्टिकोण उत्पन्न हो गया?

~***~

संविधान सभा में और कुछ नहीं बल्कि अराजकता ही होनी थी; किंतु इस सभा में कांग्रेस पार्टी की उपस्थिति ने इस संभावना को पूरी तरह खत्म कर दिया। कांग्रेस ने इसकी कार्रवाइयों में अनुशासन तथा नियम की भावना प्रदर्शित की। यह कांग्रेस पार्टी के अनुशासन का ही कमाल था कि प्रारूप समिति इस पूर्ण विश्वास के साथ संविधान सभा में संविधान का प्रारूप प्रस्तुत कर सकी कि प्रत्येक लेख तथा प्रत्येक सुधार का क्या भविष्य होगा। संविधान सभा में समय समिति के सुचारु ढंग से पारित हो जाने का पूरा श्रेय कांग्रेस पार्टी को दिया जाना चाहिए।

~***~

मैंने गांधीजी से कुछ भी ग्रहण नहीं किया और न ही उन्होंने मेरे आध्यात्मिक, नैतिक अथवा सामाजिक दृष्टिकोण को प्रभावित किया। गांधीजी इस देश के लिए एक रचनात्मक खतरा बन गए थे। उन्होंने सभी विचारों के लिए रास्ते बंद कर दिए थे।

~***~

चूँकि कांग्रेस सुधारवादी पार्टी नहीं है, अत: कांग्रेस पर इस बात का भरोसा नहीं किया जा सकता कि वह सामाजिक सुधारों अथवा आर्थिक पुनर्रचना के कार्यक्रम चलाएगी, जिनके बगैर दलितों का कभी भला नहीं हो सकता। इसलिए दलितों का ऐसी पार्टी में शामिल होना निरर्थक और मूर्खतापूर्ण कार्य होगा। कांग्रेस कभी उनकी भलाई के लिए कुछ नहीं करेगी, बल्कि वह हमेशा की तरह उनका इस्तेमाल ही करती रहेगी।

~***~

जो दलित आज कांग्रेस में हैं, वे भी हाशिए पर पड़े हुए हैं तथा यह हाशिया इतना विशाल है कि यह खत्म नहीं किया जा सकता। ये ही वे नुकसान हैं, जिन्हें दलितों को कांग्रेस में शामिल होकर उठाना होगा। ये केवल नुकसान नहीं हैं, मैं इन्हें 'भयानक दुष्परिणाम' कहूँगा। आज सभी सामाजिक आंदोलन मृतप्राय हैं। सारी राजनीतिक शक्ति दूर जा चुकी है। यह कहना ठीक नहीं होगा कि नए संविधान से दलितों को सबकुछ मिल जाएगा। वास्तविकता यह है कि वे जंजीरों में जकड़े हुए हैं।

~***~

कुछ दलित लोग कांग्रेस के टिकट पर चुनाव लड़ रहे हैं। वे महत्त्वाकांक्षी व्यक्ति

मात्र हैं, जो विधायिका की सीढ़ी पर इस आशा से चढ़ना चाहते हैं कि रेवड़ियाँ बाँटे जाते समय उन्हें भी कुछ-न-कुछ जरूर हासिल होगा। उन्हें इस बात की चिंता नहीं है कि वे किसकी सीढ़ी से चढ़ना चाहते हैं। ये महत्त्वाकांक्षी दलित लोग भी कांग्रेस में शामिल कभी नहीं होते, अगर उनके लिए कांग्रेस के बिना निर्दलीय रूप से चुनाव जीतना संभव होता।

***

जो कोई भी अपने दिमाग का इस्तेमाल करेगा, वह इस बात का अहसास करेगा कि कांग्रेस सूरज और चाँद की तरह अमर नहीं है। एक दिन इसका भी अंत होना है।

***

मेरा दृढ़ विश्वास है कि दलितों का कांग्रेस अथवा किसी भी बड़े राजनीतिक दल में शामिल होना उनके लिए घातक सिद्ध होगा।

***

कोई भूखा व्यक्ति उस आलोचक को पसंद नहीं करेगा, जो उससे कहे कि मेरे प्यारे दोस्त, यद्यपि मैं सत्ता में हूँ, यद्यपि मैं अधिकार धारण करता हूँ, यद्यपि मुझे व्यवस्था ठीक करने का वैधानिक अधिकार है; फिर भी तुम मुझसे किसी चमत्कार की उम्मीद मत करो, क्योंकि मुझे उत्तराधिकार में ही ये सब समस्याएँ प्राप्त हुई हैं। यदि सरकार एक निश्चित समय सीमा के अंदर परिणाम नहीं देती तो लोग इतने निराश, इतने दुखी हो जाते हैं कि वे यह मानने लगते हैं कि सरकार नाम की कोई चीज ही नहीं है। यदि हम संसद् में बैठकर इस विषय में अपनी जिम्मेदारियाँ नहीं समझेंगे और एक निश्चित समय के अंदर लोगों के हित तथा कल्याण के कार्यों को नहीं करेंगे तो मैं समझता हूँ कि एक दिन ऐसा अवश्य आएगा। मेरे मन में इस बात की जरा भी शंका नहीं है कि ऐसा होने पर संसद् के बाहर लोग इसे घृणा की दृष्टि से देखने लगेंगे और ऐसी नौबत कभी भी नहीं आने देना चाहिए।

***

संसद् की सदस्यता के लिए शिक्षा एकमात्र योग्यता नहीं हो सकती। यदि मैं इसे बुद्ध के शब्दों में कहूँ तो उन्होंने कहा था कि मनुष्य को दो गुणों की आवश्यकता है—पहला है ज्ञान और दूसरा है शील। शील के बगैर ज्ञान बड़ा ही खतरनाक होता है। इसके साथ शील होना अत्यंत आवश्यक है, जिसके द्वारा हममें चरित्र, साहस, प्रलोभन से मुक्त होने जैसे सद्गुण विकसित होते हैं। मैं यह देखने को उत्सुक हूँ कि इस महान् सदन में शील से रहित किसी व्यक्ति का प्रवेश नहीं होना चाहिए।

***

अगर राजनीतिक दल अपने निहित स्वार्थों के कारण इस मसले पर ध्यान नहीं देंगे तो समय के साथ लोग स्वयं इस बात का ध्यान रखने लगेंगे। लोग ऐसे किसी भी व्यक्ति को इस सदन में प्रवेश की अनुमति नहीं देंगे, जो इस सदन में बैठकर अपने दायित्वों का सही ढंग से निर्वाह न कर सके, न ही वे उसे दुबारा वापस ही आने देंगे। वे परिणाम चाहते हैं। वे चाहते हैं कि उनकी समस्याओं पर ध्यान दिया जाए और मुझे पूरा विश्वास है कि वे यह लक्ष्य तभी प्राप्त कर सकेंगे जब वे अच्छे लोगों को ही इस सदन में भेजेंगे।

***

अब समय बदल चुका है। हम सभी अब नए आदर्शों की ओर मुड़ रहे हैं। वे अब जनता के लिए सरकार चाहते हैं और उन्हें इस बात से कोई फर्क नहीं पड़ता कि वह जनता की सरकार है अथवा जनता द्वारा सरकार है। अगर हम इस संविधान को अक्षुण्ण रखना चाहते हैं, जिसमें हमने जनता की सरकार, जनता के लिए सरकार तथा जनता द्वारा सरकार का सिद्धांत समाहित किया है तो हमें अपने पथ में आनेवाली बुराइयों के प्रति लापरवाह न होने का संकल्प लेना होगा। ये बुराइयाँ ही जनता को जनता द्वारा सरकार के स्थान पर जनता के लिए सरकार को पसंद करने के लिए बाध्य करेंगी।

***

यदि हम लोकतंत्र को रूप में ही नहीं अपितु यथार्थ में भी बरकरार रखना चाहते हैं तो हमें क्या करना चाहिए? मेरे विचार से, सबसे पहले हमें संवैधानिक तरीकों से अपने सामाजिक व आर्थिक लक्ष्यों की ओर तेजी से आगे बढ़ना चाहिए। इसका अभिप्राय यह है कि हमें क्रांति के खूनी तौर-तरीके छोड़ने होंगे। हमें सविनय अवज्ञा, असहयोग तथा सत्याग्रह जैसे रास्ते भी छोड़ने होंगे। जब आर्थिक और सामाजिक लक्ष्यों को प्राप्त करने के लिए संवैधानिक तरीके शेष न बचें तो ही असंवैधानिक तरीकों का प्रयोग सही ठहराया जा सकता है; लेकिन जब तक संवैधानिक उपायों के रास्ते बचे हैं तब तक इन असंवैधानिक तरीकों को किसी भी आधार पर उचित नहीं ठहराया जा सकता। ये तौर-तरीके और कुछ नहीं हैं, बल्कि अराजकता का व्याकरण हैं तथा जितनी जल्दी हम इनसे मुक्ति पा लेंगे, हमारे लिए उतना ही अच्छा होगा।

***

दूसरी बात जो हमें करनी चाहिए वह है हमें जॉन स्टुअर्ट की उस चेतावनी को ध्यान में रखना है, जो उन्होंने आजादी को बरकरार रखने के इच्छुक व्यक्तियों को दी थी। इसके अनुसार, हमें अपनी आजादी को किसी के नियंत्रण में नहीं देना चाहिए, यहाँ तक कि उस महान् पुरुष के हाथों में भी नहीं, जिसको सत्ता सौंपते समय हम

उस पर विश्वास करते हों; क्योंकि वह भी हमारी आजादी की प्रतीक संस्थाओं को नष्ट कर सकता है। उन महापुरुषों के प्रति ऋणी अथवा कृतज्ञ होना गलत नहीं है, जिन्होंने देश के लिए जीवन-पर्यंत अपनी सेवाएँ दीं; किंतु कृतज्ञता की भी एक सीमा होती है। जैसाकि आयरिश देशभक्त डैनियल और कोलोनेल ने लिखा भी है—कोई पुरुष अपने सम्मान की कीमत पर ऋणी नहीं हो सकता, कोई महिला अपने सतीत्व की कीमत पर ऋणी नहीं हो सकती तथा कोई देश अपनी आजादी की कीमत पर ऋणी नहीं हो सकता। यह चेतावनी किसी अन्य देश की तुलना में भारत के मामले में तो और भी अधिक आवश्यक है, क्योंकि दुनिया के किसी अन्य देश की तुलना में भारत की राजनीति में नायक-पूजा तथा भक्ति की परंपरा कुछ अधिक ही है। धर्म में भक्ति मार्ग आत्मा को मुक्ति की ओर ले जाता है, किंतु राजनीति में भक्ति अथवा नायक-पूजा का मार्ग निश्चय ही अवनति तथा अवश्यंभावी तानाशाही की ओर ले जाता है।

***

दुर्भाग्यवश आज भारत में सामान्यत: लोग तथा विशेषकर राजनेता, चाहे वे जिस राजनीतिक दल के हों, अकसर सत्याग्रह करते तथा बंद कराते देखे जा सकते हैं। उन्हें इस बात की कोई परवाह नहीं कि बिना कोई गलती किए इससे लोग कितना कष्ट भुगतते हैं।

***

आप लोग पिछले पंद्रह वर्षों से मेरा जन्मदिन मना रहे हैं। मैं इन समारोहों में कभी शामिल नहीं हुआ। मैंने हमेशा इनका विरोध किया है। अब आपने मेरी स्वर्ण जयंती मना ली है। बस, इतना ही पर्याप्त है। अब आगे से ऐसे किसी समारोह के आयोजन की आवश्यकता नहीं है।

***

मेरा मानना है कि कोई संविधान कितना भी अच्छा क्यों न हो, वह एक बुरे संविधान में निश्चय ही बदला जा सकता है। यदि इसका प्रयोग अच्छे ढंग से किया जाए तो यह संविधान अच्छा साबित हो सकता है। संविधान की कार्य-प्रणाली संविधान की प्रकृति पर ही पूर्णत: निर्भर नहीं होती। संविधान किसी राष्ट्र को विधायिका, कार्यपालिका और न्यायपालिका जैसे अंग ही दे सकता है। राष्ट्र के ये अंग किस प्रकार कार्य करते हैं, यह उन लोगों तथा राजनीतिक दलों पर निर्भर है, जो लोगों की इच्छाओं तथा नीतियों के कार्यान्वयन हेतु सत्ता में हैं, जो लोगों की इच्छाओं तथा नीतियों के कार्यान्वयन हेतु सत्ता में रहकर कार्य करते हैं। कौन कह सकता है कि भारत के लोग तथा उनके राजनीतिक दल किस प्रकार भविष्य में कार्य करेंगे? क्या वे अपने उद्‌देश्यों

की पूर्ति के लिए संवैधानिक तरीकों का इस्तेमाल करेंगे या वे क्रांतिकारी तौर-तरीकों को पसंद करेंगे? संविधान के अच्छे होने से इस बात पर कोई असर नहीं पड़नेवाला। लोगों तथा उनके दलों की भावी भूमिका के विषय में जाने बिना संविधान के विषय में कोई भी धारणा बनाना निरर्थक होगा।

***

सामाजिक व आर्थिक लोकतंत्र राजनीतिक लोकतंत्र के प्रमुख तंतु हैं। एक वर्ग तथा दूसरे वर्ग के बीच समानता, एक लिंग और दूसरे लिंग के बीच समानता की उपेक्षा करना, जो हिंदू समाज की मूल जड़ में हैं, को छुए बगैर आर्थिक समस्याओं पर कानून पारित करना हमारे संविधान का मजाक उड़ाना है और गोबर के ढेर पर महल बनाने का ख्वाब देखना है।

***

इस संघीय राज्य में अनेक छेद हैं और ये दरक सकते हैं। हम अपने समाज में दरारें डाल रहे हैं। हममें एकता नहीं है। हममें एकता न होने के कारण यह पूरी संरचना कभी भी बिखर सकती है। अत: हमें समय रहते इसे बिखरने से बचाने के लिए कुछ कदम उठाने चाहिए। मैंने सुझाव दिया था कि उत्तरी प्रांतों को छोटे-छोटे क्षेत्रों में बाँटकर छोटा कर दिया जाए, ताकि दक्षिण भारत के लोग किसी प्रकार के भारी दबाव का अनुभव न करें। मेरा यह भी सुझाव था कि इसका एक अन्य समाधान है इस देश की दो राजधानियाँ बनाया जाना। मेरा सुझाव है कि हैदराबाद को भारत की दूसरी राजधानी बनाया जाना चाहिए।

***

अंग्रेजों के भारत से जाने के बाद भारत के पास केवल एक राजधानी है, जो दिल्ली है। अंग्रेजों से पहले हमेशा से भारत की दो राजधानियाँ रही हैं। मुगलकाल में भारत की पहली राजधानी दिल्ली तथा कश्मीर में श्रीनगर दूसरी राजधानी हुआ करती थी। जब अंग्रेज आए तो उन्होंने भी दो राजधानियाँ बनाईं। पहली कलकत्ता थी और दूसरी शिमला थी। जब उन्होंने कलकत्ता के बजाय दिल्ली को राजधानी बनाया तो भी उन्होंने शिमला को ग्रीष्मकालीन राजधानी बनाए रखा। यहाँ तक कि जलवायुविक स्थितियों के कारण ही अंग्रेजों और मुगलों ने दो राजधानियाँ बनाई थीं। न तो अंग्रेज और न ही मुगल पूरे बारह महीनों तक लगातार दिल्ली या कलकत्ता में रह सकते थे। दिल्ली के गरमी के महीने मुगलों के लिए काटने मुश्किल थे, इसी कारण उन्होंने श्रीनगर को गरमी के महीनों के लिए दूसरी राजधानी बनाया। इसी प्रकार अंग्रेजों के लिए कलकत्ता की गरमी भी सहन करने योग्य नहीं थी, इसलिए उन्होंने भी दो

राजधानियाँ स्थापित कीं। इन जलवायुविक परिस्थितियों के अलावा तीन अन्य स्थितियाँ भी अब देखी जा सकती हैं। मुगलों तथा अंग्रेजों के शासनकाल में लोकप्रिय सरकार जैसी कोई चीज नहीं थी। अब हमारे पास लोकप्रिय सरकार है और जनता की सुविधाओं का ध्यान रखना हमारा प्रथम दायित्व है।

***

दक्षिण भारत के लोगों के लिए दिल्ली आना अत्यंत असुविधाजनक है। वे यहाँ की दूरी तथा ठंड दोनों से परेशान होते हैं। यहाँ तक कि उत्तरी भारत के लोगों को भी यहाँ की गरमी से परेशानी होती है। वे शिकायत नहीं करते, क्योंकि वे घर के नजदीक हैं तथा सत्ता की कुरसी के बहुत पास हैं। दूसरी बात दक्षिण भारत के निवासियों की भावनाओं का ध्यान रखना है तथा तीसरी बात सुरक्षा की है।

***

दक्षिण भारतीयों का सोचना है कि उनके देश की राजधानी उनसे बहुत दूर है तथा उत्तर भारत के लोग शासन कर रहे हैं। तीसरी बात निस्संदेह अधिक महत्त्वपूर्ण है। वह यह है कि दिल्ली अभेद्य स्थान नहीं है। यह हमारे पड़ोसी देशों की बम वर्षा की सीमा में आसानी से आ सकती है। यद्यपि भारत अपने पड़ोसी देशों के साथ शांति के साथ रहने का प्रयास कर रहा है, परंतु यह नहीं माना जा सकता कि भारत को कभी किसी देश के साथ युद्ध नहीं करना पड़ेगा और यदि ऐसी स्थिति आती है तो भारत सरकार को दिल्ली छोड़नी पड़ेगी और अपना शासन चलाने के लिए किसी अन्य स्थान की खोज करनी पड़ेगी। वह कौन सा स्थान है, जहाँ भारत सरकार अपना शासन चलाने के लिए आ सकती है? एक बार कलकत्ता के विषय में भी सोचा जा सकता है; परंतु कलकत्ता तिब्बत से बम-वर्षा की जद में है। यद्यपि भारत और चीन आज अच्छे मित्र हैं, किंतु कोई भी यह विश्वासपूर्वक नहीं कह सकता कि कब तक हमारे बीच यह दोस्ती कायम रहेगी। इस स्थिति में भारत और चीन के बीच संघर्ष की संभावना विद्यमान है। ऐसे मौके पर कलकत्ता का राजधानी होना निरर्थक होगा। एक अन्य विकल्प के रूप में अगला शहर बंबई को माना जा सकता है। लेकिन बंबई एक बंदरगाह है और हमारी भारत सरकार के बंबई आ जाने पर हमारी नौ सेना उसे सुरक्षा प्रदान कर पाने में अक्षम सिद्ध होगी। क्या कोई चौथी जगह भी इस बारे में सोची जा सकती है? मेरे विचार से, हैदराबाद एक ऐसा ही स्थान हो सकता है। हैदराबाद, सिकंदराबाद तथा बोलराम को मुख्य आयुक्त के प्रांतों में परिवर्तित किया जा सकता है तथा इसे भारत की दूसरी राजधानी बनाया जा सकता है। हैदराबाद भारत की दूसरी राजधानी बनने की सभी शर्तें पूरी करता है। हैदराबाद सभी राज्यों से समान दूरी पर है। सुरक्षा के

दृष्टिकोण से यह शहर केंद्र सरकार को पर्याप्त सुरक्षा देगा। यह भारत के सभी भागों से समान दूरी पर है। यह दक्षिण भारत के लोगों को इस बात का संतोष देगा कि उनकी सरकार कभी-कभी उनके साथ रहती है।

***

सर्दियों के मौसम में सरकार दिल्ली में रहे तथा अन्य महीनों में यह हैदराबाद में रह सकती है। हैदराबाद के पास वे सभी सुविधाएँ हैं, जो दिल्ली में हैं तथा यह दिल्ली से कहीं बेहतर शहर है। इसमें दिल्ली की भाँति भव्यता भी है। यहाँ की इमारतें सस्ती हैं तथा वास्तव में दिल्ली की तुलना में कहीं अधिक सुंदर हैं। इसलिए हैदराबाद, सिकंदराबाद तथा बोलराम को भारत की दूसरी राजधानी बनाया जाना चाहिए। सौभाग्यवश यह कार्य बहुत सुगमतापूर्वक किया जा सकता है तथा इससे संपूर्ण दक्षिण भारत, महाराष्ट्र एवं आंध्र के लोगों को भी संतुष्टि मिलेगी।

***

मुझे प्रसन्नता है कि भारत पाकिस्तान से अलग हो गया। मैं पाकिस्तान के निर्माण का प्रबल समर्थक था। मैंने विभाजन की वकालत की थी, क्योंकि मेरा मानना है कि केवल विभाजन से ही हम न सिर्फ स्वतंत्र होंगे, बल्कि मुक्त भी होंगे। अगर भारत और पाकिस्तान एक रहते तो एक समय वह आता जब हिंदू स्वतंत्र होने के बावजूद मुसलिमों की दया पर आश्रित रहते। केवल स्वतंत्र भारत हिंदुओं की दृष्टि से एक मुक्त भारत नहीं होता। यहाँ एक देश में दो राष्ट्रों की एक सरकार होती। हिंदू महासभा और जन संघ के होते हुए भी निस्संदेह मुसलिम ही यहाँ राज कर रहे होते।

***

जब देश का विभाजन हो रहा था, तब मैंने सोचा था कि ईश्वर अपने शाप को वापस ले रहे हैं तथा भारत को अब एक महान् व समृद्ध देश बनाएँगे, किंतु मुझे भय है कि शाप दोबारा भी वापस आ सकता है, क्योंकि मैं देख रहा हूँ कि जो भाषाई आधार पर राज्यों के गठन की वकालत कर रहे हैं, वे दिल से अपनी क्षेत्रीय भाषाओं को उन राज्यों की राजभाषा बनाना चाहते हैं। यह विचार अखंड भारत के आदर्श के लिए मौत की घंटी के समान है। क्षेत्रीय भाषाओं की राजभाषाएँ बनाने से भारत को अखंड राष्ट्र बनाने और भारतीयों में पहले भारतीय होने की भावना जगाने की धारणा नष्ट हो जाएगी।

***

उत्तर तथा दक्षिण के मध्य बड़ी भिन्नताएँ हैं। उत्तर परंपरावादी है, दक्षिण प्रगतिवादी है। उत्तर अंधविश्वासी है, दक्षिण तर्क से परिपूर्ण है। दक्षिण शिक्षा की दृष्टि

से काफी आगे है, जबकि उत्तर शिक्षा की दृष्टि से पिछड़ा हुआ है। दक्षिण की संस्कृति आधुनिक है, उत्तर की संस्कृति पुरातन है; किस प्रकार उत्तर का शासन दक्षिण को मान्य हो सकता है? पहले से ही इस बात के संकेत हैं कि दक्षिण उत्तर से टूटकर अलग होना चाहता है।

***

श्री राजगोपालाचारी ने राज्य पुनर्गठन आयोग में एक वक्तव्य दिया है, जो 27 नवंबर, 1950 के 'टाइम्स ऑफ इंडिया' में प्रकाशित हुआ है। उन्होंने अपने वक्तव्य में यह कहा है, 'यदि राज्य पुनर्गठन की योजना को अगले 15 वर्षों तक के लिए ठंडे बस्ते में डालना संभव न हो तो एकमात्र विकल्प यह है कि केंद्र सरकार ही एकल राज्य के रूप में पूरे भारत का संचालन करे तथा जिले के अधिकारियों एवं जिले के बोर्डों से सीधे संपर्क रखे, जिन पर क्षेत्रीय आयुक्तों का नियंत्रण हो। ड्राइंग रूम में बैठकर इतिहास के परिणामों को समझने और खींची गई सीमाओं तथा अनुभागों के लिए लड़ने में राष्ट्र की ऊर्जा व्यय करना समय की बरबादी होगी।' राजगोपालाचारी ने आगे कहा, 'मेरा दृढ़ विश्वास है कि देश के दक्षिणी हिस्से का राजनीतिक महत्त्व बरकरार रखने के लिए एक बड़े दक्षिणी राज्य का अस्तित्व में बने रहना आवश्यक है। दक्षिण को तमिल, मलयालम तथा अन्य छोटे राज्यों में बाँटना प्रत्येक की महत्ता को पूर्णत: समाप्त करना होगा तथा इसका अंतिम परिणाम यह होगा कि इससे भारत और अधिक निर्धन बनेगा।'

***

श्री राजगोपालाचारी ने अपने संपूर्ण दृष्टिकोण को यहाँ पर अभिव्यक्त नहीं किया है। ऐसा उन्होंने मेरे समक्ष खुलकर प्रस्तुत किया था, जब वे राष्ट्र के प्रधान थे तथा मैं संविधान का प्रारूप तैयार करते समय प्रभारी कानून मंत्री था। एक बार मैं नित्य के इंटरव्यू के लिए श्री राजगोपालाचारी के पास गया। ऐसे ही एक इंटरव्यू में श्री राजगोपालाचारी ने संविधान सभा द्वारा तैयार किए जा रहे संविधान के विषय में अपने विचार रखे थे। उन्होंने उस समय मुझसे कहा था, 'आप एक बड़ी गलती करने जा रहे हैं। सभी क्षेत्रों के समान प्रतिनिधित्व के साथ संपूर्ण भारत के लिए एक संघ का विचार अनुचित है। ऐसे संघ में भारत का प्रधानमंत्री और राष्ट्रपति हमेशा हिंदी भाषी क्षेत्र से ही होगा। आपको दो संघ बनाने चाहिए। एक उत्तर का संघ हो तथा एक दक्षिण का संघ हो और उत्तर एवं दक्षिण के संघों को मिलाकर महासंघ बने, जिसमें महासंघ के पास विधायन हेतु तीन विषय हों तथा दोनों संघों का समान प्रतिनिधित्व हो।'

***

ये श्री राजगोपालाचारी के वास्तविक विचार हैं। उनके ये विचार मेरे पास एक

रहस्योद्घाटन के रूप में आए, जो उन्होंने एक कांग्रेसी के रूप में अपने दिल से निकालकर रखे थे। अब मैं श्री राजगोपालाचारी को भारत के उत्तर तथा दक्षिण के रूप में दो टुकड़े करने की कोशिश करनेवाले व्यक्ति के रूप में देखने लगा था।

यह नहीं भूलना चाहिए कि संयुक्त राज्य अमेरिका में उत्तर तथा दक्षिण के बीच एक गृहयुद्ध हुआ था। भारत में भी उत्तर तथा दक्षिण के बीच गृहयुद्ध हो सकता है। समय ऐसे संघर्ष के लिए जमीन तैयार करेगा। यह भी नहीं भूलना चाहिए कि उत्तर तथा दक्षिण के बीच बड़ी सांस्कृतिक भिन्नताएँ हैं और सांस्कृतिक भिन्नताएँ सुलगती हुई तीली की भाँति होती हैं।

***

मुझे वे दिन याद हैं, जब भारत के राजनीतिक विचारक 'हम भारत के लोग' शब्दों से क्रुद्ध हो जाया करते थे। इसके स्थान पर वे 'भारत राष्ट्र' की बात कहना अधिक पसंद करते थे। मेरा मानना है कि हम एक राष्ट्र हैं, यह मानकर हम एक बड़े छल-छद्म की सृष्टि कर रहे हैं। कई हजार जातियों में बँटे हम लोग कैसे एक राष्ट्र हो सकते हैं? जितनी जल्दी हम यह समझ लें कि 'राष्ट्र' शब्द के सामाजिक व मनोवैज्ञानिक अर्थ की दृष्टि से हम अभी तक एक राष्ट्र नहीं बन पाए हैं, उतना ही अच्छा होगा। तभी हम एक राष्ट्र होने की आवश्यकता को महसूस कर सकेंगे तथा इस लक्ष्य को प्राप्त करने के रास्तों तथा तरीकों के विषय में गंभीरतापूर्वक सोच सकेंगे। इस लक्ष्य को प्राप्त करना अत्यंत कठिन कार्य है। भारत में यह अमेरिका में किए गए प्रयासों से कहीं अधिक कठिन है। ये जातियाँ राष्ट्र-विरोधी हैं, क्योंकि सर्वप्रथम ये सामाजिक जीवन में पृथकता लाती हैं। ये जातियों के बीच ईर्ष्या तथा दुर्भावना को जन्म देती हैं। हमें इन सभी समस्याओं का सामना करना होगा, तभी हम वास्तव में एक राष्ट्र बन सकेंगे। राष्ट्र होने की दशा में भाईचारा ही एकमात्र कारक होगा। भाईचारे के बगैर समानता तथा स्वतंत्रता रगों की परतों से अधिक गहरी नहीं होगी।

***

राष्ट्रीय तथा सामाजिक संस्था के रूप में जाति के विषय में कोई भी दृष्टिकोण अपनाया जाए, परंतु इसकी उपेक्षा कर पाना संभव नहीं है और जब तक जाति किसी व्यक्ति की सरकारी अथवा सामाजिक पहचान के लिए प्रयुक्त की जाती रहेगी, तब तक यह दावा नहीं किया जा सकता है कि अवांछित संस्था को स्थायी बनाने में दशकीय गणना सहायक सिद्ध होती है।

***

यदि भारत राष्ट्रीयता की भावना जाग्रत् करने में सफल नहीं हो पाया है तो इसका

एकमात्र कारण यहाँ की ग्रामीण व्यवस्था है। इससे यहाँ के सभी लोग स्थानीय विशिष्टता या स्थानीय देशभक्ति में उलझकर रह गए हैं। इससे अपेक्षाकृत बृहत् नागरिक भावनाओं के लिए कोई जगह नहीं बची है। प्राचीन ग्राम पंचायतों के अंतर्गत भारत एकता के सूत्र में बँधे लोगों के एक देश के स्थान पर ग्रामीण समुदायों का एक ढीला-ढाला पुंज बनकर रह गया है। इसमें एक शासन में जुड़ाव के अतिरिक्त अन्य कोई बंधन नहीं है। इन दिनों जब आप राष्ट्रीय भावना लाने के लिए प्रयत्न कर रहे हैं, जब आप एक समान राष्ट्रीयता तथा एक जैसी भारतीय नागरिकता को स्थापित करने का प्रयास कर रहे हैं तो हमें ऐसा कोई काम नहीं करना चाहिए, जिससे इन प्रयासों को आघात पहुँचे।

****

यदि ग्राम पंचायतें अस्तित्व में आती हैं तो इनमें अल्पसंख्यकों के लिए विशेष प्रतिनिधित्व दिया जाना जरूरी है। इसी प्रकार इनमें किसी भी कीमत पर दलित जातियों के लिए विशेष प्रतिनिधित्व होना चाहिए।

****

इस सदन में एक ऐसा वर्ग है, जो एकदम से उछल पड़ेगा और कहेगा कि यह सांप्रदायिकता है। मैं स्वीकार करता हूँ कि मेरी नीति सांप्रदायिक है। मैं इससे शर्मिंदा नहीं होता। मैं तो कहूँगा कि सांप्रदायिकता के बगैर भारत राजनीतिक प्रगति के रास्ते पर आगे नहीं बढ़ सकता। सांप्रदायिकता के बिना भारत में कोई अपनो सरकार नहीं हो सकती। यह वह बात है, जिसे मैं चुनौती के भय के बगैर स्पष्ट रूप से कह सकता हूँ।

****

विरोधाभासों की यह जिंदगी हम कब तक जीते रहेंगे? कब तक हम अपने सामाजिक तथा आर्थिक जीवन में समानता की उपेक्षा करते रहेंगे? यदि हम लंबे समय तक इसकी अनदेखी करेंगे, तो हम ऐसा अपने राजनीतिक लोकतंत्र की क्षति की कीमत पर ही करेंगे। हमें जल्द-से-जल्द इस विरोधाभास को दूर कर देना चाहिए, अन्यथा असमानता के शिकार लोग उठ खड़े होंगे और राजनीतिक लोकतंत्र की इस संरचना को नष्ट कर देंगे, जिसे इस सभा ने अत्यंत परिश्रम के साथ बनाया है।

****

अब मैं संविधान के इस संशोधन के विषय में सरकार के रुख पर रोशनी डालना चाहूँगा। अभी संविधान केवल चार वर्ष पुराना है और अभी तक यह परिपक्व नहीं हुआ है। इसके चार वर्ष सात माह के जीवन में संविधान का यह तीसरा संशोधन है। मुझे नहीं पता कि दुनिया के किसी भी संविधान को इतनी तेजी से संशोधित किया

गया है और यदि मैं ऐसा कहता हूँ तो यह कार्य वर्तमान सरकार द्वारा जल्दबाजी में किया जा रहा है।

~***~

क्या हमारी सरकार इन आधारभूत नियमों को देख रही है? यह सच है कि हमारा संविधान अत्यधिक लचीला है। यह न तो अमेरिकी संविधान की तरह है और न ही ऑस्ट्रेलिया के संविधान जैसा ही है। संविधान तैयार करनेवाले व्यक्ति इस तथ्य से भली-भाँति अवगत थे कि भविष्य में ऐसी परिस्थितियाँ आएँगी जब संविधान में संशोधन किया जाना आवश्यक होगा और आप ज्वलंत सामाजिक समस्याओं को संविधान की बाध्यताओं के कारण स्थगित नहीं रख सकते। इसी कारण यह प्रस्ताव किया गया कि अनुच्छेद 368 में उल्लिखित प्रावधान यथेष्ट हैं।

~***~

किंतु मेरे मन में किंचित् भी संदेह नहीं है कि जिन्हें संविधान के प्रारूप से कोई लेना-देना नहीं है, उन्होंने भी यह कभी नहीं सोचा होगा कि सरकार मतदाताओं को सूचित किए बगैर अत्यंत जल्दबाजी में संविधान में संशोधन करने लगेगी।

~***~

मैं मानता हूँ कि मतदाताओं को सूचना देना राजनीतिक जीवन तथा पार्टी के जीवन का एक सामान्य सिद्धांत है। प्रत्येक पार्टी को कुछ कार्य करने के लिए जनादेश मिलता है। जनादेश के बगैर एक पार्टी कुछ भी नहीं कर सकती। प्राय: मतदाताओं को धोखे में नहीं रख सकते तथा आप निर्वाचित होकर आए हैं और आपके पास संविधान को संशोधित करने का पूरा अधिकार है, आप यह नहीं सोच सकते। यही वह कार्य है, जो हमारी सरकार कर रही है। मात्र इसलिए क्योंकि उसके पास बहुमत है और वे सोचते हैं कि उनके पास न सिर्फ किसी भी उस कानून को बनाने की ताकत है, जिसके माध्यम से वे सत्ता प्राप्त कर चुके हैं, बल्कि मात्र निर्वाचित हो जाने से ही वे संविधान को बदलने की ताकत भी रखते हैं। उन्हें अब अपने कृत्यों के विषय में जनता को सूचित करने की भी आवश्यकता नहीं रह गई है, यहाँ तक कि उन्हें संविधान का संशोधन करते समय भी किसी को सूचित करने की जरूरत नहीं है।

~***~

क्या संविधान किसी साधारण कानून से किसी भी प्रकार भिन्न नहीं है? क्या यह कागज का टुकड़ा मात्र है, जो किसी व्यक्ति के चाहने मात्र से संशोधित किया जा सकता है? मराठी में एक कहावत है, जिसमें हम कहते हैं कि यदि कोई बूढ़ी स्त्री मरती है तो इससे अधिक प्रभाव नहीं पड़ता; मगर हमें डर लगने लगता है कि कहीं यम देवता बार-बार यहाँ आने के आदी न हो जाएँ और हम यम देव का बार-

बार आना रोकना चाहने लगते हैं। इससे कोई फर्क नहीं पड़ता कि बूढ़ी महिला मरी है या यमदेव उसे झपट्टा मारकर ले गए हैं। यहीं वह बात है जो यहाँ पर हो रही है तथा मैं यह देख रहा हूँ कि सरकार के मन में संविधान के लिए बहुत कम सम्मान शेष बचा है और उसमें संविधान की अवमानना करने का भाव जाग्रत् होता जा रहा है।

***

मुझे नहीं लगता कि सरकार को किसी बात की कोई परवाह है। मैं वही चेतावनी दोबारा देना चाहता हूँ और कुछ वजहें बताना चाहता हूँ कि क्यों संविधान को बार-बार संशोधित नहीं किया जाना चाहिए और इससे बेवजह छेड़छाड़ किया जाना बंद किया जाना चाहिए। इसी कारण मैं कहता हूँ कि संविधान को इतने हलके ढंग से तथा बार-बार नहीं सुधारा जाना चाहिए। मैं नहीं जानता कि सरकार मेरी बात सुनेगी, शायद नहीं सुनेगी।

***

दूसरे देशों में यदि न्यायपालिका सरकार की इच्छा के विपरीत संविधान के किसी खंड का अभिप्राय निकालती है तो सरकार को उससे सहमत होना पड़ता है। वहाँ सरकारें अदालतों के निर्णयों को नहीं उलटतीं। यहाँ हमारे देश में हमने एक अलग किस्म की मानसिकता विकसित कर ली है। हमारी मानसिकता है कि यदि उच्चतम न्यायालय के न्यायाधीश हमारी इच्छा के अनुरूप निर्णय नहीं देते हैं तो हम उन्हें हटा सकते हैं। यही वह बात है। इसके मुकाबले मुझे अपने उच्चतम न्यायालय के व्यवहार से प्रसन्नता होती है। अपने अस्तित्व में आने के थोड़े से समय में ही उच्चतम न्यायालय के अनेक दौर मुझे दिखाई दिए हैं। मुझे याद है कि अपनी शक्ति की पहली चमक दिखाते हुए उच्चतम न्यायालय ने यह घोषित किया अथवा यह घोषित करने का साहस किया कि भारतीय दंड संहिता का कुछ अंश दोषपूर्ण है। हमारी सरकार इससे तुरंत क्षुब्ध हो गई और वह एक संशोधन ले आई कि न्यायपालिका की यह टिप्पणी गलत थी।

***

मुझे आशा है कि सरकार द्वारा लगातार किए जा रहे संविधान संशोधनों के बावजूद उच्चतम न्यायालय अपने निष्पक्ष निर्णय देता रहेगा और उसे सरकार के कहे की चिंता नहीं होगी। मुझे नहीं लगता कि कोई भी व्यक्ति यह कहेगा कि उच्चतम न्यायालय ने कोई निर्णय संविधान की मूल भावना के विरुद्ध दिया है।

***

सरकार संविधान में संशोधन करने जा रही है। उसे संविधान को संशोधित करने

का चस्का लग गया है। यदि आप संविधान को हर शनिवार को संशोधित करेंगे तो ऐसे संविधान का क्या मतलब है?

***

जो व्यक्ति संविधान से संतुष्ट नहीं हैं, उन्हें मात्र दो-तिहाई बहुमत लाने की आवश्यकता है और यदि वे वयस्क मताधिकार के आधार पर दो-तिहाई बहुमत नहीं ला पा रहे हैं तो यह माना जाएगा कि संविधान के प्रति उनकी असंतुष्टि को साधारण जनता का समर्थन हासिल नहीं है।

***

प्रत्येक देश में यदि बुद्धिजीवी वर्ग सत्ता में नहीं है तो भी सर्वाधिक प्रभावशाली वर्ग है। बुद्धिजीवी वह वर्ग है, जो भविष्यद्रष्टा होता है, जो सुझाव दे सकता है तथा नेतृत्व प्रदान कर सकता है। कोई भी ऐसा देश नहीं है, जहाँ पर कुछ लोगों का समूह विद्वान् विचारों तथा कृत्यों से परिपूर्ण न होता हो। यह प्रेरणादायक समूह होता है तथा बुद्धिजीवी वर्ग का प्रतिनिधित्व करता है। यह कहना अतिशयोक्ति नहीं होगी कि एक देश की संपूर्ण नियति उसके बुद्धिजीवी वर्ग पर निर्भर करती है। यदि बुद्धिजीवी वर्ग ईमानदार है, स्वतंत्र है तथा निष्पक्ष है तो इसके द्वारा उठाए गए कदमों पर विश्वास किया जा सकता है तथा किसी समस्या के उपस्थित होने पर उसे नेतृत्व प्रदान करने हेतु कहा जा सकता है। यह सत्य है कि विद्वत्ता स्वयं में कोई गुण नहीं है। यह बुद्धिजीवी वर्ग द्वारा इस्तेमाल किए गए तौर-तरीकों अथवा उन तरीकों के प्रयोग में निहित है, जिनसे सफलता पाई जाती है।

***

एक बुद्धिजीवी व्यक्ति एक अच्छा व्यक्ति हो सकता है, किंतु वह आसानी से एक धूर्त व्यक्ति में भी परिवर्तित हो सकता है। इसी प्रकार एक बुद्धिजीवी वर्ग ऐसी उच्च कोटि की आत्माओं का समूह हो सकता है, जो सहायता करने के लिए सदैव तैयार रहते हों, मानवता के कष्टों को दूर करने के लिए तत्पर रहते हों अथवा वे सरलतापूर्वक ठगों के गिरोह हो सकते हैं या संकीर्ण विचारों के लोगों का गुट हो सकते हैं, जिनसे उन्हें समर्थन मिलता हो।

□

भारत में शूद्र

यदि चातुर्वर्ण्य व्यवस्था का आशय समाज का चार भागों में विभागीकरण ही होता तो कुछ भी आपत्तिजनक न होता। दुर्भाग्यवश, चातुर्वर्ण्य व्यवस्था रहस्यमयी है। इस व्यवस्था के अंतर्गत समाज को चार भागों में बाँटने के साथ-साथ चार वर्णों का जीवन-स्तर निर्धारित करने के लिए क्रमिक असमानता का प्रावधान रखा गया है।

~***~

क्रमिक असमानता की प्रणाली मौखिक न होकर वैधानिक और दंडनीय है। चातुर्वर्ण्य-संरचना में शूद्र निम्नतम स्तर पर हैं। इसके साथ ही उन पर अनेक प्रतिबंध लगाए गए हैं, जिससे वे अपने स्तर से उबर न सकें।

~***~

मेरा स्पष्ट मत है कि ब्राह्मण-विहित विधान से पीड़ित प्राचीन आर्यों के शूद्र कालांतर में इतने पतित हो गए कि जन-जीवन में उनका कोई स्तर ही नहीं रह गया अर्थात् उनका स्तर समाज में निम्नतर हो गया। इसके दो परिणाम निकले—प्रथम तो यह कि 'शूद्र' शब्द का अर्थ ही बदल गया। इसका अर्थ एक विशेष जाति का नाम न रहकर सामान्यत: असभ्य, असंस्कृत एवं अधम जातियों का संयुक्त नाम हो गया। दूसरा परिणाम यह हुआ कि 'शूद्र' शब्द का अर्थ व्यापक होने के कारण दंड-विधान का भी विस्तार हुआ। इस प्रकार तथाकथित आधुनिक शूद्र भी उस दंड-विधान से शासित होने लगे, जिससे आदि शूद्र होते थे। कुछ भी हो, तथ्य यह है कि अपराधियों के लिए निर्धारित दंड-विधान निर्दोषों पर भी क्रूरता से लागू होने लगे। यदि हिंदू विधि निर्माता इस ऐतिहासिक वास्तविकता से विमुख न होते कि आधुनिक निम्न वर्णीय जन आदि शूद्रों से भिन्न हैं तो यह अंधेर नहीं होता। दुर्भाग्य

से ये दंड-विधान आदि शूद्रों की ही भाँति आधुनिक शूद्रों पर बर्बरता से लागू हैं।

***

शूद्र कौन थे और वे प्राचीन आर्यों के समाज का चौथा वर्ग क्यों और कैसे बने? संक्षेप में मेरा उत्तर इस प्रकार है—

(1) शूद्र आर्यों की जातियों में सूर्यवंशीय हैं।

(2) एक समय था, जब आर्यों में केवल तीन ही वर्ण थे—ब्राह्मण, क्षत्रिय और वैश्य।

(3) शूद्रों का प्रथम वर्ण न था। वे आर्यों के द्वितीय वर्ण—क्षत्रिय वर्ण—के ही एक अंग थे।

(4) शूद्र राजाओं और ब्राह्मणों में निरंतर संघर्ष चला और ब्राह्मणों को शूद्रों के बीभत्स अत्याचार सहने पड़े।

(5) शूद्रों के दमन से आक्रांत ब्राह्मणों ने घृणावश शूद्रों का उपनयन बंद कर दिया।

(6) उपनयन-विरोध से शूद्रों का सामाजिक पराभव हुआ। वे सामाजिक स्तर पर इतने पतित हुए कि उन्हें वैश्यों से भी नीचे एक और वर्ण चौथा वर्ण बनाना पड़ा।

***

आज का हिंदू समाज पाँच श्रेणियों में विभाजित है। पहला वर्ग पुरातनपंथी लोगों का है, जो हिंदू सामाजिक प्रणाली में कोई दोष नहीं मानते। दूसरा वर्ग आर्य समाजियों का है। वे वेदों में विश्वास व्यक्त करते हैं और किसी ऐसी बात को नहीं मानते, जो वेद-निहित नहीं है। तीसरा वर्ग इस व्यवस्था को मृतप्राय कहता है, क्योंकि कानून उसे मान्यता नहीं देता। चौथा वर्ग राजनेताओं का है—वे इन सब बातों से उदासीन हैं। उनके लिए सामाजिक व्यवस्था में सुधार से भी अधिक महत्त्व सुधार का है। पाँचवाँ वर्ग शांत चिंतकों का है, जो सामाजिक सुधार को समाज से भी अधिक महत्त्वपूर्ण मानता है।

***

'कानून वर्ण-व्यवस्था को नहीं मानता' का गलत अर्थ नहीं लगाना चाहिए। संक्षेप में कहा जा सकता है—

(1) इसका अर्थ यह नहीं कि जाति-व्यवस्था को मानना अपराध है।

(2) इसका अर्थ यह नहीं है कि वर्ण-व्यवस्था समाप्त हो गई है।

(3) इसका अर्थ यह भी नहीं कि नागरिकों के अधिकारों की प्राप्ति के निमित्त किसी का वर्ण-निर्धारण आवश्यक होने पर ऐसा निर्णय नहीं किया जाएगा।

(4) इसका अर्थ केवल यह है कि कानून जाति-व्यवस्था को मान्यता प्रदान नहीं करता।

***

सामाजिक संगठन कानून के बल पर नहीं चलते, कुछ अन्य बातों के बल पर चला करते हैं। इनमें सामाजिक व धार्मिक बल प्रमुख हैं। वर्ण-व्यवस्था धार्मिक आधार पर आधारित है और यही कारण है कि हिंदुओं का सामाजिक बल इसका साथ देता है। वैधानिक निषेध न होने के कारण ही धार्मिक मान्यता के बल पर जाति-वर्ण-व्यवस्था फलती-फूलती रही है। इसका सबसे पुष्ट प्रमाण यह ही है कि शूद्रों और अछूतों की स्थिति आज भी पहले जैसी ही है।

***

मेरे मत का आर्य समाजियों से दो आधारों पर मतभेद है। आर्यसमाजियों का विश्वास है कि आर्यों में प्राचीन काल से ही चार वर्ण हैं। इसके विपरीत, प्राचीन ग्रंथों से स्पष्ट है कि प्राचीन काल में आर्यों में केवल तीन ही वर्ण थे।

***

हिंदू समाज का एकमात्र वर्ग सामाजिक सुधारों की अनिवार्यता को मानता है। उनके विचार से, इस समस्या के समाधान में एक लंबा सनय और अनेक पीढ़ियों का प्रयास अपेक्षित है। इसीलिए इसका अनुसंधान स्थगित करना अनुचित है। एक उत्साही हिंदू राजनीतिज्ञ भी यह मानता है कि सामाजिक संकीर्णता हर मोड़ पर संकट खड़ा कर देती है। ये कठिनाइयाँ सामाजिक न होकर हर क्षण की कठिनाइयाँ हैं। मुझे प्रसन्नता है कि हिंदू समाज में एक ऐसा भी वर्ग है। यद्यपि ऐसे लोगों की संख्या कम ही होगी, फिर भी मैं अपना पक्ष उन्हीं के समक्ष प्रस्तुत करता हूँ।

***

यह कहा जा सकता है कि मैंने हिंदू धर्म ग्रंथों को अपेक्षित सम्मान नहीं दिया। इसके दो कारण हैं। प्रथम तो यह कि मैंने इतिहासवेत्ता की दृष्टि से गवेषणा की है। इतिहासवेत्ता सत्य की खोज के लिए धार्मिक ग्रंथों एवं इतर ग्रंथों में कोई भेदभाव नहीं करता। उसका मुख्य ध्येय तो सत्य की खोज रहता है। इस कार्य में मैं हिंदू ग्रंथों के प्रति सम्मान प्रदर्शित नहीं कर सका तो वह सत्य है। दूसरी बात यह है

कि धार्मिक ग्रंथों में श्रद्धा कराई नहीं जाती, वह तो स्वाभाविक स्थितियों में स्वयं पनपती या भंग होती है।

***

हिंदू ग्रंथों के प्रति ब्राह्मण विद्वान् की श्रद्धा तो स्वाभाविक है, किंतु एक अब्राह्मण शोधकर्ता के लिए अस्वाभाविक। इस भेद को सरलता से समझा जा सकता है। ये पवित्र ग्रंथ ब्राह्मणों द्वारा लिखित हैं और इनमें ब्राह्मणों की श्रेष्ठता व विशेषाधिकारों का वर्णन है। अत: ब्राह्मण इन ग्रंथों की पवित्रता-पावनता को क्यों न अक्षुण्ण बनाए रखें।

***

शूद्र के विषय में खोज करते समय मेरे मस्तिष्क में पूर्ण ऐतिहासिकता रही है। यह सर्वविदित है कि इस देश में अब्राह्मणों का—शूद्रों का—एक आंदोलन चल रहा है और उससे मेरा निकट का संबंध है।

***

मैं 'महाभारत' के चालीसवें अध्याय के 'शांतिपर्व' के रचनाकार का आभारी हूँ। हालाँकि यह कहना कठिन है कि वह कौन था—व्यास, वैशंपायन, सूत, लोमहर्ष अथवा भृगु। इनमें से कोई भी क्यों न हो, उसने वैजवन का पूर्ण वृत्तांत देकर अनुग्रह ही किया है। यदि वह वैजवन को शूद्र न कहता तो शूद्रों के उद्‌भव का मूल स्रोत ही विलुप्त हो जाता। भावी पीढ़ी के लिए इतनी महत्त्वपूर्ण सूचना सामग्री को सुरक्षित रखे जानेवाले रचनाकार का यह अपार उपकार है।

***

यह सभी जानते हैं कि 'शूद्र' भारतीय हिंदू समाज का चौथा वर्ण है; किंतु किसी ने यह जानने का प्रयास नहीं किया कि शूद्र कौन थे और वे हिंदू समाज के चतुर्थ वर्ण कैसे बन गए? यह एक महत्त्वपूर्ण प्रश्न है। यह जानना अत्यंत आवश्यक है कि वे समाज की विकास परंपरा अथवा किसी परिवर्तन के फलस्वरूप समाज का चौथा वर्ण बन गए या बनाए गए।

***

शूद्रों की खोज के लिए हमें सर्वप्रथम भारतीय आर्यसमाज की चातुर्वर्ण्य व्यवस्था की जानकारी प्राप्त कर लेना आवश्यक है। चातुर्वर्ण्य की व्यवस्था की जानकारी हमें 'ऋग्वेद' के दसवें मंडल के 90वें मंत्र 'पुरुष-सूक्त' से मिलती है। मंत्र में कहा गया है—

(1) पुरुष के सहस्र सिर, सहस्र नेत्र और सहस्र पैर हैं। संपूर्ण ब्रह्मांड उसने

दस अंगुल से आच्छादित किया हुआ है।

(2) पुरुष स्वयं ही ब्रह्मांड है—जो हो गया है और जो भवितव्य है। वह अमरत्व का स्वामी है और भोजन से वृद्धि पाता है।

(3) यह उसकी महानता है। पुरुष उससे भी उत्तम है। समस्त ब्रह्मांड उसका चतुर्थांश है। शेष तीन-चौथाई शून्य में अनर है।

(4) तीन-चौथाई के साथ पुरुष ऊपर चढ़ा। उसका एक चतुर्थांश यहाँ उत्पन्न हुआ। उसने विस्तृत रूप धारण कर खाद्य-अखाद्य सभी पदार्थों को आच्छादित कर लिया।

(5) उससे वीर्य उत्पन्न हुआ और वीर्य से उत्पन्न पुरुष संपूर्ण पृथ्वी पर फैल गया।

(6) जब देवताओं ने पुरुष की बलि से यज्ञ किया तो वसंत का घृत बना, ग्रीष्म की लकड़ी और पतझड़ की समिधा।

(7) पुरुष का दूर्वा घास पर वध किया गया, जिससे देवताओं, साध्यों और ऋषियों ने यज्ञ किया।

(8) इस विश्व यज्ञ से दही और घी बना, जंगली और पालतू पशु पैदा हुए।

(9) उस यज्ञ से ऋक् और साम की ऋचाएँ, छंद और यजुर्वेद के मंत्र पैदा हुए।

(10) उससे घोड़े, दोनों ओर दाँतवाले पशु, गाय, बकरी और भेड़ पैदा हुए।

(11) जब देवों ने पुरुष का विभाजन किया, उसे कितने भागों में बाँटा? उसका मुख क्या था? बाँह क्या थे? पैर क्या थे?

(12) ब्राह्मण उसका मुख, क्षत्रिय बाहु, वैश्य और शूद्र उसके पैरों से उत्पन्न हुए।

□

संदर्भ ग्रंथ

डॉ. बाबा साहब भीमराव आंबेडकर, राइटिंग्स एंड स्पीचेज, (महाराष्ट्र सरकार का प्रकाशन) खंड 1 से खंड 14 तक।

सोर्स मेटेरियल ऑफ डॉ. बाबा साहब अंबडेकर एंड द मूवमेंट ऑफ अनटचेबल्स, (महाराष्ट्र सरकार का प्रकाशन) खंड—1

डॉ. आंबेडकर : लाइफ एंड मिशन, धनंजय कीट।

आंबेडकर : द टोटल रिवोल्यूशनरी : डी. के. बैसंतरी।

द चमार्स : जी. डब्ल्यू. ब्रिग्स।

रेनीनिसेंसज एंड रिमेंबरेंसेज ऑफ डॉ. बी. आर. आंबेडकर, नानक चंद रत्तू।